Pulp and Paper Chemistry and Technology   Volume 3
Paper Chemistry and Technology

Edited by Monica Ek, Göran Gellerstedt, Gunnar Henriksson

Pulp and Paper Chemistry and Technology Volume 3

This project was supported by a generous grant by the Ljungberg Foundation (Stiftelsen Erik Johan Ljungbergs Utbildningsfond) and originally published by the KTH Royal Institute of Technology as the "Ljungberg Textbook".

Paper Chemistry and Technology

Edited by Monica Ek, Göran Gellerstedt,
Gunnar Henriksson

DE GRUYTER

Editors

Dr. Monica Ek
Professor (em.) Dr. Göran Gellerstedt
Professor Dr. Gunnar Henriksson
Wood Chemistry and Pulp Technology
Fibre and Polymer Technology
School of Chemical Science and Engineering
KTH − Royal Institute of Technology
100 44 Stockholm
Sweden

ISBN 978-3-11-048343-7

Bibliographic information published by the Deutsche Nationalbibliothek

The Deutsche Nationalbibliothek lists this publication in the Deutsche Nationalbibliografie; detailed bibliographic data are available in the Internet at http://dnb.d-nb.de.

Typesetting: WGV Verlagsdienstleistungen GmbH, Weinheim, Germany.
Printing and binding: Hubert & Co. GmbH & Co. KG, Göttingen, Germany.
Cover design: Martin Zech, Bremen, Germany.

Foreword

The production of pulp and paper is of major importance in Sweden and the forestry industry has a profound influence on the economy of the country. The technical development of the industry and its ability to compete globally is closely connected with the combination of high-class education, research and development that has taken place at universities, institutes and industry over many years. In many cases, Swedish companies have been regarded as the initiator of new technology which has started here and successively found a general world-wide acceptance. This leadership in knowledge and technology must continue and be developed around the globe in order for the pulp and paper industry to compete with high value-added forestry products adopted to a modern sustainable society.

The production of forestry products is based on a complex chain of knowledge in which the biological material wood with all its natural variability is converted into a variety of fibre-based products, each one with its detailed and specific quality requirements. In order to make such products, knowledge about the starting material, as well as the processes and products including the market demands must constitute an integrated base. The possibilities of satisfying the demand of knowledge requirements from the industry are intimately associated with the ability of the universities to attract students and to provide them with a modern and progressive education of high quality.

In 2000, a generous grant was awarded the Department of Fibre and Polymer Technology at KTH Royal Institute of Technology from the Ljungberg Foundation (Stiftelsen Erik Johan Ljungbergs Utbildningsfond), located at StoraEnso in Falun. A major share of the grant was devoted to the development of a series of modern books covering the whole knowledge-chain from tree to paper and converted products. This challenge has been accomplished as a national four-year project involving a total of 30 authors from universities, Innventia and industry and resulting in a four volume set covering wood chemistry and biotechnology, pulping and paper chemistry and paper physics. The target reader is a graduate level university student or researcher in chemistry / renewable resources / biotechnology with no prior knowledge in the fields of pulp and paper. For the benefit of pulp and paper engineers and other people with an interest in this fascinating industry, we hope that the availability of this material as printed books will provide an understanding of all the fundamentals involved in pulp and paper-making.

For continuous and encouraging support during the course of this project, we are much indebted to Yngve Stade, Sr Ex Vice President StoraEnso, and to Börje Steen and Jan Moritz, Stiftelsen Erik Johan Ljungbergs Utbildningsfond.

Stockholm, August 2009   Göran Gellerstedt, Monica Ek, Gunnar Henriksson

List of Contributing Authors

Göran Annergren
Granbacken 14
85634 Sundsvall, Sweden
jag.consulting@sundsvall.mail.telia.com

Gunnar Engström
Rådjursstigen 38, 5tr
170 76 Solna, Sweden
gengstrom@bredband.net

Nils Hagen
SCA Graphic Research
Box 846
85 123 Sundsvall, Sweden

Tom Lindström
KTH Royal Institute of Technology
Dept. of Solid Mechanics,
School of Engineering Sciences
100 44 Stockholm, Sweden
toml@fpirc.kth.se

Bo Norman
Innventia AB
Box 5604
114 86 Stockholm, Sweden
bo.norman@innventia.com

Lennart Salmén
Innventia AB
Box 5604
114 86 Stockholm, Sweden
lennart.salmen@innventia.com

Stig Stenström
Center for Chemistry and
Chemical Engineering
Department of Chemical Engineering
P.O. Box 124
221 00 Lund, Sweden
Stig.Stenstrom@chemeng.lth.se

Lars Wågberg
KTH Royal Institute of Technology
Chemical Science and Engineering
Fibre and Polymer Technology
100 44 Stockholm, Sweden
wagberg@pmt.kth.se

Magnus Wikström
Billerud AB
P.O. Box 703
169 27 Solna, Sweden
magnus.wikstrom@billerud.com

Contents

1 Structure of the Fibre Wall

Lars Wågberg
Department of Fibre and Polymer Technology, KTH

1.1 Background

When the structure and the swelling of the cell wall of the fibres are discussed, there is often a focus on how the water is accommodated in the fibre wall. The water in the cell wall is usually divided into water in a gel phase and water in voids in the fibre wall. However, some authors also claim that the concept of the fibre wall as a gel is an unnecessary exercise, since most of the effects achieved by changing the chemical environment around the fibres can be explained by a swollen surface layer of the fibres, see e.g. Pelton (1993). Without entering this debate, it is a fact that the discussion about how the water is accommodated in the fibre wall is very dependent on the structure of the wall and how this fibre wall is changed by different process conditions. Therefore it is necessary to start by discussing the current understanding of the structure of the fibre wall and then go on to discuss how recent developments have changed our understanding of the structure of the fibre wall and how it is changed upon for example lignin and hemicellulose removal.

1.1.1 Current Understanding of the Structure of the Fibre Wall

Our current understanding of the structure of the cell wall of papermaking fibres is dominated by the work of Stone and Scallan, see e.g. Lindström (1986), and their work will therefore be reviewed in some detail. In an early paper, Stone and Scallan (1965a) introduced the concept of the multilamellar structure of the fibre wall with the lamellae arranged concentrically with the cell wall axis. The number of lamellae was dependent on the degree of swelling and upon drying the lamellae join together. According to their measurements, the water-swollen cell wall consisted of several hundred lamellae each less than 100 Å thick and separated by an average distance of about 35 Å. These results were obtained through nitrogen gas adsorption and the lamellar structure was "visualised" through a methacrylate embedding followed by scanning electron microscope analysis. This latter technique results in an unnatural swelling of the fibre wall and the results can at best be used for qualitative discussions. The nitrogen gas adsorption, however, is very accurate and these results also showed that the saturation point of the cell wall was 0.3 cm^3/g.

The basis behind the nitrogen adsorption method is an adsorption isotherm of nitrogen onto the fibres (Haselton 1954). The fibres are exposed to an increasing concentration, i.e increasing pressure, of N_2 gas and the amount of nitrogen adsorbed onto the fibres is determined at the different N_2 pressures. Up to a monolayer coverage of nitrogen on the fibres the adsorption can be described by the BET equation, i.e. equation (1.1).

$$p / v(p_0 - p) = 1 / v_m c + [(c-1) / v_m c] \cdot p / p_0 \tag{1.1}$$

where
    p = Pressure in mm Hg of the gas at a certain temperature
    p_0 = Vapour pressure of the pure gas at the same temperature
    v = Volume of adsorbed gas in ml (STP)
    v_m = Volume of the gas in ml (STP) to form a monolayer
    c = Constant

By applying this equation to the adsorption of nitrogen onto the fibres the volume of gas needed to form a monolayer, v_m, on the fibres can be determined and by using this entity and the size of the molecules the specific surface area of the fibres can be determined according to equation (1.2).

$$\text{Area} = \frac{N v_m L}{22,400 \cdot 10^{20}} \ \text{m}^2/\text{g} \tag{1.2}$$

where
    N = Avogadros number
    v_m = Volume of the gas in ml (STP) to form a monolayer (ml/g adsorbent)
    L = Molecular crossectional area in Ångström^2

A combination of the specific surface area of the fibres and the specific volume of the material in the fibre wall the thickness of lamellae in the fibre wall can then be calculated from equation (1.3) by assuming a cylindrical lamellae arrangement in the fibre wall

$$t = \frac{2V}{A} \tag{1.3}$$

where
    V = Specific volume of the material in the fibre wall
    A = Specific surface area of the fibre wall
    T = Thickness of the concentric lamellae in the fibre wall

For air dried chemically delignified fibres a specific surface area of 1 m^2/g can usually be measured and assuming a specific volume of the material to be 0.63 cm^3/g, equation (1.3) leads to a thickness of the lamellae of 1.28 µm whereas a similar calculation for a solvent exchanged fibre with a specific surface area of around 100 m^2/g leads to a lamellae thickness of 128 µm.

From the adsorption of nitrogen on the fibres it is also possible to determine the total void volume inside side the fibre wall by performing the adsorption measurements at higher nitrogen pressures.

Stone and Scallan (1965b) also introduced a pulp preparation technique, where the wood chips could be delignified under extremely well controlled conditions and the first results with this technique were virtually the same as the earlier published results regarding the multilamellar structure of the fibre wall. They were also able to show that the specific surface area and the volume of voids in the fibre wall, from nitrogen gas adsorption, increased as the degree of delignification increased. The surface area increased from $13.3\,m^2/g$ at a yield of 95.4 % for kraft pulps to $274\,m^2/g$ at a yield of 47.6 %. Corresponding figures for the void volume in the fibre wall were 0.019 and $0.578\,cm^3/g$.

By applying the same technique for delignification, Stone and Scallan (1967) prepared a number of kraft pulps with different yields and they then characterised these pulps with N_2 adsorption, pressure plate analysis and a method where they determined the non-solvent water in the fibre wall. In this latter technique, macromolecules with a molecular mass too high to allow for penetration into the fibre wall were used and by knowing the total amount of water in the system and the concentration of the macromolecules in the water phase, it was possible to determine the volume of the fibre wall inaccessible to the polymer. In the pressure plate technique, a water-saturated sample is subjected to an increasing air pressure and the volume of liquid remaining in the sample is measured at different pressures. From this it is then possible to estimate a total void volume of the fibre wall. The results of this comparison are summarised in *Table 1.1*.

Table 1.1. Summary of the results from the void volume estimation of the fibre from laboratory prepared kraft pulps as detected by different methods. From Stone and Scallan (1967).

Pulp yield (%)	Lignin (%)	Volume (cm³/g) from N₂ ads.	Volume (cm³/g) from pressure plate	Volume (cm³/g) from non-solvent water
100	27.0	0.01	0.4	0.42
92.4	27.3	0.04	0.67	0.70
89.0	27.6	0.08	0.74	(0.66)
80.0	28.5	0.27	0.86	0.92
77.8	28.2	0.33	0.94	0.94
70.4	25.2	0.54	1.06	1.08
61.6	19.3	0.63	1.16	1.22
53.4	12.3	0.55	1.21	1.28
48.7	6.5	0.57	1.14	1.21

It can be seen in the table that there is a large difference in the void volumes as determined by the nitrogen gas adsorption and by the other techniques. From measurements on water-saturated samples, which had been subjected to different degrees of water removal before solvent exchange drying and analysis with N_2 adsorption, the authors concluded that the value obtained with the pressure plate method and the non-solvent water method was "correct". This term was then called the fibre saturation point (FSP) since it included water in the gel phase of the fibre wall and in the void volume of the fibre wall.

4

When the yield of the pulp was decreased, the number of fibres/g increased and, in order to compensates for this, Stone and Scallan (1967) multiplied the FSP with the yield and were hence able to compare the water associated with the same number of fibres for the different yields. Assuming a density of the components in the fibre wall they were also able to calculate how the solid material decreased with decreasing yield. They also made a further division of the pores in the fibre wall into macro-pores, i.e. the pores that could be determined with nitrogen adsorption, and micro-pores, which were defined as the difference between the FSP-value and the nitrogen adsorption value. A representation of this is given in *Figure 1.1.*

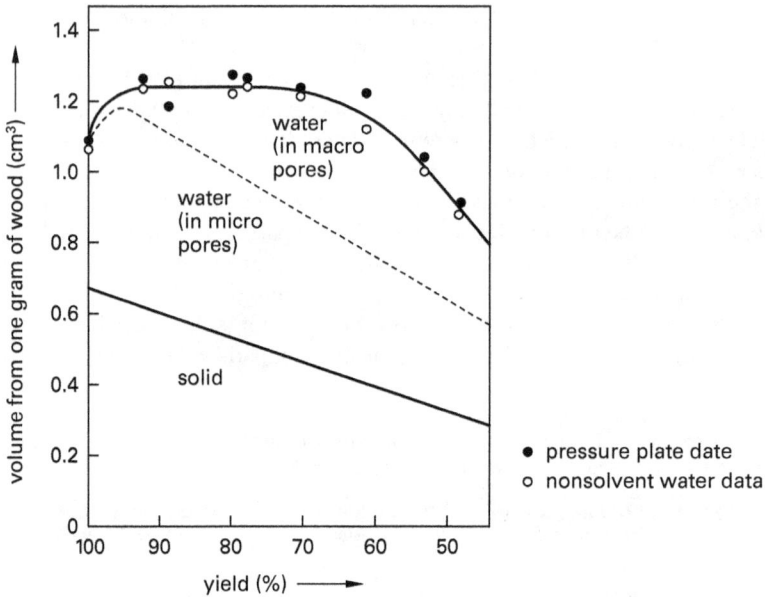

Figure 1.1. The different volumetric changes in the fibre wall occurring when the degree of delignification is increased. The graph shows the volumetric changes from one gram of wood, i.e. the yield times the FSP-value, and the volume of solid is calculated from an assumed fibre wall material specific volume of 0.667 cm^3/g. The macro-pore water is defined as the volume of N_2 adsorption and the micro-pore water is defined as the difference between the FSP-value and the macro-pore volume. From Stone and Scallan (1967) and the data are collected from experiments with a kraft pulp.

There is a small increase in the volume of the fibre wall at the beginning of the cook when the yield decreases but the fibre wall thickness is constant between 95 % and 70 % yield and the obvious interpretation of this is that the volume of material removed is filled with water. Below a 70 % yield, the situation is different and the fibre wall volume decreases as the yield decreases.

In a continuation of this series of papers (Stone and Scallan 1968) the authors used the solute exclusion technique to determine how the size of the pores of the fibre wall changes as the yield of the pulp decreases. By using macromolecules with known dimensions and measuring how the volume of non-solvent water changed with the molecular mass, it was possible to estimate the size distribution of the pores of the fibre wall. However, the interpretation of the data is not straightforward since the shape of the pore is very important for the partitioning of the molecules between the fibre-wall and the solution when the size of the molecule and the size of the pore are similar, Lindström (1986) and Alince (1991). It can nevertheless be concluded that Stone and Scallan (1968) showed that the size of the pores, i.e. the pore width, in the fibre wall increases as the yield decreases, that an approximate size of the pores in the wood is between 5 and 40 Å and that the size of the pores in the fibre wall of a pulp with a yield of 44.6 % is between 10 and 100 Å.

This was definitely a large step in our understanding of the structure of the fibre wall and this picture is still very dominating. It should however be stressed that these figures must be taken as relative. Recent experimental evidence (van de Ven 1997, Li et. al 1993, Häggqvist et al 1998, Andreasson et al 2003 and Maloney et al 1999) has shown that the absolute value of the pore size may be different from the values given by Stone and Scallan (1968) and this will discussed later on in this chapter.

Stone and Scallan (1968) also used the solute exclusion technique to suggest a new structural model of the fibre wall where the earlier defined macro-pores were found to be of the order of 25–300 Å. This was defined as the interlamellar pore width and the micropore was defined as intralamellar pores in the size range of 5–25 Å, although the authors stated that the upper limit was somewhat arbitrarily determined. Based on these results and on electron micrographs, Scallan (1974) introduced the now very well known picture of how the internal fibrillation takes place with increasing swelling of the fibre wall. This figure is shown in *Figure 1.2.*

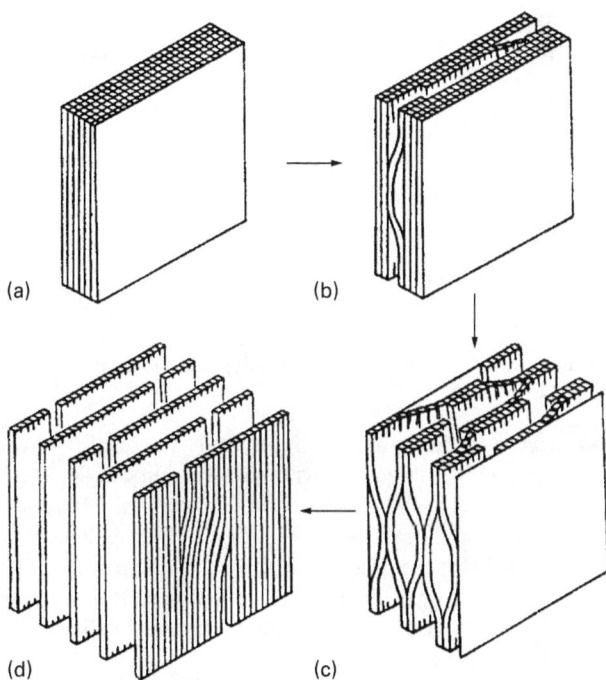

Figure 1.2. Schematic view of how the fibrillation of the fibre wall takes place when the swelling is increased. From Scallan (1974).

It should be pointed out that the forces causing this fibrillation can not be solely the swelling forces within the fibre wall. There is also a need for a considerable mechanical action on the fibre wall in order to make this fibrillation occur.

In the models presented by Stone and Scallan the arrangement of different chemical components in the fibre wall is not discussed, even though some comments were made regarding the arrangement of pores around the cellulosic microfibrils (Stone and Scallan 1968). In order to fill this gap in knowledge, Kerr and Goring (1975) conducted work in which they studied ultra-thin microtome sections of permanganate-stained fibres from black spruce with transmission electron microscope. They showed that cellulose was not stained by the permanganate and that only a minor part of the hemicellulose was stained, which means that it was possible to see the distribution of the lignin in the fibre wall. From cross-sections both along the fibres and across the fibres, they concluded that the ultrastructural arrangement of the different components of the fibre wall could be represented by the picture shown in *Figure 1.3*.

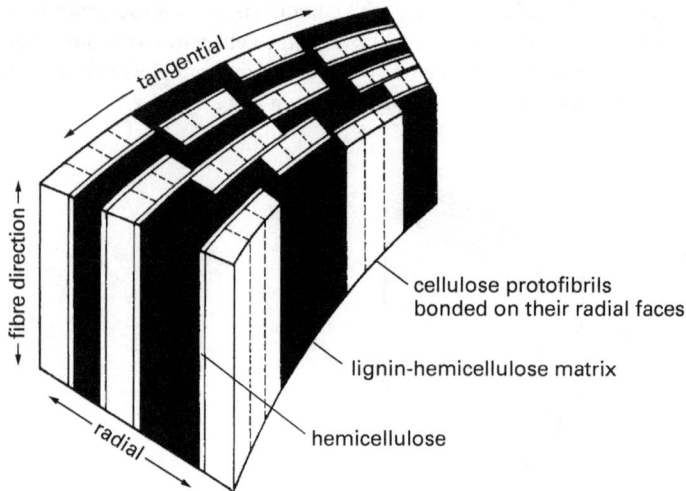

Figure 1.3. Schematic representation of the ultrastructural arrangement of cellulose, lignin and hemicellulose in the fibre wall of black spruce tracheids. From Kerr and Goring (1975).

In this picture, 1–2 fibrils are associated in the radial direction, with a size of 35–70 Å, whereas 3–4 fibrils are associated in the tangential direction, with a dimension of 104–140 Å. When the authors started to compare the amounts of the different components in the fibre wall, they found literature values corresponding to 0.26:0.45:0.29 for the lignin: cellulose: hemicellulose and in the microtome sections they found that the proportion stained by the permanganate was equivalent to 0.45. This means that a fraction of the hemicellulose had been stained by the permanganate and might therefore be assumed to be associated with the lignin. The rest of the hemicellulose was assumed to be associated with the cellulose microfibrils.

1.1.2 Recent Developments in the Evaluation of the Pore Structure of Fibres

Since the properties of the fibre wall will be determined both by the type of material in the fibre wall and how it is organised inside the fibre wall large research efforts have recently been conducted to find a more elaborate structure determination of the fibre wall (van de Ven 1997, Li et. al 1993, Häggqvist et al 1998, Andreasson et al 2003, Duchesne et al 1999 and Maloney et al 1999).

First of all the use of new techniques for determination of fibre structure via cryofixation and deep etching of the fibres followed by Field Emission Scanning Electron Microscopy (FE-SEM) should be mentioned (Duchesne 1999). This technique has revealed a new very open structure of the fibre wall, for most commercial chemical pulps in their never dried form, where the fibrils are clearly separated. The openings between the fibrils are of the order of 10–50 nm, naturally depending on the degree of delignification, and the lateral dimension of fibril aggregates are shown to be of the order 10–20 nm. This is demonstrated in *Figure 1.4* where a Cryo-FE-SEM image of the swollen fibrillar structure of the fibre wall is demonstrated for delignified fibres from spruce. In *Figure 1.5*, a similar section from an air dried spruce fibre is shown. It should be stressed that the scale in the figures is the same and the size of the bar is 100 nm in both diagrams.

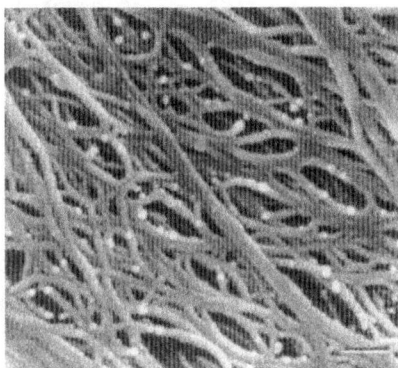

Figure 1.4. High resolution Cryo-FE-SEM micrograph showing the surface structure of a frozen hydrated kraft pulp fibre from spruce. The bar in the figure corresponds to 100 nm. (Duchesne 1999).

From these two figures it is clear that the openings in the fibre wall are larger then the pore size as estimated with the solute exclusion technique but this might not be so unexpected since this method is known not to be able to determine the correct pore size distribution of the fibres eventhough it is excellent for a determination of the Fibre Saturation Point (FSP) as defined earlier. It is also clear from these figures that drying in air lads to a collapse of the pores in the fibre wall most likely caused by the capillary forces between the fibrils during drying.

Despite the elegance of these types of micrographs it is very difficult to quantify the pore size and pore size distribution from these figure based on single fibre evaluations. Methods based on some type of average property is the fibres is therefore of very large interest and Li et al 1999 used ^{2}H and ^{1}H NMR relaxation measurements to determine the pore size distribution within pulp fibres. With this technique the relaxation of water molecules within the fibres are determined from NMR spin lattice relaxation profiles using D_2O and H_2O as probe liquids. The principal behind the method is schematically shown in *Figure 1.6*.

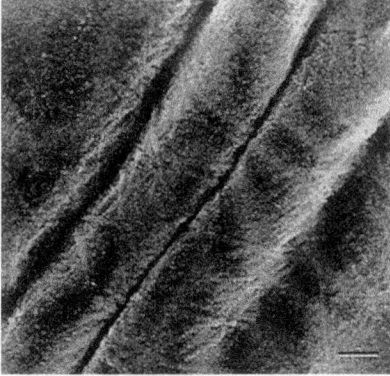

Figure 1.5. FE-SEM micrograph of an air dried kraft pulp fibre from spruce. The bar in the figure corresponds to 100 nm (Duchesne 1999).

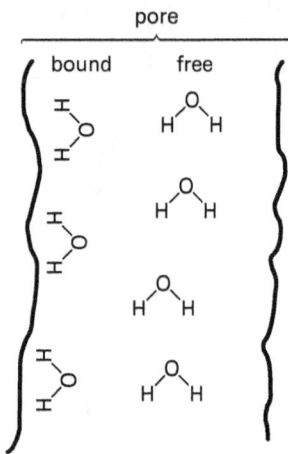

Figure 1.6. Schematic representation of the principal behind the NMR relaxation method. It is assumed that two types of water exist in the fibre wall, bound water and free water.

As indicated in the figure the method is based on a number of assumptions. First of all it is assumed that two types of water molecules exists within the fibre wall, bound water that will be affected by the surface of the fibres and free water with the same properties as bulk water. This means that the method will determine a surface to volume ratio of the water inside the fibre wall. In order to calculate and average pore radius of the fibre wall it is also assumed that the pores within the fibre wall has a cylindrical arrangement, similar to that shown in *Figure 1.2*. In mathematical terms this can be summarised by equation (1.4) and (1.5)

$$M(\tau) = M_0\left[1 - 2\exp\left(-\frac{\tau}{T_1}\right)\right] \tag{1.4}$$

where

$M(\tau)$ = Magnetization at time τ

M_0 = Magnetization at time 0

T_1 = The population-averaged relaxation time

T_1 will be determined by the relaxation time of the free water, T_{1F} and the relaxation time of bound water, T_{1B}, and the amount of water in the different states. Assuming a cylindrical geometry the relation between T_1 and T_{1F} and T_{1B} can be written

$$\frac{1}{T_1}(r) = P_B \frac{1}{T_{1B}} + (1 - P_B)\frac{1}{T_{1F}} = \frac{1}{T_{1F}} + \frac{Sn}{V\rho}\left(\frac{1}{T_{1B}} - \frac{1}{T_{1F}}\right) = \frac{1}{T_{1F}} + \frac{2n}{r\rho}\left(\frac{1}{T_{1B}} - \frac{1}{T_{1F}}\right) = \alpha + \frac{\beta}{r} \tag{1.5}$$

where

T_1 = Population averaged relaxation time in a pore with a uniform pore size

T_{1B} = Relaxation time of bound water

T_{1F} = Relaxation time of free water

P_B = Fraction of water dynamically perturbed by the surface

S = Specific surface area of the a pore in the solid

V = Volume of a pore in the solid

n = Amount of "bound" water

ρ = Density of the liquid

r = radius of the pore

$$\alpha = \frac{1}{T_{1F}}$$

$$\beta = \frac{2n}{\rho}\left(\frac{1}{T_{1B}} - \frac{1}{T_{1F}}\right)$$

A crucial parameter in order to convert the magnetization data to a pore size distribution data is the b parameter in equation 5 and details about the determination of this parameter is given in (Li et al 1993). However, as can be seen in this set of equations it is possible to calculate a pore size distribution from NMR relaxation measurements after some assumptions about the geometry of the fibre wall.

Andreasson et al (2003) conducted an investigation similar to the investigation by Stone and Scallan (1965b) where a set a kraft pulps from carefully selected chips were carefully delignified and then subjected to NMR relaxtion measurements and a typical result from these measurements are shown in *Figure 1.7*.

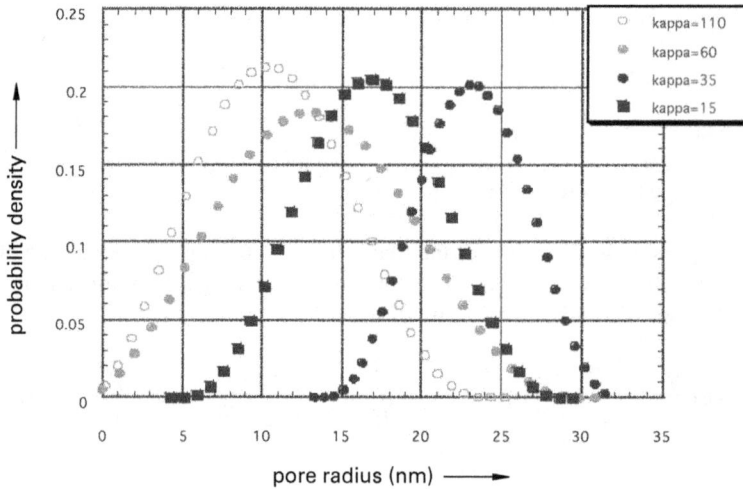

Figure 1.7. Pore size distribution of pores within the fibre for a set of kraft pulps from spruce, delignified to different yield, as determined with NMR relaxation measurements (Data from Andreasson 2003).

As can be seen from this figure the average pore radius increases from 10 nm to around 23 nm as the kappa number is changed from 110 to 35 and then again decreases to around 17 nm as further lignin and hemicellulose is removed to a kappa number of 15. Simultaneous measurements of the total volume of liquid held by the fibre wall, for the different fibres shown in *Figure 1.6*, also showed that as the hemicellulose an the lignin is removed from the fibre wall the volume of the fibre wall is not dramatically changed which is agreement with the data shown in *Figure 1.1*. This also supports the hypothesis that as material is removed from the fibre during kraft pulping empty spaces are created within the fibre wall but that the total volume of the fibre wall is fairly constant. The pore sizes as determined with the NMR relaxation technique is also in very good agreement with the size of the pores as determined with the cryo-FE-SEM technique as shown in *Figure 1.4*. The discrepancy between the NMR/cryo-FE-SEM and the pore sizes as determined with the Solute Exclusion Methods (Stone and Scallan 1968) might naturally have several causes. However, when examining the structure of the fibre wall as revealed in *Figure 1.4* it seems plausible to suggest that the different methods are sensitive to/measures different parts of the fibre wall. In order to be able to characterise the pores inside the fibre wall the polymers used in the solute exclusion technique have to enter through the pores in the external part of the fibre wall. As can be seen in figure the S1 layer has a considerable amount of small pores that will restrict the polymers, used in the SEC method, from entering the fibre wall and therefore the SEC method will most probably be sensitive to the pores in the outermost layers of the fibre wall. As was mentioned earlier the NMR relaxation method will measure the surface to volume ratio of water inside the fibre wall and therefore this method will give an average of all pores within the fibre wall. With this in mind it might be suggested that the different methods are sensitive to different parts of the fibre wall and therefore they are complementary to each other. It should also, once again be stated, that the SEC method can not be used to determine the pores size distribution of the fibres unless the shape of the pores inside the fibre wall are used to correct the SEC data. Since this shape of the pores is not known the method is hard to use more than for a qualitative discussion of the pore size distribution.

Another method based on polymer adsorption has also been used to estimate the pore size of the pores within the fibre wall (van de Ven 1997). In this method, different cationic polyelectrolytes, i.e. cationic polyethyleneimines, are adsorbed to fibres and compared with the adsorption to a microcrystalline cellulose (MCC) with virtually no internal pores. The results are then presented as adsorbed amount as a function of the size of the adsorbing molecules and at a certain size there will be a decrease in the adsorbed amount for the porous fibres, compared with the MCC, since these molecule will have no access to the internal parts of the fibre wall. Typical results from these measurements are shown in *Figure 1.8* and as can be seen there is a break in the curves for the fibres at a molecular size of 13 nm and assuming that the pores have to be 3–5 times larger for the molecules to enter the average pore size would be 40–65 nm which is of the same order of magnitude as determined with the NMR technique and the cryo-FE-SEM.

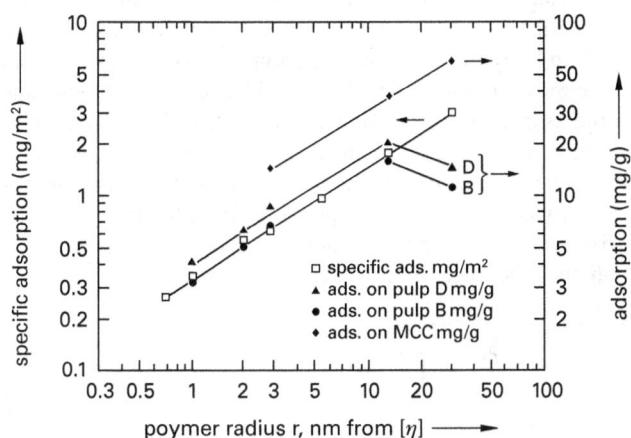

Figure 1.8. Adsorption of PEI, cationic polyetyleneimine, of different molecular mass, size, to two different types of pulp fibres and to MCC at pH=10. Adsorption capacity in mg/g is compared with specific adsorption in mg/m² (van de Ven 1997).

It is naturally difficult to exactly explain the difference between the SEC method and the polyeletrolyte adsorption method and van de Ven (1997) tried to explain the difference partly by the difference in the driving force for interaction with the fibres between the dextrans used in the SEC and the PEI used in the adsorption method. The dextrans have no interaction at all with the fibre surface and might be very sensitive to depletion effects in small pores whereas this is not applicable for the PEI. It can nevertheless be summarised that the PEI adsorption results in pores somewhat larger than the NMR relaxation method but they are of the same order of magnitude and it should be kept in mind that the PEI molecules are by no means monodisperse and therefore the agreement between the two methods is very good.

The final method that recently has been developed for pore size measurements is the thermoporosimetry method (Maloney 1999). The basis behind this method is that water in small cavities will have a lower freezing point compared with free water. This can mathematically be described with Gibbs-Thompson equation, equation (1.6).

$$D = \frac{4\, V\, T_0\, \sigma_{ls}}{\Delta H_m \ln \dfrac{T}{T_0}}$$

(1.6)

where

D = diameter of the pore

V = molar volume of solvent

σ_{ls} = surface tension between solid and liquid

ΔH_m = latent heat of melting

T = Melting temperature

T_0 = Melting temparetaure of pure water

With this method a sample of wet fibres is first frozen and then the temperature is increased and the latent heat of melting is recorded and transformed to a melted amount of water per gram of fibres. From these data a pore size distribution curve can then be calculated up to rather large pores which is unique for this method and a typical pore size distribution from measurements with water is shown in *Figure 1.9*. The maximum pore size that could be determined, with water as a solvent, was around 500 nm and as indicated in the figure there is still a big difference in pore volume at this pore size as compared with FSP measurements. It was postulated (Maloney 1999) that the difference between the FSP and Thermoporosimetry data was due to macropores in the fibre wall and the pores as determined with the thermoporosimetry method was termed micropores. These micropores consist of pores larger than 1 nm and pores with non-freezing water, i.e. pores smaller than 1 nm which is the lower detection limit for water.

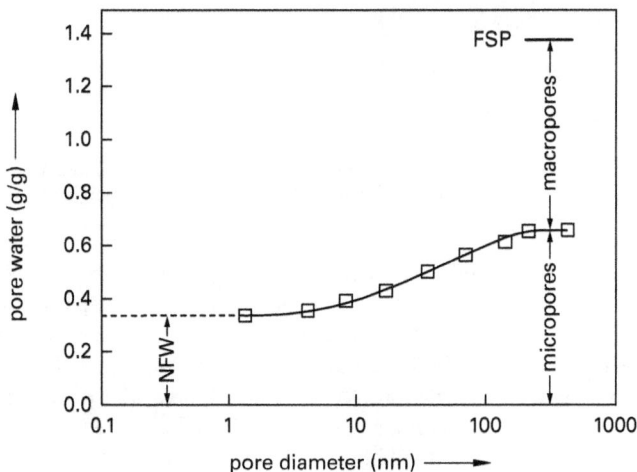

Figure 1.9. Pore size distribution of bleached softwood kraft fibres in their Na$^+$ form as determined with thermoporosimetry in water. The classification of the pores in micropores and macropores is indicated in the figure and the location of the FSP bar is arbitrary and should not be confused with a certain pore size (Maloney et al 1999).

There are however several limitations with this method. Due to limitations in the equipment smaller pores than 1 nm in water can not be determined with this method and for another solvent, cyclohexane it was found that the minimum pore size was around 7 nm. Cyclohexane is recommended for use instead of water since it is known not to cause any disturbances of the fibre structure upon freezing, which was found for water, and water has to be exchanged to cyclohexane via a solvent exchange procedure before testing. Another advantage with using cyclohexane is that larger pores can be determined with this solvent. It should though be kept in mind that the solvent exchange might cause a structure change of the fibre wall that is very hard to quantify.

The liquid contained in pores that can not be measured with the thermoporosimetry method was assumed (Maloney et al 2001) to have a log-normal size distribution and together with the measurable pores a bimodal pore size distribution curve can be presented from these types of measurements. A typical pore size distribution from measurements with never dried bleached softwood fibres, both beaten and unbeaten, are shown in *Figure 1.10*.

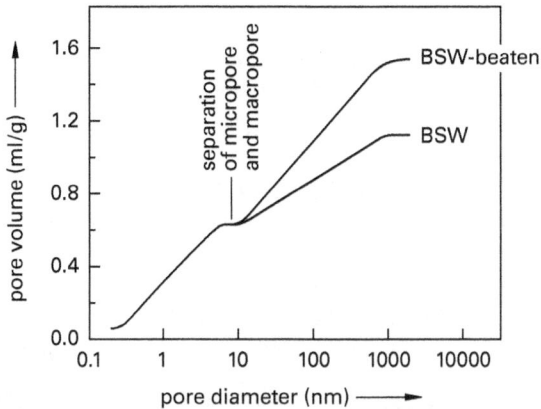

Figure 1.10. Pore size distribution for a never dried bleached softwood pulp where the pores smaller than 10 nm have been assumed to have a log-normal size distribution. Data for both beaten and unbeaten fibres are shown. The bimodal size distribution is a result of the assumed log-normal size distribution of the pores smaller than 10 nm but might be supported by literature data (Maloney et al 2001).

As seen in this figure pore sizes up to 3 μm can be measured with this method and it is also clear that a mechanical beating changes the larger pores in the fibre wall.

1.2 Concluding Remarks

As has been made obvious from this chapter a lot of hard work and a lot of advanced equipment have been applied to determine the pore size of the pores within the fibre wall and the pore size distribution. It is also clear that different techniques will measure different parts of the fibre wall and a lot of effort have over the years been spent on debating which method that is correct and which method that is not. In the opinion of the author all methods have drawbacks and advantages and it is absolutely essential that the user of the different methods should have a deep in-

14

sight into the principals of the methods before starting to interpret the results from the measurements. It is the aim of this chapter to give an insight into the knowledge about the structure of the fibre wall and into the basics behind the different measuring principles.

It seems as if there is an agreement about that the radius of the pores in the fibre wall is around 10 to 80 nm, naturally depending on which type of fibres that are studied and how they are treated, and that there is a difference in the size in the exterior part of the fibre wall and in the interior part of the fibre wall. This will naturally be very important when depositing some functional additive within the fibre wall but it will naturally also be very important for the flexibility of the fibres and the way the fibres will conform towards each other during drying and consolidation of paper. These latter correlations are not clear but there are many new and advanced methods available for characterisation of the pores within the fibre wall.

1.3 References

Alince, B. (1991) *Tappi J.* 11: 200.
Andreasson, B., Forsström, J. and Wågberg, L. (2003) *Cellulose* 10: 111.
Duchesne, I. and Daniel, G. (1999) *Nordic Pulp Paper Res. J.* 14, 2: 129.
Häggqvist, M., Li, T.-Q. and Ödberg, L. (1998) *Cellulose* 5, 1: 33.
Haselton, W.R. (1954) *Tappi* 37, 9: 404.
Li, T.-Q., Henriksson, U. and Ödberg, L. (1993) *Nordic Pulp Paper Res.* J. 8, 3: 326.
Lindström, T. (1986) In: A. Bristow and P. Kolseth (Eds.) *Paper – Structure and Performance.* New York, Basel: Marcel Dekker, Inc., p.75 and p.99.
Maloney, T.C. and Paulapuro, H. (1999) *J. Pulp Paper Sci.* 25, 12: 430.
Maloney, T.C. and Paulapuro, H. (2001) In: C.F. Baker (Ed.) *The Science of Papermaking. Trans. 12th Fundamental Res. Symposium held at Oxford.* The Pulp & Paper Fundamental Res. Soc., p. 897.
Pelton, R. (1993) *Nordic Pulp Paper Res. J.* 8, 1: 113.
Scallan, A.M. (1974) *Wood Sci.* 6, 3: 266.
Stone, J.E. and Scallan, A.M. (1965a) *J. Polym. Sci.. Part C* 11: 13.
Stone, J.E. and Scallan, A.M. (1965b) *Pulp Paper Mag. Canada* 66, 8: T407.
Stone, J.E. and Scallan, A.M. (1965c) In: F. Bolam (Ed.) *Consolidation of the Paper Web, Transact. Symp. Cambridge, Vol. 1.* London, UK: Tech. Sect. Brit. Paper Board Makers Assoc., p.1453.
Stone, J.E. and Scallan, A.M. (1967) *Tappi* 50: 10496.
Stone, J.E. and Scallan, A.M. (1968a) *Cellulose Chem. Techn.* 2, 3: 343.
Stone, J.E. and Scallan, A.M. (1968b) *Pulp Paper Mag. Canada* 69, 6: T288.
van de Ven, T.G.M. and Alince, B. (1997) In: C.F. Baker (Ed.) *The Fundamentals of Papermaking, Trans. 11th Fundamental Res. Symp. Cambridge.* Pira International, p. 771.

2 Structure and Properties of Fibres

Lennart Salmén
Innventia AB

2.1 Introduction

The physical properties of plant fibres like stiffness, strength and swelling due to moisture uptake, are highly dependent on the structure or morphology of the cell wall. In plants nature has evolved an optimal utilization of the wood polymers in the structural arrangement of these within the cell wall. Thus trees are provided by their tracheids, the fibres, with a strength to rise over a 100m, to withstand hurricane winds swinging the crown in all directions, to be downloaded with tons of snow and ice and still being able to transport water up to the crown. Thus the fibre construction as a hollow, laminated fibre reinforced tube is optimised both with regard to its design and to strength criteria in both its transverse and longitudinal direction and as well with regard to compressive and tensile forces.

2.2 Cell Wall Structure

The structural organisation of the cell wall regulating its mechanical and physical properties may be grouped into the following hierarchical levels:
* the layered structure of the fibre wall
* the fibre reinforced structure of each cell wall layer
* the orthortropic organisation of the matrix components
* the polymer chemistry of the constituents

2.2.1 Cell Wall Layers

With its organisation into a layered structure of plies, cell wall layers, with the reinforcements of cellulose fibrils in different directions the fibre resembles in a way that of a man made pipe for pressure applications. When there are demands for high strength properties in different directions a laminated oriented ply construction generally offers a good solution. This is what is often used in the fibre reinforced composite field for complicated constructions as for instance rotor blades for wind turbines and has long been used for example as plywood for the demand of strong, lightweight boxes. In the case of the fibre construction essentially three layers may be distinguished, see *Figure 2.1*. The thinner outer and inner layers, which are the S_1 and S_3 layers, have the reinforcing cellulose fibrils arranged at a high angle to the fibre direction. These layers provide the fibre with its transverse stiffness allowing it to be used as pipes for the water transport within the stem. In the middle, between these layers lies the S_2 layer which is distinguished by its low fibril angle. This layer makes up about 70 to 80 % of the material of the wood fibre and it is thus dominating the mechanical properties of the fibre. Due to its low fibril angle the S_2 layer provides the fibre with a very high stiffness and strength in the longitudinal fibre direction thus providing the tree with its strength properties for supporting the load of itself as well as for withstanding the bending forces acting on the stem caused by the blowing winds.

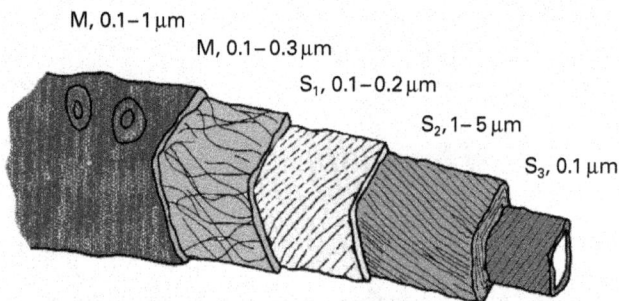

Figure 2.1. Schematic picture of a wood fibre in this case liberated intact from the wood so that the middle lamella, M, remains on the fibre surface as from a low refined high temperature thermomechanical pulp. P is the primary wall with irregular structure of the cellulose fibrils while in the secondary wall layers, the S_1, S_2, S_3, the cellulose fibrils are ordered at a specific angle.

Morphologically the softwood trees creates thin walled earlywood fibres in the spring mainly providing for water transport and thick-walled latewood fibres in late summer mainly providing the tree with the strength properties. Thus latewood fibres have a more or less rectangular cross section with a large thickness of the wall whereas the earlywood fibres are square to hexagonal in cross section. For the cell wall layers the largest difference is in the thickness of the secondary S_2–wall so that the relative proportion between the layers differs somewhat between the two fibre types, see *Table 2.1* (Fengel and Stoll 1973).

Table 2.1. The difference in thickness of the cell wall layers between earlywood and latewood fibres of Norway spruce (*Picea abies*) (Fengel and Stoll 1973).

Cell wall layer	Average thickness (µm)		Relative contribution (%)	
	Earlywood	Latewood	Earlywood	Latewood
ML/2	0.09	0.09	4.2	2.1
S$_1$	0.26	0.38	12.5	9.0
S$_2$	1.66	3.69	78.7	85.4
S$_3$	0.09	0.14	4.5	3.3
S	2.10	4.30	99.9	99.8

2.2.2 Fibre Reinforced Structure

Each of the cell wall layers of the fibre is composed of a mixture of the main wood polymers cellulose, lignin and hemicelluloses. The distribution between these polymers differs somewhat between the cell wall layers, see *Figure 2.2*, which to some extent affects properties but mainly the dissolution of lignin during pulping and bleaching processes. For the polysaccharides there is also a difference in the composition among the different cell wall layers, see *Table 2.2* (Meier 1961). As seen the outermost layers, the primary wall and middle lamella differs in particular in their high content of pectic substances such as arabinan and galactan. Within the secondary wall there is a higher relative content of xylan in the S$_1$- and S$_3$-walls than in the S$_2$-wall in comparison with glucomannan.

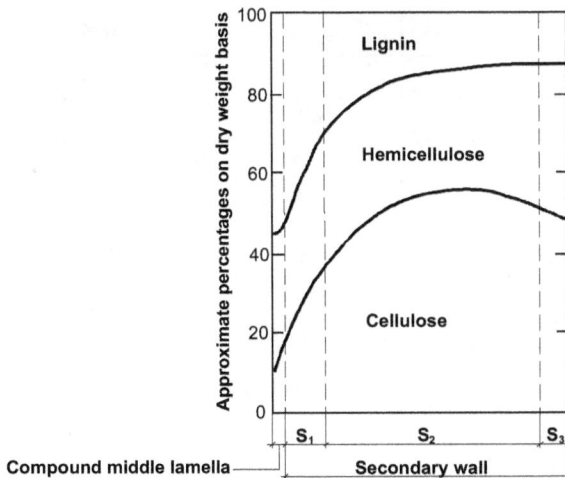

Figure 2.2. Distribution of the principal chemical constituents within the cell wall layers of softwoods. The diagram is based upon the one given by Panshin (Panshin and Zeeuw 1980) but modified for the lignin content in the compound middle lamella (ML+P) according to Westermark et al. (Westermark et al 1988).

The arrangement of the wood polymers within the cell wall is that of a fibre reinforced structure with partly crystalline cellulose fibrils as the reinforcing materials in a matrix of the other, amorphous wood polymers; lignin and hemicelluloses. The present conception of the structure favours that the cellulose microfibrils, 3 to 4 nm in cross section (Larsson et al 1999) are aggregated into larger entities, the cellulose fibrils or cellulose aggregates, of an average cross section of 20 nm (Duchesne and Daniel 2000, Fengel 1970) in which most probably some of the glucomannan is incorporated (Fengel 1970, Atalla et al 1993, Salmén 2000). Adjacent to these aggregates mostly glucomannan is arranged in parallel with the fibrils (Åkerholm and Salmén 2001). In the space in-between these aggregates, occupying about 3 to 4 nm (Terashima et al 1993), it is likely that the lignin matrix is interspaced with xylan entities (Salmén and Olsson 1998), see *Figure 2.3*.

Table 2.2. Percentages of polysaccharides in the different cell wall layers of the fibre wall to Meier (Meier 1961).

Polysaccharide	ML+P[a]	S$_1$	S$_2$ outer part	S$_2$ inner part + S$_3$
Spruce				
Galactan	16.4	8.0	0.0	0.0
Arabinan	29.3	1.1	0.8	0.0
Cellulose	33.4	55.2	64.3	63.6
Glucomannan	7.9	18.1	24.4	23.7
Glucuronoarabinoxylan	13.0	17.6	10.7	12.7
Pine				
Galactan	20.1	5.2	1.6	3.2
Arabinan	29.4	0.6	0.0	2.4
Cellulose	35.5	61.5	66.5	47.5
Glucomannan	7.7	16.9	24.6	27.2
Glucuronoarabinoxylan	7.3	15.7	7.4	19.4

[a] Contains a high percentage of pectic acid as well

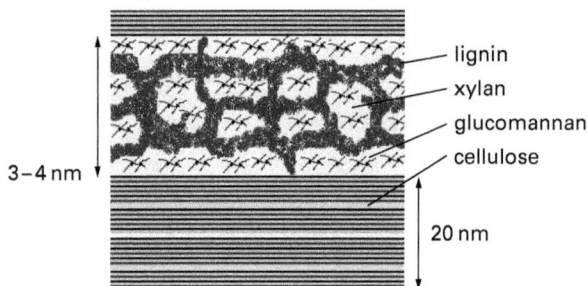

Figure 2.3. Schematic illustration of the arrangement of lignin and different hemicelluloses within the space between cellulose fibrils (Salmén 2000).

2.2.3 Matrix Organisation

There are strong indications that the arrangement of the cellulose aggregates and the lignin rich matrix shows a pattern of a tangential lamellation when viewing the fibre in a transverse cross-section (Ker and Goring 1975). This is also evident when observing the structure of the cross section of the secondary wall by AFM, *Figure 2.4*, where a structural arrangement of the fibril aggregates in the tangential direction is appearing (Fahlén and Salmén 2003). The spacing in-between the fibril aggregates are also of such small dimensions that it is clearly necessary on a volumetric basis that some of the hemicelluloses will be arranged in-between the cellulose microfibrils within the cellulose aggregates (Salmén 2000).

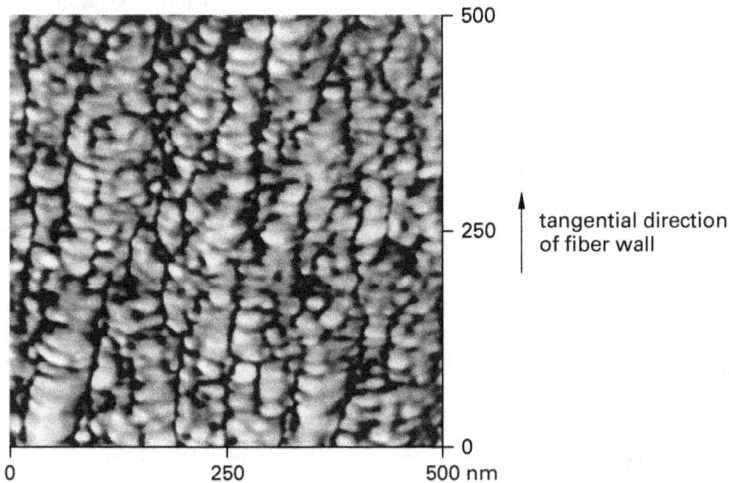

Figure 2.4. Atomic Force Microscopic, AFM- image of a transverse cross section of the secondary wall of a spruce fibre. Cellulose aggregates averaging 18 nm arranged in a pattern of tangential lamellation is evident (Fahlén and Salmén 2003).

Such a structure is also what may be concluded from the change in shape of the fibre due to dissolution of mainly lignin but also some hemicelluloses during pulping. With a radial arrangement of lignin lamellas, the fibre would after removal of lignin had shrunk but with maintained thickness. On the other hand with a tangential lamelation the thickness would have decreased keeping the fibre diameter essentially constant, see *Figure 2.5*. That the latter case is what happens was shown in the nineteen seventies by Stone et al. (Stone et al 1971).

The dissolution of lignin and hemicelluloses during pulping give rise to wholes in the fibre wall, generally termed pores. This lignin diffused out of the cell wall has been found to have a molecular weight corresponding to a spherical diameter of between 3.5 to 10 nm (Willis et al 1987, Favis et al 1984). This corresponds well with the size of the pores created in the fibre wall, which with indirect methods have been indicated to be between 5 to 15–20 nm in diameter (Berthold and Salmén 1997, Stone and Scallan 1967) as seen from *Figure2.6*. In TMP which correspond to wood, i.e. chemically intact fibres, the water uptake in the material occurs as designed to apparent pore sizes below 2 nm and is due to a swelling of the matrix wood polymers. In the chemical pulps water uptake occurs to a larger extent in pores above 2 nm i.e. where no

20

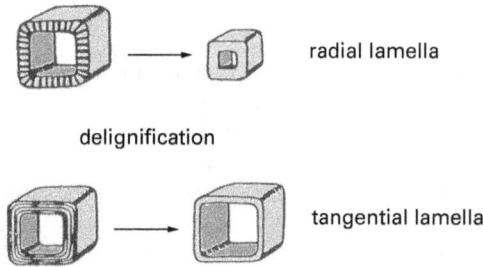

Figure 2.5. Schematic illustration of the possible ways of fibre wall shrinkage due to delignification followed by drying. In reality pulping of fibres results in a change in the fibre form in such a way that the fibre form is maintained while the fibre wall thickness decreased thus essentially the behaviour for a tangential lamella organization (Stone et al 1971).

pores exists for the TMP, thus in holes created by the dissolution of lignin and hemicelluloses. The creation of these pores is essential for the development of a fibre that is flexible and compressible and therefore this is essential for the improvement of the fibre bonding potential of the fibres.

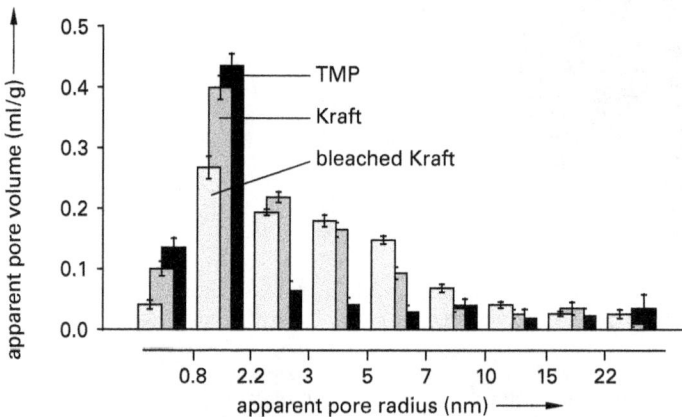

Figure 2.6. The apparent pore size distribution for a TMP, a Kraft softwood pulp and a bleached softwood pulp determined by Inverse Size Exclusion Chromatography, ISEC. Error bars represents standard deviation (Berthold and Salmén 1997).

The size of the lignin molecules diffused from the cell wall are much larger than the corresponding tangential and radial space s in the fibre wall between cellulose aggregates pointing to that the lignin has a lager spread in the length direction of the fibre as also indicated by the schematic illustration of lignin dissolution by Favis and Goring (Favis and Goring 1984) as developed from the well-known structure of the Kerr-Goring model (Kerr and Goring 1975) of the fibre wall shown in *Figure 2.7*. This size of the lignin molecules that are dissolved also points to the fact that the lignin in the fibre wall behaves as a mechanically heterogeneous material. For this to occur a size of the polymer in the order of 5 to 15 nm in diameter is required (Kaplan 1976, Schick and Donth 1981). This means that the lignin in the fibre wall may show a single softening behaviour, i.e. a single glass transition temperature, which also has been determined to occur at 90 °C (1 Hz) for water saturated conditions of the fibre material (Salmén 1984).

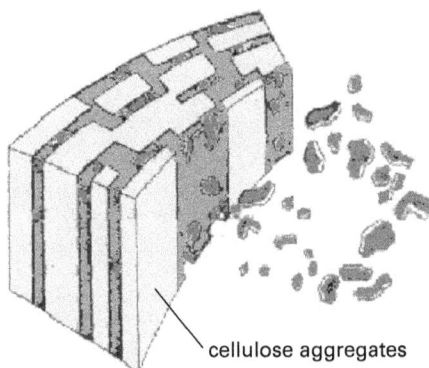

cellulose aggregates

Figure 2.7. Schematic illustration of the dissolution of lignin out of the fibre wall (Favis and Goring 1984) developed from the well-known model picture of the secondary cell wall the Kerr-Goring model (Kerr and Goring 1975). The lignin entities have been shown to appear as flakes, which may with an orientation in the tangential/length direction be fitted into the spacing between the cellulose aggregates.

2.2.4 Wood Polymer Properties

From the structure of the fibre wall it is clear that the cellulose aggregates, which are highly oriented in the direction of the length of the fibre gives the fibre its highly anisotropic properties. This however means that it will be the properties of the other wood polymers, the hemicelluloses and the lignin, which will have a large influence on properties in the transverse direction (Bergander and Salmén 2001). Of the wood polymers, cellulose, hemicelluloses and lignin most attention has been given to the cellulose when it comes to determining mechanical properties of the wood constituents. The rationale is of course the cellulose being the stiff reinforcing material in the composite and therefore the component that to a high degree determines the properties of the fibre in the most important direction, its length. More and more evidence have though been established suggesting an orientation of both hemicelluloses and lignin in the direction of the cellulose fibrils, mainly based on spectroscopic data (Åkerholm and Salmén 2001, Åkerholm and Salmén 2003). It is suggested that the cellulose and hemicelluloses acts as templates for the lignin deposition thus giving also the lignin an orientation in relation to the cellulose fibrillar direction (Atalla 1998). An orientation in the plane of the cell wall has also been indicated from Raman spectroscopy (Agarwal and Atalla 1986) as also depicted in *Figure 2.8.*

Cellulose

For cellulose there are both theoretical estimates as well as experimental determinations of its elastic properties. Most recent analysis tends to favour values in the longitudinal direction for the cellulose crystalline region between 120 to 170 GPa (Tashiro and Kobayashi 1991, Kroon-Batenburg et al 1986) although there are some older theoretical estimates of such high values as 246 and 319 GPa (Gillis 1969). When comparing theoretical estimates with experiments both

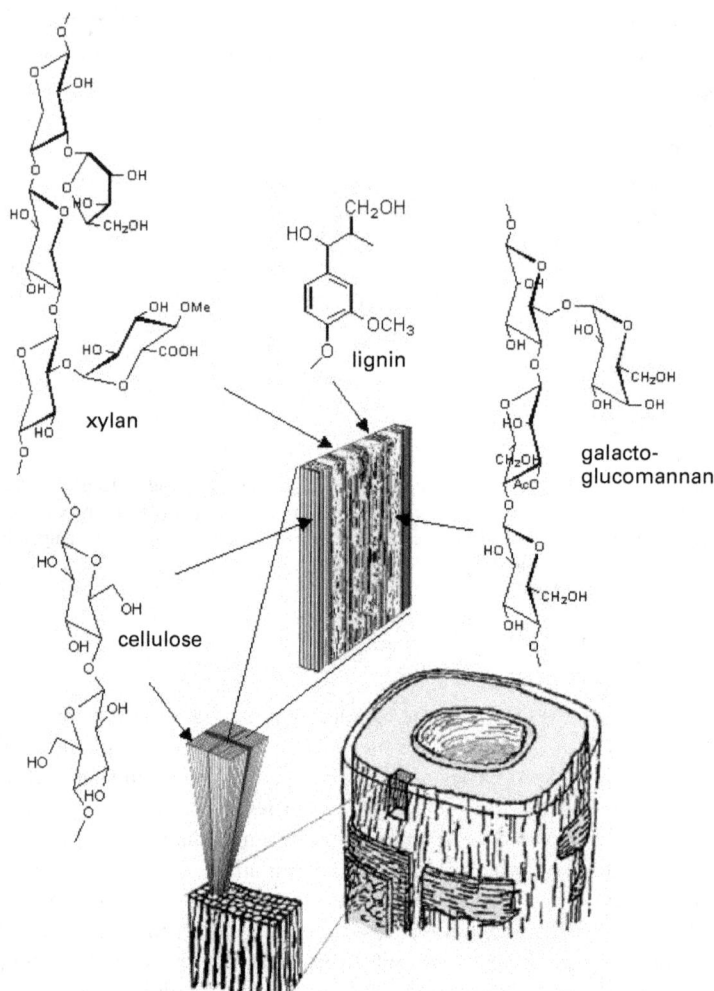

Figure 2.8. Schematic picture of the arrangement of the wood polymers, cellulose, lignin and hemicelluloses in the secondary wall structure of wood fibres. All of the wood polymers have an orientation with the main axis in the direction of the fibre, i.e. following the direction of the cellulose fibrils. The lignin is also oriented with the plane of the molecule in a transverse direction to the fibre wall (Åkerholm and Salmén 2003).

for cellulose I, as well as the experimentally easier material cellulose II, a reasonable value of a modulus seems to be ca 134–136 GPa for the stiffness of cellulose I (Kroon-Batenburg et al 1986, Sakurada et al 1962).

Hemicellulose

For hemicelluloses material data are extremely scarce. Cousins' data on extracted xylan fractions (Cousins 1978), which might be considered as representative indicate a random value of the elastic modulus for the dry material of 8 GPa. However it is known that hemicelluloses will crystallize when isolated which would affect the modulus determinations. It is also known that the hemicelluloses at least the glucomannan is oriented along the cellulose fibrils within the fibre wall (Liang et al 1960, Keegstra et al 1973, Åkerholm and Salmén 2001). This means that the in-situ hemicelluloses might have a bulk modulus that is lower than that determined by Cousins but that the orientation will mean that it is stiffer in the fibre direction and much weaker in its transverse direction.

Lignin

Also when it comes to lignin data only exists for extracted materials, which then have been severely modified compared to the original material inside the cell wall. Under dry conditions values for an isotropic lignin between 4 to 7 GPa have been reported (Cousins 1976, Cousins et al 1975). In the native cell wall it is though highly probable that also the lignin is orthotropic with a preferred orientation in the direction of the cellulose fibrils (Åkerholm and Salmén 2003). Data given in *Table 2.3* of the elastic properties of the wood polymers may be seen as reasonable estimates. Where the values for cellulose are highly probable while the values for lignin and hemicellulose should only be considered as in the right range indicating their orthotropic nature as well as their moisture sensitivity.

Table 2.3. Estimates of mechanical data for wood polymers.

Polymer		Dry conditions	Moist conditions
Cellulose			
	E_x	134 GPa	134 GPa
	E_y	27.2 GPa	27.2 GPa
	G	4.4 GPa	4.4 GPa
	u	0.1	0.1
Hemicellulose			
	E_x	4.0 GPa	40 MPa
	E_y	0.8 GPa	8 MPa
	G	1.0 GPa	10 MPa
	u	0.2	0.2
Lignin$_x$			
	E_x	2.0 GPa	2.0 GPa
	E_y	1.0 GPa	1.0 GPa
	G	0.6 GPa	0.6 GPa
	u	0.3	0.3

Moisture sorption characteristics

All of the wood polymers are hygroscopic and thus readily absorbs moisture from the surrounding air in relation to its relative humidity, RH. The extent of moisture sorbed is directly associated with the number of hydrophilic sites, i.e. the hydroxyl, OH-groups and the carboxylic COOH-acid groups. For sulphonated fibres also sulphonic SO_3H-acid groups do adsorb water. The acidic groups adsorb considerably more water around them as compared to the hydroxyl groups. Also, for these acidic groups the number of water molecules adsorbed depend on the conterion so that for the carboxylic acid groups sodium (a monovalent ion) sorbs more than calcium (a divalent ion) which sorbs more than the acidic proton form while for the sulphonic acid groups the acidic proton form sorbs more than the sodium which sorbs more than the calcium (Berthold et al 1996). For the normal wood fibre the content of acidic groups are though rather low so that only at the highest relative humidities effects are visible. Of more importance is the relative composition of the fibre as can be seen from *Figure 2.9* showing the absorption for the different wood polymers as well as for TMP-fibres (Takamura 1968). The lignin has a lower content of hydroxyl groups as compared to the hemicelluloses while the lower adsorption for the cellulose is due to its crystalline/paracrystalline regions, which are not adsorbing any water. The adsorbed amount of moisture for the pulp fibre may as seen be accurately calculated from the sum of adsorption for its components, cellulose, hemicellulose and lignin, knowing the relative amount of these in the fibre (Takamura 1968). It is also obvious, the parallel sorption curves, that it is the same mechanisms, i.e. that of the sorption by the hydroxyl which are dominating the behaviour for the wood polymers.

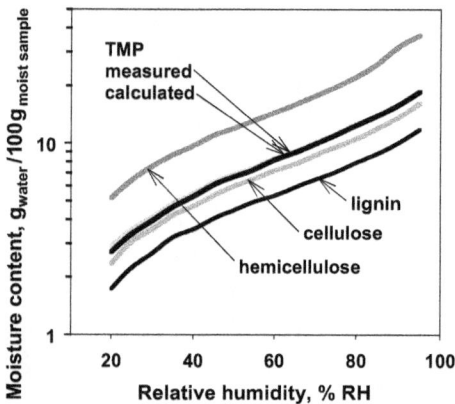

Figure 2.9. Moisture content (gram water per 100 gram of moist sample) as a function of RH for the wood polymers; lignin, cellulose, hemicelluloses as well as for a TMP of pine. The calculated relationship is based on the relative amount of the wood polymers in the TMP and the absorption curves for these. Redrawn from data of Takamura (Takamura 1968).

2.2.6 Softening

As the constituents of the wood fibre are polymeric components their properties are highly dependable on their softening temperature, the glass transition temperature. In completely dry conditions all of the wood polymers softens at very high temperatures (Back and Salmén 1982), 230 °C for cellulose, 205 °C for lignin and around 170 to 180 °C for hemicelluloses as illustrated in *Figure 2.10* by the drop in elastic modulus for papers of different composition (Salmén 1979). This is for the carbohydrates partly a reflection of the high hydrogen bonding capacity.

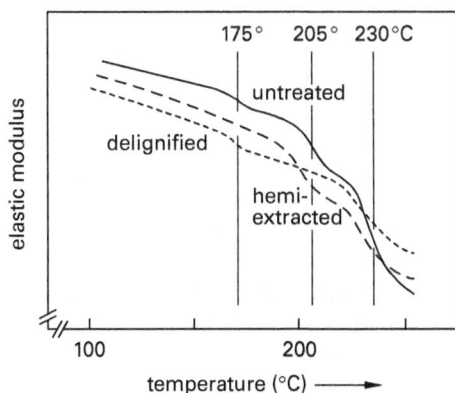

Figure 2.10. The identification of separate glass transition temperatures for hemicellulose at 175 °C, for lignin at 205 °C and for disordered cellulose at 230 °C in the elastic modulus –temperature curves from papers selectively extracted of either lignin or hemicelluloses respectively (Salmén 1979).

This fact makes the wood polymers also highly sensitive to their moisture content as moisture functions as a softener, a low molecular weight component lowering the softening temperature, for them, *Figure 2.11* (Salmén 1990, Salmén 1982). Thus for the hemicelluloses the softening temperature is lowered even below zero at high moisture contents, approaching 100 % RH,

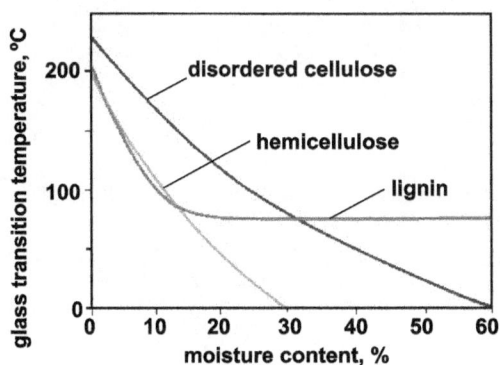

Figure 2.11. Glass transition temperature of wood polymers as a function of moisture content (Salmén 1990). Lignin has a limiting reduction of the glass transition due its three-dimensional structure. For cellulose the crystalline restrictions shifts the transition in the way that higher moisture contents are needed for the required mobility to result in a softening.

while for lignin the branched structure limits the transition temperature to about 70 °C under fully wet conditions (Salmén 1984). The cellulose disordered regions makes the softening of these not particularly pronounced and for conditions where it may influence properties its transition is most likely shifted to higher moisture contents due to the restrictions from crystalline regions as schematically shown in *Figure 2.11*.

2.3 Fibre Properties

The mechanical properties of wood fibres are as discussed in section 2.2 highly dependent on the stiffness of the cellulose fibrils and the arrangement of these in the thickest cell wall layer the S_2-layer. When isolating the fibres, especially the earlywood fibres of chemical pulps made from softwood they will tend to collapse, as illustrated in *Figure 2.12* while latewood fibres and thick-walled hardwood fibres more or less tend to maintain their shape. In softwoods the earlywood fibres dominate in terms of number of fibres and the collapse favours a structure giving much higher flexibility of the fibres a phenomena that is improved with the extent of the pulping and beating processes. In the wet, swollen state, the differences in fibre flexibility is not that pronounced, *Figure 2.13*, although earlywood fibres are more flexible than latewood fibres (Hattula and Niemi 1988). The most significant factor influencing the wet fibre flexibility is the yield of the pulp. With the dissolution of lignin and the creation of pores in the cell wall these will make the fibre to appear as a delaminated structure under wet conditions, which renders its high flexibility. With beating this delamination of the fibre wall is further enhanced but the effect is much smaller than the yield effect (Hattula and Niemi 1988).

latewood fiber   earlywood fiber

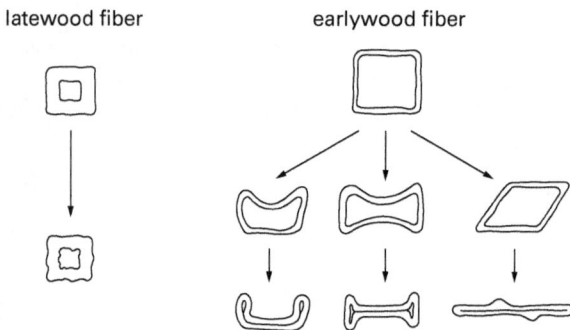

Figure 2.12. Schematic drawings of latewood and earlywood fibres and their collapse modes as a result of drying: after Jayme and Hungher (Jayme and Hunger 1961).

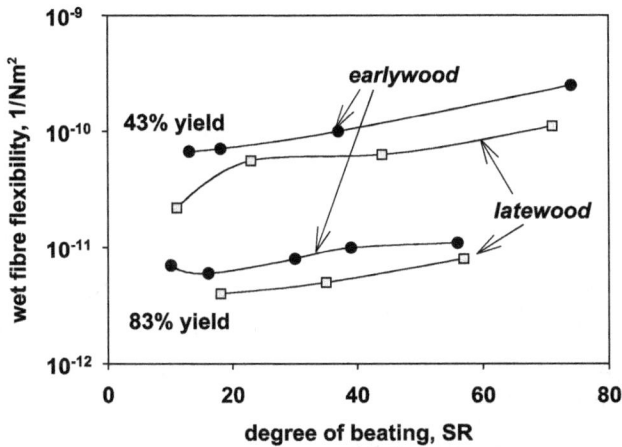

Figure 2.13. Wet fibre flexibility of softwood spruce Kraft fibres as a function of beating degree for two pulps of different yield as determined by the method of Tam Doo and Kerekes (Tam Doo and Kerekes 1981). Earlywood fibres shows a higher wet fibre flexibility than latewood fibres (Hattula and Niemi 1988).

2.3.1 Stress Strain Properties

The tensile stress-strain properties of wood fibres are highly dependent on the angle of the cellulose fibrils within the S_2-layer, the fibril angle a shown in *Figure 2.14* for data on chemical pulp fibres of black spruce. For small angles less than 10 degrees the fibres are highly elastic,

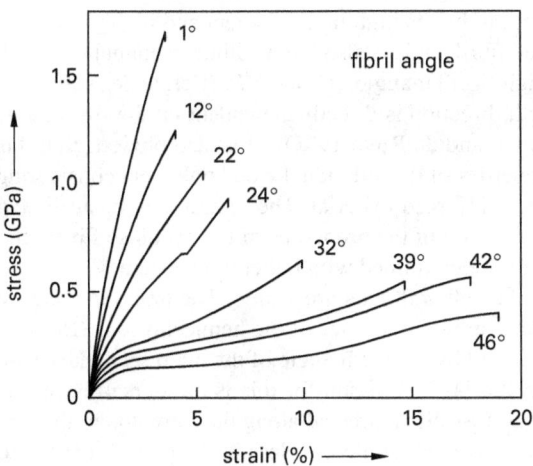

Figure 2.14. Stress-strain curves of single chemical pulp fibres of black spruce. Redrawn from Page and El-Hosseiny (Page and El-Hosseiny 1983). At low fibril angles the cellulose fibrils gives the high stiffness and strength of the fibre while at the high fibril angles the shearing of the matrix and an unwinding of the structure gives the extremely high extensibility of the fibre.

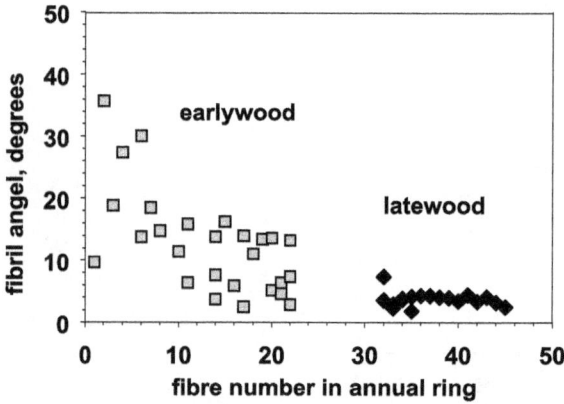

Figure 2.15. The fibril angle for individual fibres in an annual ring in the mature wood of a spruce wood tree measured with confocal laser scanning microscopy (Bergander 2001). In the earlywood the fibril angle various considerably while it in the latewood fibres is rather constant.

have high strength values and show a more or less brittle behaviour. For high fibril angles the behaviour is more like that of a ductile material exhibiting very high straining levels. Due to the high fibril angle for these fibres most of the deformation probably occurs due to a flow of the hemicellulose/hemicellulose-lignin matrix resulting in an orientation of the cellulose fibrils in the longitudinal fibre direction, i.e. in the direction of straining. Most of the fibres from soft-wood that grows in the Nordic countries have though a fibril angle below 15° thus being more or less elastic. Juvenile wood fibres probably show somewhat of a ductile behaviour with a fibril angle in the range of 15 to 30°, while the behaviour of compression wood fibres with a fibril angle reaching up to 30 to 45° may be different due its high lignin content and highly different structure of the cell wall. The variation of the fibril angle is also large within an annual ring with the first earlywood fibres having a rather high fibrillar angle, *Figure 2.15* (Bergander 2001).

The elastic modulus of fibres in its length direction is directly dependent on the fibril angle of the S_2-layer as seen in *Figure 2.16* (Salmén and de Ruvo 1985). It is also obvious that this modulus is very well predicted from the properties of the individual wood polymers considering the cell wall as a laminated composite, fibre reinforced material. The reason may be attributed to the high dependence in the longitudinal direction of the properties of the cellulose fibrils and that the elastic properties of cellulose have been determined with rather high accuracy.

For the properties in other directions of the cell wall data are scarce. The transverse elastic modulus of the fibre wall is highly dependent on the properties of the hemicelluloses (Bergander and Salmen 2002) but is hardly at all affected by the fibril angle of the main secondary wall the S_2-wall as seen in *Figure 2.17* (Salmén 2001). The reason for this is of course that in the transverse direction the cellulose fibrils of the S_2-wall is directed along the fibre due to the low fibril angle implying that the transverse forces are acting more or less in the cross direction to the cellulose fibrils.

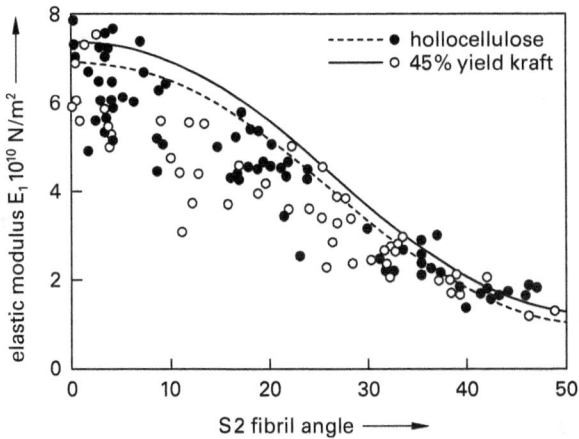

Figure 2.16. The elastic modulus of chemically pulped softwood fibres in the longitudinal direction as a function of the fibril angle in the S_2 wall (Page et al 1977). The curves in the figure are based on model calculations considering the cell wall as a laminate built up of cellulose reinforcements in a matrix of lignin and hemicelluloses using the mechanical properties of the wood polymers as established for isolated materials (Salmén and de Ruvo 1985).

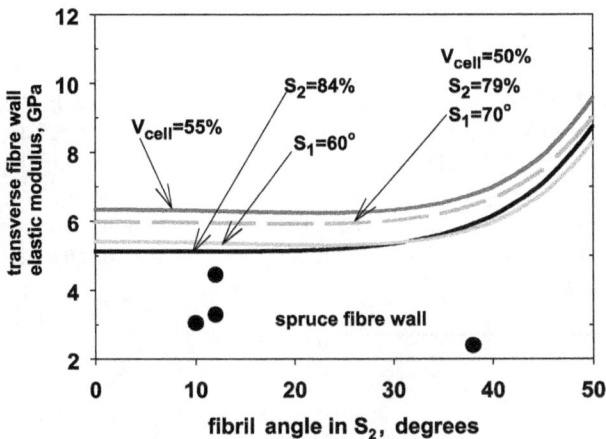

Figure 2.17. The transverse elastic modulus of softwood fibres as calculated considering the cell wall as a laminate built up of cellulose reinforcements in a matrix of lignin and hemicelluloses using the mechanical properties of the wood polymers as established for isolated materials plotted as a function of the fibril angle in the S_2-layer for some cases of defined composition of the cell wall (Salmén 2001). V_{cell} refers to the relative amount of cellulose in the fibre, S_2 to the relative volume of this cell wall and S_1 to the fibril angle of this layer. Also data from measurements on radial sections of the cell wall tested in the transverse direction on a micro-tensile stage in an ESEM are included (Bergander and Salmén 2000).

The morphological differences between earlywood and latewood fibres are of course most obviously related to the differences in cell wall thickness. Thus latewood fibres have a higher tensile stiffness and a higher breaking load than earlywood fibres. Generally also the elastic modulus and the breaking stress (quantities which take the cross sectional area in consideration) are higher for the latewood fibres as a consequence of a smaller fibril angle and a higher percentage of the S_2-layer. As a consequence the elastic modulus in the transverse direction is lower for latewood fibres than for earlywood fibres.

When straining a fibre to high degrees it will tend to twist due to its helical fibrillar structure. This is a well-known phenomenon for tubular helical structures where the axis of the loading in its length direction is not coinciding with the main symmetry axis of the material. Thus a free fibre subjected to high tensile forces will twist and buckle to a shape as illustrated in *Figure 2.18*. Within the paper sheet structure such deformations are not possible but instead forces are built up from this tendency of the fibre structure to twist.

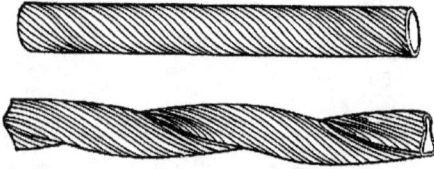

Figure 2.18. A helically wound tube, a fibre, subjected to high tensile forces will tend to deform by buckling and twisting as the material symmetry axis do not coincide with the loading direction.

2.3.2 Effects of Moisture and Temperature

With increasing temperature the wood polymers softens and this is of course reflected in the mechanical properties of the fibres as seen for the decrease of the torsional stiffness of Kraft fibres in *Figure 2.19* (Kolseth and Ehrnrooth 1986). The overall decreasing stiffness reflects the loss in hydrogen bond and van der Waals forces between the molecules with the increasing temperature and in relation to those the effect of the hemicellulose softening in the region of 150 to 200 °C is very small. This however reflects the fact that in the wood fibre it is the cellulose fibrils that are the load taking components.

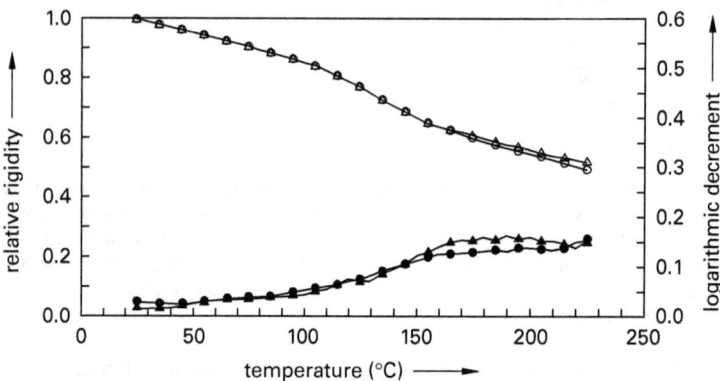

Figure 2.19. Relative rigidity, relative torsional stiffness, (unfilled symbols) and logarithmic decrement (filled symbols) versus temperature for two high-yield Kraft pulp fibres; triangles = 50 % yield, 8 % lignin content, circles = 63.5 % yield, 21.4 % lignin content (Kolseth and Ehrnrooth 1986).

As moisture functions as a softener for the wood polymers it is not surprising that increased moisture content, or relative humidity leads to a decreasing stiffness of the wood fibre, *Figure 2.20* (Kolseth 1983). The magnitude of the loss in stiffness is very much what can be expected as an effect of the loss of stiffness of the hemicelluloses.

It is noteworthy that the magnitude of the decrease of the fibre stiffness caused by the softening of the hemicelluloses is within reasonable limits independent of the fibre composition as both shown experimentally and by theoretical calculations (Kolseth and Ehrnrooth 1986; Salmén et al 1985). The increased damping (logarithmic decrement) reflecting the viscous contribution from hemicellulose softening shows only a moderate increase at the higher relative humidities which also is a consequence of the strong dependence of the cellulose fibrils in the wood cell wall structure. The relative modulus decrease can anyhow very well be modelled from the data on the drop of stiffness of the hemicellulose polymer, a drop amounting to several decades in stiffness, with consideration of the laminar, reinforced structure of the fibre wall. As seen in *Figure2.21* the decrease in stiffness is much higher in the transverse direction of the fibre than in its longitudinal direction, which reflects the orientation of the cellulose fibrils. Along the fibre the cellulose fibrils have a stronger reinforcing capability than in the transverse direction of course as a reflection of the orientation of the fibrils, the low fibril angle in the S_2-layer.

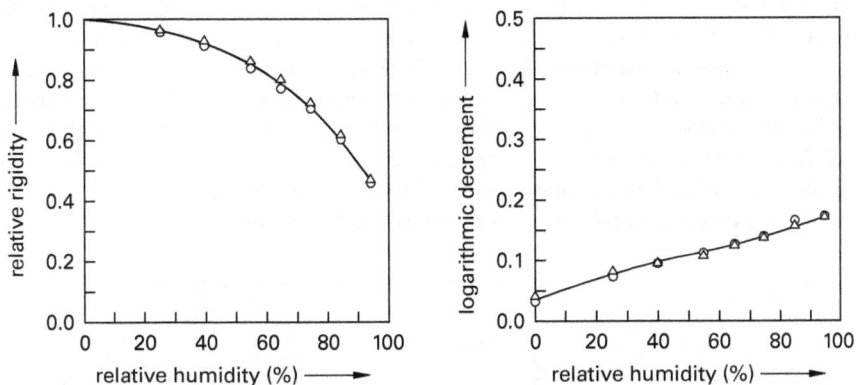

Figure 2.20. The relative rigidity (torsional stiffness) and logarithmic decrement (damping) versus relative humidity for two high-yield Kraft pulp fibres; triangles = 50% yield, 8% lignin content, circles = 63.5% yield, 21.4 % lignin content (Kolseth 1983).

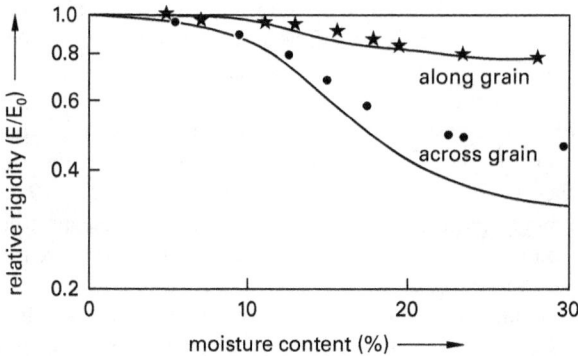

Figure 2.21. The relative decrease in elastic modulus due to an increase in relative humidity along fibres (along grain) and transverse to fibres (across grain) reflecting the softening of the hemicelluloses (Salmén 1986). Data of the relative modulus of modulus of spruce according to data of Carrington (Carrington 1922):. The lines show the calculated changes assuming hemicellulose softening according to data of Cousins (Cousins 1978).

Due to the angle that the fibrils in the secondary wall S_2, the main structure of the cell wall, makes with the fibre axis there will be torsional forces created during moisture changes due to the contraction/expansion within the hemicellulose rich matrix. This is how an asymmetric laminate behaves, which the fibre resembles as soon as the fibre form diverts from being cylindrical, i.e. when it has collapsed to some extent. With increasing relative humidity the expansion of the hemicelluloses results thus in a twist of the fibre amounting to several turns for an entire fibre, *Figure 2.22* (Kolseth 1983). Such a twisting of fibres is of special concern when flash dying pulps. However it also implies that in the paper sheet stresses are built up due to the restraining of the twisting during drying of the paper from the network structure.

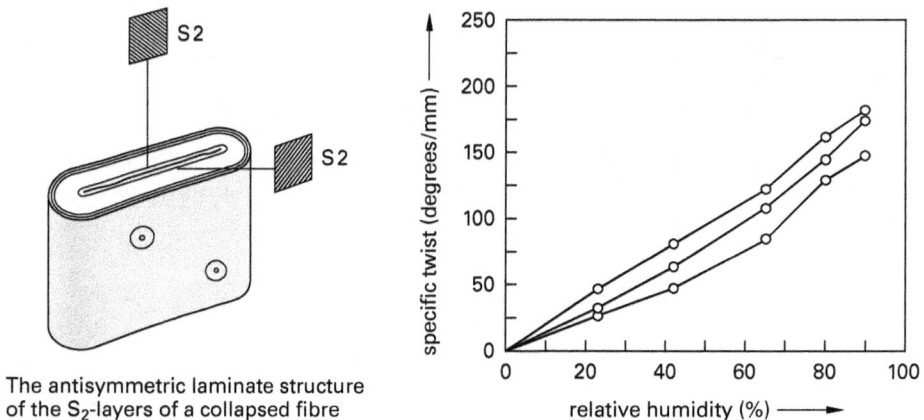

The antisymmetric laminate structure of the S_2-layers of a collapsed fibre

Figure 2.22. The fibre structure of a collapsed fibre exposing cellulose fibres in opposite angles in the two secondary walls thus composing an antisymmeric laminate, a structure that will cause twisting due to expansion of the matrix material between the reinforcing fibrils. This twisting is given as the specific twist in degrees/mm length for a Kraft fibre as a function of relative humidity (Kolseth 1983).

2.3.3 Effects of Drying Stresses

When fibres are loaded under axial tension during drying the normal shrinkage of the fibre will be hindered resulting in an orientation of the structure as illustrated in *Figure 2.23*. This will result in a fibre exhibiting higher stiffness and higher breaking load, *Figure 2.24* (Jentzen 1964). At the same time the strain to failure decreases up to 40 %. The phenomenon has been referred to as an effect of "drying stresses" or "dried-in stresses".

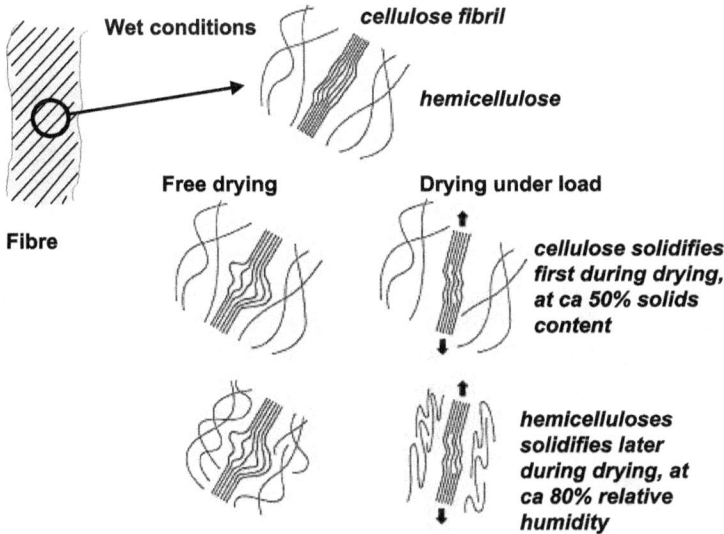

Figure 2.23. Schematic picture of the effect of the load on a fibre during drying on the structural arrangement of the cellulose and hemicelluloses. When subjected to a load during drying both the cellulose disordered regions and the hemicelluloses will be more aligned in the direction of stress. On the macroscopic scale the fibrils will also align so that the fibril angle will diminish.

If the fibre on the other hand is compressed during the drying the stiffness and the strength decreases while the strain to failure increases considerably, *Figure 2.25* (Dumbleton 1972). When straining the fibre it will be permanently deformed, much more so for a fibre prior subjected to compressive forces, *Figure 2.25*. With each subsequent cycle also the stiffness of the fibre is increased. What happens is a form of strain hardening (or strain weakening in compression). At the molecular level the tensile strain will result in a rearrangement of the cellulose disordered regions and of the hemicellulose molecules so that they will be more aligned with the stress, see *Figure 2.23*. The opposite of course applies to compressive forces. At the macroscopic level the straining results in an orientation of the cellulose fibrils in the direction of the straining. For the rearranging to take place the fibre matrix material within the fibre has to be somewhat pliable. This is why so large effects may occur due to loads during drying but also that the permanent deformation is larger at higher relative humidities. To a high degree the phenomenon is related to the solidification of the wood polymers that locks in the deformation. If the fibre was unloaded under wet conditions most of the deformation would be lost due to the shrinkage of the fibre.

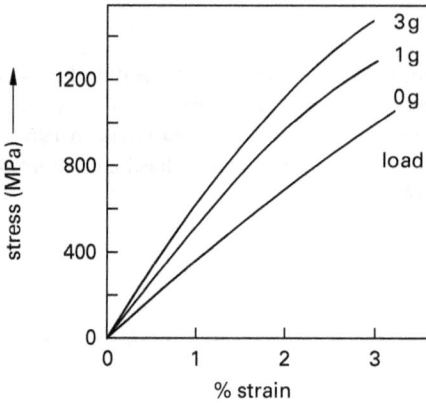

Figure 2.24. Effect of drying restraints on the stress-strain curve of fibres from never-dried latewood holocellu-lose fibres of long leaf pine (Jentzen 1964). The higher the load on the fibre the higher is the stiffness and the breaking load of the fibre.

Therefore whenever a fibre is subjected to higher humidity dried in stresses will be released the more so the higher the humidity (most stresses will of course be released for a fibre that is completely wetted). The development of dried in stresses is an essential part of the papermaking operation to obtain the desired strength properties of the paper product.

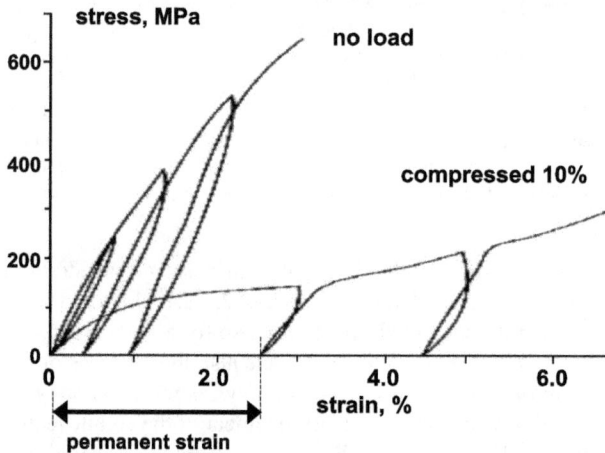

Figure 2.25. Load – unloading cycles for holocellulose latewood fibres dried either freely or compressed 10%, after Dumbleton (Dumbleton 1972).

2.4 References

Agarwal, U.P. and Atalla, R.H. (1986) In-situ Raman microprobe studies of plant cell walls: Macromolecular organization and compositional variability in the secondary wall of Picea mariana (Mill.). *B.S.P. Planta* 169: 325–332.

Åkerholm, M. and Salmén, L. (2001) Interactions between wood polymers studied by dynamic FT-IR spectroscopy. *Polymer.* 42(3): 963–969.

Åkerholm, M. and Salmén, L. (2003) The mechanical role of lignin in the cell wall of spruce tracheids. *Holzforsch.*

Atalla, R.H. (1998) Cellulose and the hemicelluloses: patterns for the assembly of lignin. In: N.G. Lewis and S. Sarkanen (Eds.) *Lignin and lignan biosynthesis*. Washington DC: Am. Chem. Soc., p. 172–179.

Atalla, R.H., Hackney, J.M., Uhlin, I., and Thompson, N.S. (1993) Hemicelluloses as structure regulators in the aggregation of nativ cellulose. *Int. J. Biol. Macromol.* 15: 109–112.

Back, E.L. and Salmén, N.L. (1982) Glass transitions of wood components hold implications for moulding and pulping processes. *Tappi* 65(7): 107–110.

Bergander, A. (2001) Local variability in chemical and physical properties of spruce wood fibers. In: *Pulp and paper chemistry and technology.* Stockholm: KTH.

Bergander, A. and Salmén, L. (2000) Variations in transverse fiber wall properties: Relations between elastic properties and structure. *Holzforschung* 54(6): 654–660.

Bergander, A. and Salmén, L. (2001) Cell wall properties and their effects on the mechanical properties of fibers. *J. Material Sci.*

Bergander, A. and Salmen, L. (2002) Cell wall properties and their effects on mechanical properties of fibers. *J. Material. Sci.* 37(1): 151–156.

Berthold, J. and Salmén, L. (1997) Effects of mechanical and chemical treatments on the pore size distribution in wood pulps examined by Inverse Size Exclusion Chromatography (ISEC). *J. Pulp Paper Sci.* 23(6): J245–J253.

Berthold, J., Rinaudo, M., and Salmen, L. (1996) Association of water to polar groups; estimations by an adsorption model for ligno-cellulosic materials. *Colloids Surfaces.* 112(2/3): 117–129.

Carrington, H. (1922) The elastic constants of spruce as influenced by moisture. *Aeronaut. J.* 26: 462–471.

Cousins, W.J. (1976) Elastic modulus of lignin as related to moisture content. *Wood Sci. Technol.* 10: 42979.

Cousins, W.J. (1978) Young's modulus of hemicellulose as related to moisture content. *Wood Sci. Technol.* 12: 161–167.

Cousins, W.J., Armstrong, R.W., and Robinson, W.H. (1975) Young's modulus of lignin from a continuous indentation test. *J. Materials Sci.* 10: 1655–1658.

Duchesne, I. and Daniel, G. (2000) Changes in surface ultrastructure of Norway spruce fibres during kraft pulping - visualisation by field emission-SEM. *Nordic Pulp Paper Res. J.* 15(1): 54–61.

Dumbleton, D.F. (1972) Longitudinal compression of individual pulp fibres. *Tappi.* 55(1): 127–135.

Fahlén, J. and Salmén, L. (2003) Cross-sectional structure of the secondary wall of wood fibers as affected by processing. *J. Mater. Sci.*

Favis, B.D. and Goring, D.A.I. (1984) A model for the leaching of lignin macromolecules from pulp fibres. *J Pulp Paper Sci.* 10(5): J139–J143.

Favis, B.D., Yean, W.Q., and Goring, D.A.I. (1984) Molecular weight of lignin fractions leached from unbleached kraft pulp fibers. *J. Wood Chem. Technol.* 4(3): 313–320.

Fengel, D. (1970) Ultrastructural behavior of cell wall polysaccharides. *Tappi* 53(3): 497–503.

Fengel, D. and Stoll, M. (1973) On the variation of the cell cross area, the thickness of the cell wall and of the wall layers of sprucewood trachieds within an annual ring. *Holzforsch.* 27(1): 1–7.

Gillis, P.P. (1969) Effect of hydrogen bonds on the axial stiffness of crystalline native cellulose. *J. Polym. Sci.* A2(7): 783–794.

Hattula, T. and Niemi, H. (1988) Sulphate pulp fibre flexibility and its effect on sheet strength. *Paperi ja Puu.* (2): 356–361.

Jayme, G. and Hunger, G. (1961) Electron microscope 2- and 3-dimensional classification of fibre bonding. In: F. Bolam (Ed.) *The formation and structure of paper.* Oxford: Tech. Sec B.P. and B.M.A.

Jentzen, C.A. (1964) The effect of stress applied during drying on some of the properties of individual pulp fibres. *Tappi* 47(7): 412–418.

Kaplan, D.S. (1976) Structure-property relationships in copolymers to composites: Molecular interpretation of the glass transition phenomenon. *J. Applied Polymer Science* 20: 2615–2629.

Keegstra, K., Talmadge, K.W., Bauer, W.D., and Albersheim, P. (1973) The structure of plant cell walls III. A model of the walls of suspension-cultivated sycamore cells based on the interconnections of the macromolecular components. *Plant Physiol.* 51: 188–196.

Kerr, A.J. and Goring, D.A.I. (1975) The ultrastructural arrangements of the wood cell wall. *Cellulose Chemistry and Technology* 9: 536–573.

Kolseth, P. (1983) Torsional properties of single wood pulp fibers, Ph. D. In: *Department of Paper Technology, The Royal Institute of Technology.* Stockholm: Department of Paper Technology, The Royal Institute of Technology, p. 24.

Kolseth, P. and Ehrnrooth, E.M.L. (1986) Mechanical Softening of Single Wood Pulp Fibers. In: J.A. Bristow and P. Kolseth (Eds.) *Paper Structure and Properties.* New York: Marcel Dekker Inc., p. 27–50.

Kroon-Batenburg, L.M., Kroon, J., and Norholt, M.G. (1986) Chain modulus and intramolecular hydrogen bonding in native and regenerated cellulose fibers. *Polym. Comm.* 27: 290–292.

Larsson, P.T., Wickholm, K., and Iversen, T. (1999) Structural elements in cellulose I10 th International symposium on wood and pulping chemistry. Yokohama. *Japan TAPPI.*

Liang, C.Y., Bassett, K.H., McGinnes, E.A., and Marchessault, R.H. (1960) Infrared spectra of crystalline polysaccharides VII Thin wood sections. *Tappi* 43(12): 1017–1024.

Meier, H. (1961) The distribution of polysaccharides in wood fibers. *J. Polymer Sci.* 51: 11–18.

Page, D.H. and El-Hosseiny, F. (1983) The mechanical properties of single wood pulp fibres. Part VI, Fibril angle and the shape of the stress-strain curve. *J. Pulp Pap. Sci., Trans Techn Sect.* 9(4): TR 99–100.

Page, D.H., El-Hosseiny, F., Winkler, K., and Lancaster, A.P.S. (1977) Elastic modulus of single wood pulp fibers. *Tappi J.* 60(4): 114–117.

Panshin, A.J. and Zeeuw, C.D. (1980) *Textbook of wood technology.* New York: McGraw-Hill.

Sakurada, I., Nukushina, Y., and Ito, T. (1962) Experimental determination of the elastic modulus of crystalline regions in oriented polymers. *J. Polym. Sci.* 57: 651–660.

Salmén, L. (1982) Temperature and water induced softening behaviour of wood fiber based materials. In: *Paper technology.* Stockholm: KTH.

Salmén, L. (1984) Viscoelastic Properties of in situ Lignin under Water-Saturated Conditions. *J. Materials. Sci.* 19: 3090–3096.

Salmén, L. (1986) The Cell Wall as a Composite Structure. In: J.A. Bristow and P. Kolseth (Eds.) *Paper Structure and Properties*. New York.: Marcel Dekker Inc., p. 51–73.

Salmén, L. (1990) On the interaction between moisture and wood fiber materials. In: D.F. Caulfield, J.D. Passaretti, and S.F. Sobczynski (Eds.) *Materials Interactions Relevant to the Pulp, Paper, & Wood Ind.* Pittsburgh.: Materials Res. Soc., p. 193–201.

Salmén, L. (2000) Structure-property relations for wood; from cell-wall polymeric arrangement to the macroscopic behavior. In: H.-C. Spatz and T. Speck (Eds.) *Plant Biomechanics 2000.* Stuttgart.: Georg Thieme Verlag, p. 452–462.

Salmén, L. and de Ruvo, A. (1985) A model for the prediction of fiber elasticity. *Wood Fiber Sci.* 17(3): 336–350.

Salmén, L. and Olsson, A.-M. (1998) Interaction between hemicelluloses, lignin and cellulose: structure-property relationships. *J. Pulp Pap. Sci.* 24(3): 99–103.

Salmén, L. (2001) Micromechanics of the wood cell wall: a tool for a better understanding of its structure. In: P. Navi (Ed.) *First International Conference of the European Society for Wood Mechanics.* Lausanne, Switzerland: EPFL.

Salmén, L., Kolseth, P., and de Ruvo, A. (1985) Modeling the softening behaviour of wood fibres. *J. Pulp Paper Sci.* 11(4): J102–J107.

Salmén, N.L. (1979) Thermal softening of the components of paper: its effect on mechanical properties. *Trans. Tech. Sect. (Can. Pulp Pap. Assoc.).* 5(3): 45–50.

Schick, C. and Donth, E. (1981) Characteristic length of glass transition; experimental evidence. *Physica Scripta* 43: 423–429.

Stone, J., Scallan, A.M., and Ahlgren, P.A.V. (1971) The ultrastructural distribution of lignin in tracheid cell walls. *Tappi* 54: 1527–1530.

Stone, J.E. and Scallan, A.M. (1967) The effect of component removal upon the porous structure of the cell wall of wood. II Swelling in water and the fiber saturation point. *Tappi* 50(10): 496–501.

Takamura, N. (1968) Studies on hot pressing and drying process in the production of fiberboard. III. Softening of fibre components in hot pressing of fibre mat. *J.Japan Wood Res. Soc.* 14(2): 75–79.

Tam Doo, P.A. and Kerekes, R.J. (1981) A method to measure wet fiber flexibility. *Tappi* 64(3): 113–116.

Tashiro, K. and Kobayashi, M. (1991) Theoretical evaluation of three-dimensional elastic constants of native and regenerated celluloses: role of hydrogen bonds. *Polymer.* 32(8): 1516–1526.

Terashima, N., Fukushima, K., He, L.-F., and Takabe, K. (1993) Comprehensive model of the lignified plant cell wall. In: H.G. Jung, D.H. Buxton, R.D. Hatfield, and J. Ralph (Eds.) *Forage cell wall structure and digestibility*. Madison, WI.: ASA-CSSA-SSSA, p. 247–270.

Westermark, U., Lidbrandt, O., and Eriksson, I. (1988) Lignin distribution in spruce (Picea abies) determined by mercurization with SEM-EDXA technique. *Wood Sci. Technol.* 22: 243–250.

Willis, J.M., Yean, W.Q., and Goring, D.A.I. (1987) Molecular weights of lignosulphonate and carbohydrate leached from sulphite chemimechanical pulp. *J. Wood Chem. Technol.* 7(2): 259–268.

3 Interactions between Fibres and Water and the Influence of Water on the Pore Structure of Wood Fibres

Lars Wågberg
Department of Fibre and Polymer Technology, KTH

3.1 Thermodynamics for the Interaction between Water and Fibres

3.1.1 Background

Wood fibres are in their natural environment always surrounded by water in different forms, i.e. liquid water, ice or water vapour. The reason to this is that the water is indirectly needed for the photosynthesis since chlorophyll is water-soluble and that water-soluble nutrients are supplied from the soil. Due to this the fibres have to interact with water in such a way that the water is transported from the roots to the leaves/needles and on to the cambial layer of the stem. One result of this is that most components of the fibres, i.e. cellulose, hemicellulose and lignin, are more or less prone to interact with water and this has several consequences

a) It is possible to liberate the fibres from each other through processing in aqueous media
b) It is possible to form different types of papers from fibres in aqueous media
c) The fibres are in the final products sensitive towards moist air and liquid water

The first two items are naturally very positive since organic solvents can be totally avoided during processing but the last item is naturally a large drawback for many of the products formed from wood fibres. In order to preserve the inherent positive properties of the fibres different treatments such as sizing and lamination with different types of water repellent plastics are commonly used. These treatments are naturally more or less successful for the protection of the fibres and the fact still remains that wood fibres will imbibe water even if they are embedded in a matrix of water repellent plastics in a composite.

Usually different components of the fibre wall are described as hydrophobic or hydrophilic but these terms are very general and in fact rather useless for the description of the properties of fibres. In the present chapter a thermodynamic description of the interaction between water and fibres has instead been adopted for the quantification of the fibre/water interaction. The result is a quantification of fibre water interaction in energy terms and also in terms that can distinguish between different driving forces for the moisture adsorption.

3.1.2 Thermodynamics for Water Vapour Sorption to Fibres

The Change in Gibbs Free Energy upon Water Sorption

The water sorption to fibres will be a spontaneous process as long as the integral change in Gibbs free energy associated with the sorption process, ΔG, has a negative value. This means that Gibbs free energy after absorption must be lower than before adsorption. In mathematical terms the integral change in Gibbs free energy associated with the sorption process can be described by equation (3.1)

$$\Delta G = n_1 \overline{\Delta G_1} + n_2 \overline{\Delta G_2} \qquad (3.1)$$

where

n_1 = moles of volatile liquid
n_2 = moles of non-volatile solid

$\overline{\Delta G_1}$ and $\overline{\Delta G_2}$ = partial molar free energy of the volatile and non-volatile components respectively.

The differential energy change is the difference between the chemical potential of the liquid in the adsorbed state and the pure liquid and this can be written as, equation (3.2)

$$\overline{\Delta G_1} = \frac{\partial(\Delta G)}{\partial n_1} = \mu_1 - \mu_0 = RT \ln \frac{P}{P_0} \qquad (3.2)$$

where

μ_1 = chemical potential of the liquid in the absorbed state
μ_0 = chemical potential of the free liquid
P_0 = vapour pressure of the pure liquid
P = actual vapour pressure
T = absolute temperature in K

According to the Gibbs-Duhem equation it is also known that, equation (3.3)

$$\overline{\Delta G_2} = \frac{\partial \Delta G}{\partial n_2} = -\int \frac{n_1}{n_2} \, \mathrm{d}\overline{\Delta G_1} = \int_0^{P/P_0} \frac{RT \, n_1 P_0}{P} \, \mathrm{d}\left(\frac{P}{P_0}\right) \qquad (3.3)$$

By combining equations (3.1)–(3.3) the following expression for the integral change in Gibbs free energy as a function of the absorption isotherm, i.e. the amount of adsorbed volatile liquid (n_1) as a function of the relative vapour pressure (P/P_0), can be achieved as, equation (3.4)

$$\Delta G = n_1 \, RT \ln\left(\frac{P}{P_0}\right) - RT \int_0^{P/P_0} \frac{n_1 P_0}{P} \, \mathrm{d}\left(\frac{P}{P_0}\right) \qquad (4)$$

By plotting $n_1 \, P/P_0$ as a function of P/P_0 it is possible to graphically evaluate the integral to the right hand side of equation (3.4) since

$$-RT \int_0^{P/P_0} n_1 \, \mathrm{d}\ln\left(\frac{P}{P_0}\right) \equiv -RT \int_0^{P/P_0} n_1 \frac{P}{P_0} \, \mathrm{d}\frac{P}{P_0} \qquad (3.5)$$

This also means that the integral change in Gibbs free energy associated with the absorption process can directly be achieved from a carefully determined absorption isotherm of water to fibres. By determining water absorption isotherms at different temperatures and applying the Gibbs-Helmholz equation, (3.6), it is possible to determine both the enthalpy of adsorption and the entropy contribution to the adsorption process according to (3.7)

$$\frac{\partial\left(\Delta G/T\right)}{\partial\left(1/T\right)} = \Delta H \qquad (3.6)$$

$$\Delta S = \frac{\Delta H - \Delta G}{T} \qquad (3.7)$$

The Change in Enthalpy upon Water Sorption

Instead of performing the water adsorption at different temperatures and applying equations (3.6) and (3.7) it might be more convenient to determine the enthalpy associated with the sorption process from calorimetric measurements. The reasons to this are several. First of all the determination of a water sorption isotherm has earlier been very time consuming. Secondly, the polymeric components of the fibre wall show changes with temperature since both hemicellulose and lignin are highly amorphous. As a result, the changes in Gibbs free energy, enthalpy and entropy might be combination processes of the sorption process and changes in properties of the components constituting the fibre wall. Naturally this problem will to some extent also be present for the calorimetric measurements but at least it is then known that the moisture content

42

is the only factor that is changing during the measurements and not both the temperature and the moisture content.

In order to separate the different enthalpy changes associated with the wetting of fibres by a liquid it might be useful to use the different processes schematically as shown in *Figure 3.1*. In this figure the wetting of the solid by water, i.e. process 1), has been separated into the

2) evaporation of n_l moles of water from the liquid
3) sorption of n_l moles of liquid onto the solid
4) the wetting of this solid with adsorbed liquid into the liquid

As is easily seen in the *Figure 3.1* process 1 is simply the sum of 2 and 4.

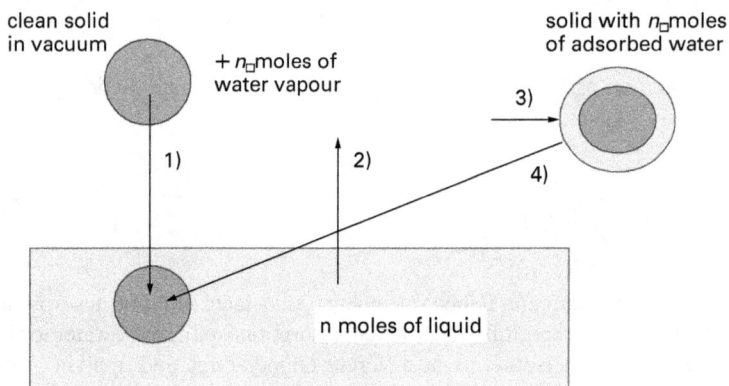

Figure 3.1. Schematic representation of the wetting of a solid by a volatile liquid. The wetting, 1), has been divided into an evaporation process, 2), a sorption process, 3), and a wetting of the water containing solid, 4). *Freely adopted from W.D. Harkins and G. Jura (1942) J.Am. Chem. Soc. 66: 919.*

In mathematical terms this can be written as, equation (3.8)

$$\Delta H_i \left(\frac{S}{L} \right) = \Delta H_i \left(\frac{S_f}{L} \right) + n_i \lambda + \Delta H_{ads} \tag{3.8}$$

where
$\Delta H_i (S/L)$ = Heat of immersion of the dry solid, i.e. process 1) in *Figure 3.1*.
$\Delta H_i (S_f/L)$ = Heat of immersion of a wet solid, i.e. process 4) in *Figure 3.1*.
$n_i \lambda$ = Heat of evaporation of n_i moles of the liquid
ΔH_{ads} = Heat of adsorption

It can also be concluded that the net integral enthalpy change upon water sorption, ΔH, is the heat change in excess of the condensation of the liquid, equation (3.9).

$$\Delta H = \Delta H_{ads} - \left(-n_i \lambda \right) \tag{3.9}$$

By combining equations (3.8) and (3.9) the following expression, equation (3.10), for the integral enthalpy change upon adsorption of vapour to the solid, fibres, can be achieved

$$\Delta H = \Delta H_i \left(S/L \right) - \Delta H_i \left(S_f/L \right) \tag{3.10}$$

This means that by measuring the heats of immersion of solids with different moisture contents it is possible to determine the integral enthalpy of adsorption over the entire range of moisture contents up to $P/P_0 = 1$. The critical issue when using this approach is that the heat of wetting of the dry solid, ΔH_i (S/L), has to be determined with high accuracy.

Once the integral enthalpy of adsorption has been determined it is possible to calculate the integral entropy of adsorption according to equation (3.7).

3.1.3 Application of the Methodology to Describe the Adsorption of Water Vapour to a Sample of Microcrystalline Cellulose

There are considerably few applications of the use of this thermodynamic description of the water sorption to natural fibres. Most probably this is due to the large experimental effort that has to be undertaken to perform these measurements. In Hollenbeck et al. (1978) a clear example of this approach was shown and this will be described in some detail to show how the methodology can be applied to a practical sample.

Initially the water sorption to microcrystalline cellulose at different humidities was determined and the following adsorption and desorption isotherm was found.

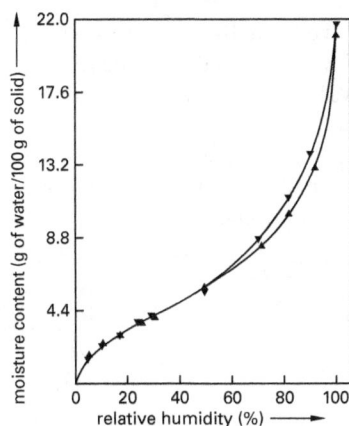

Figure 3.2. Water vapour adsorption (▲) and desorption (▼) on microcrystalline cellulose at 25 °C. Adopted from Hollenbeck et al. (1978).

This type of adsorption isotherm is typical for what is called a type II isotherm which means that the water is first adsorbed in a monolayer on the surface (in Figure 3.2 this would correspond to a relative humidity (P/P_0) of 0.3) and then in several layers and finally at higher moisture contents capillary condensation will occur ($P/P_0 > 0.8$). As can be seen there is a small hysterisis between the adsorption and desorption branches and this will be discussed further in the next paragraph where the influence of water on fibre structure is described.

Following this measurement the heats of immersion of different samples of microcrystalline cellulose was measured and the results from these measurements are shown in *Figure 3.3*.

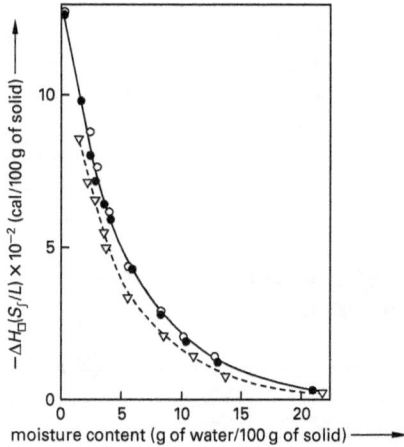

Figure 3.3. Heats of immersion of microcrystalline cellulose containing different amounts of adsorbed and desorbed water at 25 °C. The different samples showing adsorption (O, ●) and desorption (∇) had been equilibrated for different times before the measurement of heats of immersion: Samples corresponding to O had been equilibrated for 3 days,● for 7 days and ∇ for 8 days before the measurements. Adopted from Hollenbeck et al. (1978).

From the adsorption isotherm in *Figure 3.2* the integral change in Gibbs free energy was calculated utilising equation (3.4). The integral enthalpy change was calculated from the results in *Figure 3.3* using equation (3.10) and the value for $\Delta H_i(S/L)$ was estimated by extrapolating the data in *Figure 3.3* to 0 moisture content. Having determined both these entities it was then easy to calculate the integral entropy change at different moisture contents and the results from these calculations are shown in *Figure 3.4*. From these data it was then also possible to calculate the differential thermodynamic properties and these results are shown in *Figure 3.5*.

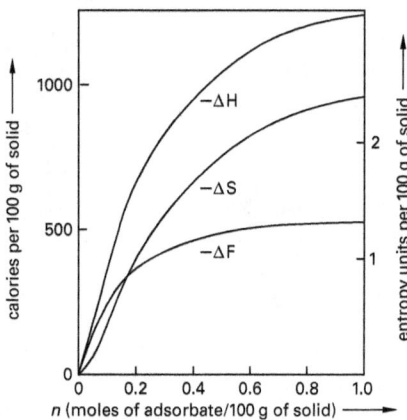

Figure 3.4. Integral thermodynamic properties describing the adsorption of water vapour onto microcrystalline cellulose at 25 °C. Adopted from Hollenbeck et al. (1978) and ΔF in the figure corresponds to ΔG in the text.

From *Figure 3.4* it can be seen that the adsorption of water onto microcrystalline cellulose is an enthalpically driven process. The integral change in Gibbs free energy starts to level off at an adsorption of about 0.6 moles per 100 g solids and at a moisture content above 0.8 moles/100 g solids the increase in the integral Gibbs free energy is approximately 0. This latter fact indicates that the interaction between the water molecules and water covered cellulose surfaces is approximately the same as the interaction between water molecules in solution.

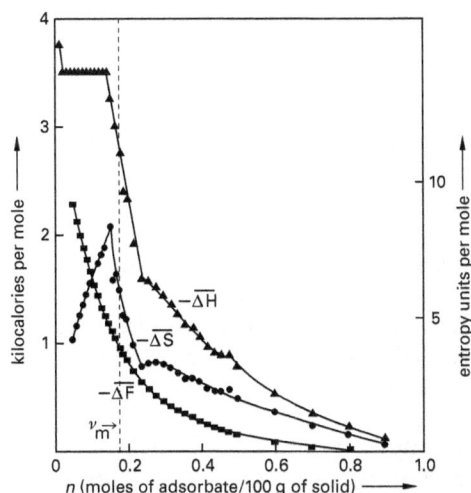

Figure 3.5. Differential thermodynamic properties describing the adsorption of water vapour onto microcrystalline cellulose at 25 °C. Adopted from Hollenbeck et al. (1978) and ΔF in the figure corresponds to ΔG in the text. The dashed line and ν_m correspond to the monolayer capacity of water molecules on the surface of the microcrystalline cellulose.

The constant differential enthalpy at water contents below 0.15 moles/100 g solid shown in *Figure 3.5* is an indication of that no swelling occurs and that the adsorption sites on the cellulose are chemically equivalent. This might not be so unexpected for a microcrystalline cellulose sample but other investigations (Morrison and Dzieciuch 1959) have shown that cotton fibres show a rapid decrease in differential enthalpy at low moisture contents, indicating a swelling of this fibrous substrate. Since the swelling of the solid will consume energy there will be a decrease in the differential enthalpy. In Hollenbeck et al. (1978) it was also suggested that the rapid decrease in differential enthalpy above an adsorption level of 0.15 moles/100 g solid is due to an increased lateral interaction between the water molecules on the surface of the cellulose. This suggestion is supported by the fact that the level coincided with the monolayer coverage level.

From the level of 3.5 kcal/mole at adsorption levels below 0.15 moles/100 g solid an average binding energy between water and cellulose was determined to 1.75 kcal/mole assuming that each water molecule forms two hydrogen bonds with cellulose. It has to be remembered that this is the energy in excess of the normal hydrogen bonds in water and a total binding energy between cellulose and water can be calculated to 6.25 kcal/mole by adding the energy from the normal hydrogen bonds in water (4.5 kcal/mole) to the experimentally determined 1.75 kcal/mole.

In *Figure 3.5* it is also shown that the entropy is decreasing over the entire adsorption range indicating that the adsorption of water molecules to cellulose leads to a net ordering of the molecules in the system. Generally there will be several terms contributing to the entropy change upon adsorption of water molecules:

1. decrease in entropy due to a change from a 3 D to a 2 D molecular orientation
2. decrease in entropy due to ordering in the surface film
3. increase in entropy due to separation of water molecules on the surface of the solid to distances larger than the average distance in solution
4. increase in entropy due to swelling of the solid

This means that at very low levels of water adsorption there will be a disordering of the system since the distance between the water molecules on the surface of the cellulose will on average be larger than the average distance between the water molecules in solution. This disordering might dominate over the 3 D to 2D change and ordering in the surface. As shown in *Figures 3.4 and Figure 3.5* this disordering of the system cannot be detected in these experiments.

Finally it was also possible to calculate the difference for the adsorption and desorption experiments and it was found that both the differential enthalpy and differential entropy was different for adsorption and desorption indicating that the material changes properties upon water vapour saturation.

As shown with this example the thermodynamic analysis of water sorption to fibres yields a lot of fundamental data that definitely can aid in the characterization of different processes occurring in the handling of fibres. Due to the rapid development of automated accurate equipment for both water sorption and calorimetric measurements this methodology is interesting for increased use in the near future.

3.1.4 Influence of Adsorbed Water on the Structure of Fibres

Reasons Behind the Hysteresis in the Adsorption/Desorption Isotherms

As shown in *Figure 3.2*, a hysteresis is detected between the adsorption and the desorption isotherm. The difference in this figure is small compared to the results shown in Morrison and Dzieciuch (1959) but the trend is the same, i.e. the desorption branch is always higher than the adsorption branch as shown in *Figure 3.6*. It is also usually found that if the fibres are repeatedly cycled at different relative humidities a steady state is found where the adsorption and desorption follows the same isotherms. However, the second desorption cycle usually shows a lower amount of adsorbed water at a certain relative humidity compared with the first desorption cycle, i.e from initially never dried fibres.

There might be several reasons to this hysteresis but the most common explanation can be traced back to the discussions earlier regarding the thermodynamic parameters for the adsorption and the desorption of water on fibres. As indicated in the previous sections the water sorption (adsorption/desorption) will change the structure of the fibre wall and this will naturally show up in the energy terms describing the water sorption. The reason to this is that energy is consumed (swelling, i.e. breaking of bonds in the fibre wall) or liberated (bond formation during water desorption) when changing the structure of the fibre wall. Very little is quantitatively known about the connection between the change in fibre wall structure and change in the free energy upon adsorption and desorption and the presented results are to some extent contradictory. In Hollenbeck et al. (1978) it was for example found that no swelling occurred upon water sorption in microcrystalline cellulose whereas in Morrison and Dzieciuch (1959) a swelling was indicated by a rapid decrease in the differential enthalpy even before the formation of a monolayer of water molecules on the fibres. Furthermore it has also been found that the pre-treatment of the fibres (specially heating) had a significant effect on the heats of wetting (Morrison and Dzieciuch 1959, Argue and Maass 1935). Generally it can be stated that the higher the drying temperature the lower the heats of wetting (Morrison and Dzieciuch 1959, Argue and Maass 1935).

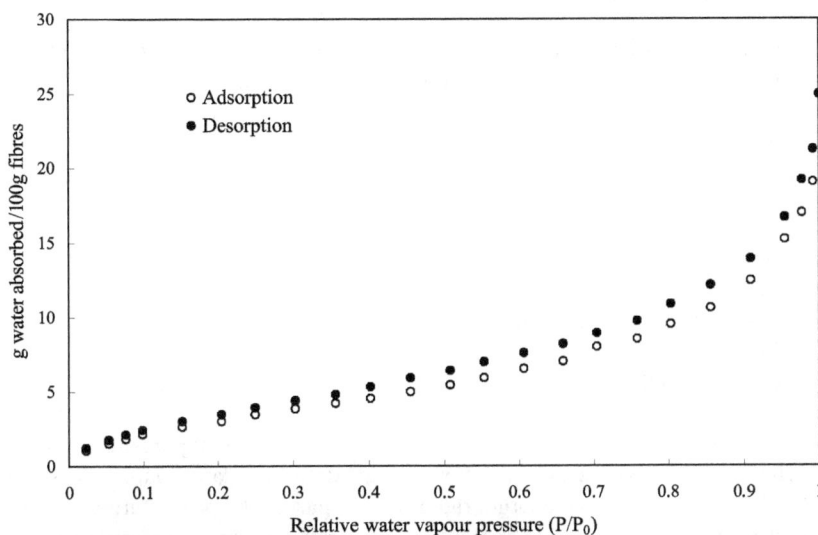

Figure 3.6. Water adsorption and desorption on cotton cellulose at 24.6 °C. Equilibrium time for each measuring point was at least 24 hours. Redrawn from Morrison and Dzieciuch (1959).

Naturally it is very difficult to compare different substrates with different chemical compositions and structures but it is quite clear that the hysterisis in the sorption isotherm is a combination of a change in the enthalpy and entropy in the interaction between water molecules and cellulose during adsorption and desorption. The exact reason to this difference in entropy and enthalpy during adsorption/desorption is not known but with the modern equipment available for both calorimetry and water sorption this would be a very interesting research area for characterisation of wood fibres pretreated in different ways.

Another explanation for this hysterisis is the existence of pores in the fibre wall with a narrow entrance and a wider central part of the pore. The existence of these kinds of pores is very likely considering the lamellar structure of the fibre wall. An idealised pore with this structure is shown in *Figure 3.7* and as indicated in this figure the entrance of these pores have a radius of R_1 whereas the central part of the pore have a radius of R_2.

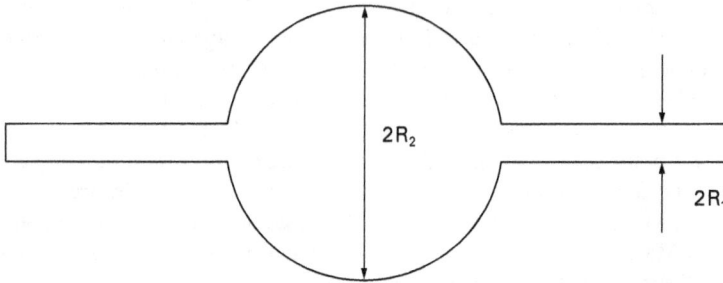

Figure 3.7. Idealised structure of a pore in the fibre wall with narrow entrances and a wider central part of the pore. R_1 and R_2 are radii characteristic of these two regions.

According to the Kelvin equation, equation (3.11), there is a relationship between the radius of a pore and the relative vapour pressure at which this pore is filled with liquid.

$$\ln(P/P_0) = \frac{-2\gamma M}{r\,d\,RT} \tag{3.11}$$

where

P/P_0 = Relative vapour pressure
γ = Surface tension of water (0.072 N/m)
M = Molecular mass of water ($18 \cdot 10^{-3}$ kg/mole)
r = radius of capillary (in m)
d = density of the water (1000 kg/m^3)

This means that during adsorption of water vapour to dry fibres the small pores will first be filled with water. The central part of the pore will not be filled until the relative vapour pressure is increased to a level corresponding to the larger radius of this part of the pore. During desorption the central part of the pore will not be emptied until the relative vapour pressure has decreased to a level where the smaller pores are emptied and then both the smaller entrances and the central part of the pore will be emptied at the same time thus causing a considerable difference between the adsorption and desorption branch.

However, there is most probably no single explanation to the detected sorption hysteresis. Most likely it is a combination of a nonregular structure of the fibre wall and a change of properties of the fibre wall showing up as a change in the thermodynamic properties of the adsorption and desorption processes.

Changes in Fibre Structure upon Drying and Wetting

As has been described in a separate chapter the fibre wall attains a porous structure as lignin and hemicellulose are removed during the liberation of the fibres from the wood. The size of the pores and the volume of pores and solid substance will vary depending on the degree of delignification. This means that the lignin and hemicellulose will be removed from the different layers of the fibre leaving cellulose fibrils in a more or less open network with water left in the open voids between the cellulose fibrils. A representation of this is shown in *Figure 3.8* where a number of SEM (Scanning Electron Microscopy) pictures (Duchesne 2001) have been combined to show the structure of the different layers in the fibre wall as the lignin/hemicellulose is removed.

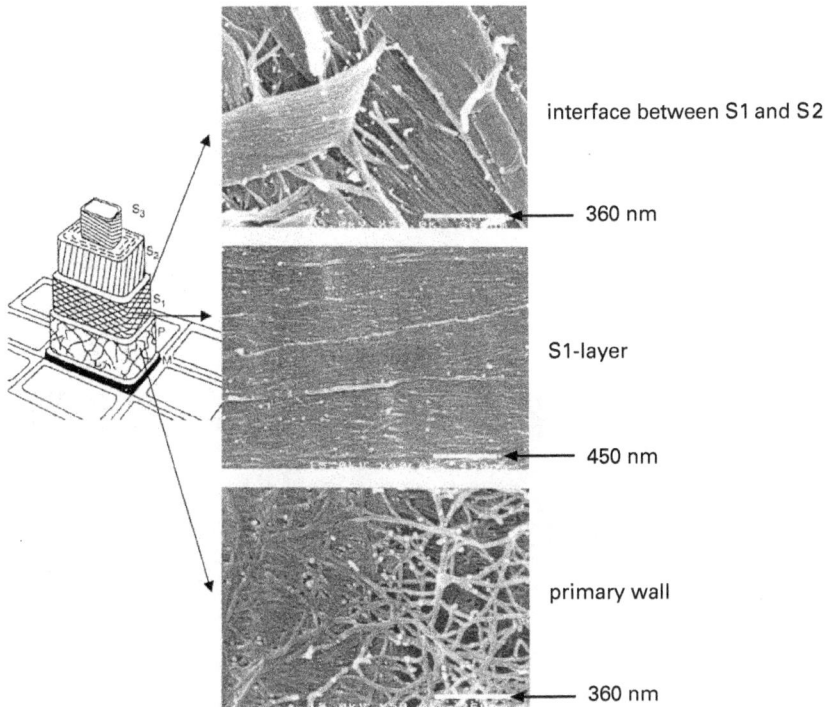

Figure 3.8. SEM micrographs showing the cellulose fibrils in the different cell wall layers after lignin removal. The white bars in the diagrams show the resolution in the diagrams. (Duchesne 2001).

As can be seen in these pictures, the delignified cell wall will be very porous and this is also in accordance with the view of the fibre wall as a porous lamellar structure as introduced by Stone and Scallan (1966) and schematically shown in *Figure 3.9*. In the figure it is also indicated that the fibre wall will change its volume as the water is removed and according to Stone and Scallan (1966) the fibre wall is almost solid when water is fully removed. Wetting and beating will reswell the fibre wall as water is imbibed in between the cellulose fibrils but as discussed before it will not be possible to fully recover the never-dried structure of the fibre wall.

From this discussion and from the earlier discussions on the reasons behind the hysteresis in the sorption isotherms it is clear that the interaction between fibres and water will lead to a change in the structure of the fibre wall. This change in structure is very important since it will influence the behaviour of the fibres during drying of the paper and in turn it will also influence the behaviour of the drying paper. Apart from a shrinking of the fibre wall the fibres will also collapse whereby the void in the lumen to a large extent disappears, at least for chemical pulps. This has also been described by Stone and Scallan (1965) and their suggestion to the change in fibre shape during processing from wood to paper is schematically shown in *Figure 3.10*.

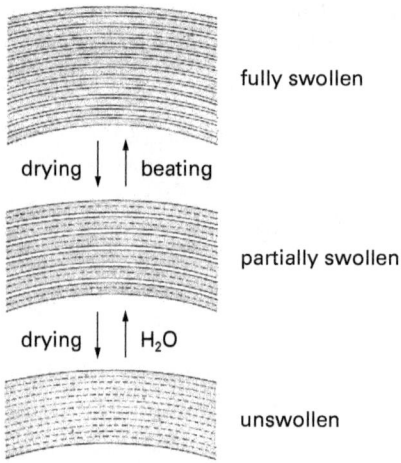

Figure 3.9. Schematic representation of the effect of water and mechanical treatment on the swelling and deswelling of the porous lamellar structure of the fibre wall. In the figure the fibre is viewed along the fibre axis. Adopted from Stone and Scallan (1966).

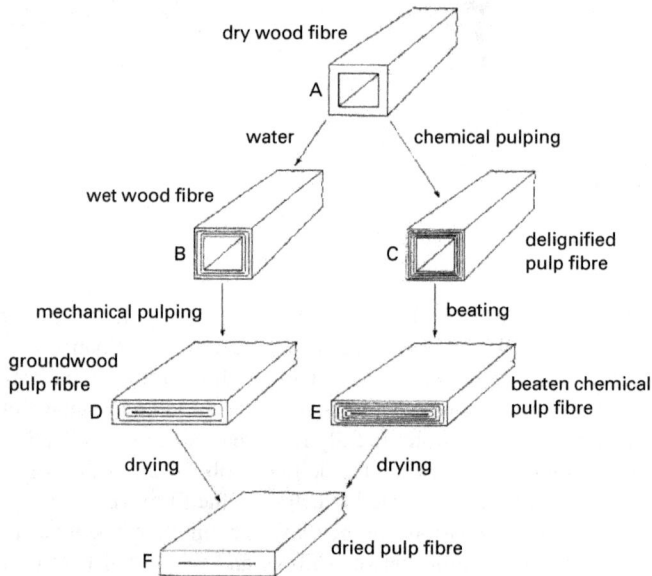

Figure 3.10. Change of the fibre wall and fibre structure as the fibre is converted from wood to paper. Adopted from Stone and Scallan (1965).

Ignoring the collapse, which is hard to quantify with techniques available today, and focusing on the change in structure of the fibre wall it is important to quantify how the dimensions of the fibres are changed as the fibre is drying. Since the fibres have a cylindrical shape it is possible for the fibre wall to shrink from the lumen and outwards or from the outside inwards towards the lumen. The first process will result in a limited change of the external dimensions of the fibres whereas the second process will lead to a considerable change of the external dimension of the fibres.

According to a careful investigation of how the fibre wall volume is changed as the yield ofthe fibres is decreasing, Stone and Scallan (1967) presented the following result, *Figure 3.11*.

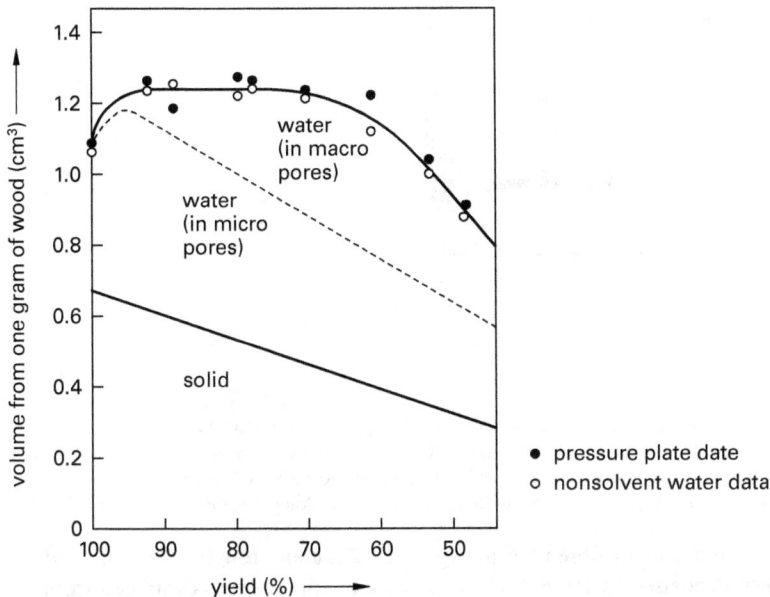

Figure 3.11. Change in the fibre wall volume and the different structural elements constituting this volume. The different volumes have been multiplied with the yield in order to compare the data for a fixed number of fibres instead for 1 g o fibres with a certain yield. Adopted from Stone and Scallan (1967).

According to this figure the volume of water that is removed when the fibres are drying is 0.69 ml/g at a yield of 80 % compared with a solid volume of 0.53 ml/g. At a yield 55 % the corresponding figures are 0.63 ml/g and 0.37 ml/g respectively. Provided that the void volume within the fibre wall is totally collapsed upon drying this would lead to a change in volume of the fibre wall with 56 % and 63 % respectively. Neglecting the change in length of the fibre upon drying this would correspond to a change in the cross -sectional area of the fibre wall of 56–63 % for these never dried fibres. In order to isolate which shrinkage process that is dominating these data can be compared with the data from Tydeman and Page (Tydeman et al. 1966) who studied how wood fibres were shrinking in the cross-direction during drying. Typical data from their investigation are shown in *Figure 3.12* where the shrinkage of wood fibres have been compared with terylene fibres used as reference fibres in their measurements since these latter fibres are unaffected by the moisture.

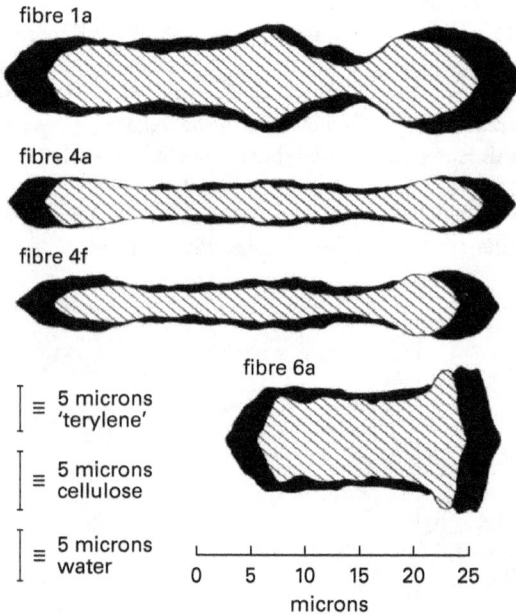

Figure 3.12. Swelling and shrinking of fibres from unbeaten sulphite pulps relative to terylene fibres used as reference fibres in the measurements. The dark area in the figure corresponds to the water saturated samples whereas the hatched areas correspond to the dry fibres. Fibres 1 and 4 were dried and reslushed whereas fibre 6 was never dried. The technique used could not show the change in lumen upon drying. Adopted from Tydeman et al (1966).

Form the change upon drying of fibre no 6 in *Figure 3.12* it could roughly be estimated that the outer area had been decreased by around 30 % compared with the 60 % estimated from water removal in *Figure 3.11*. Considering also that all pores are most probably not closed upon drying it is obvious that a significant part of the volume change of the fibres is caused by a change of the outer dimensions of the fibre wall. Nevertheless, a considerable part is caused by a shrinkage outwards from the lumen. Since the statistics behind this kind of comparison is very poor it has to be treated with caution and it is included more as a demonstration how this kind of problem could be handled. However, the fact remains that there are very few studies available where water removal from the fibre wall is combined with measurements of the shrinkage of the fibre wall despite the importance of the issue.

Irreversibility of Pore upon Drying

One large problem associated with drying of fibres before using them for paper production is that some of the pores of the fibre wall are irreversibly closed upon drying. Some of the paper-making potential of the fibres is thereby lost and has to be recovered with increased beating that in turn creates other problems such as reduced dewatering and increased consumption of paper chemicals. These problems are well known and the main reasons to the integration of pulp and papermaking, especially for papers containing high yield pulps. When fibres are dried and reslushed it is found that some water is imbibed into the fibre wall but the volume of water in the fibre wall is much smaller than the volume of water in the never-dried fibres. The fact that some of the pores will re-open and others will remain closed is both interesting and important. Today it is not resolved which types of pores that are permanently closed and which types of pores that might be re-opened and a knowledge of this and how the situation could be changed would be very important for an improved use of recycled fibres and for better treatment of broke in the paper production.

In an investigation about pore closure upon drying, Stone and Scallan (1966) prepared samples that were conditioned at different relative humidities and then split in two parts. One part was subjected to solvent exchange (to avoid further pore closure) and drying whereas the other part was reslushed, dewatered and subjected to solvent exchange before drying. Both samples were then subjected to pore volume analysis by nitrogen adsorption and the results shown in *Figure 3.13* indicate that the pore closure at "higher" amounts of water gives rise to an irreversible pore closure whereas the pore closure at lower amounts of water is more or less reversible.

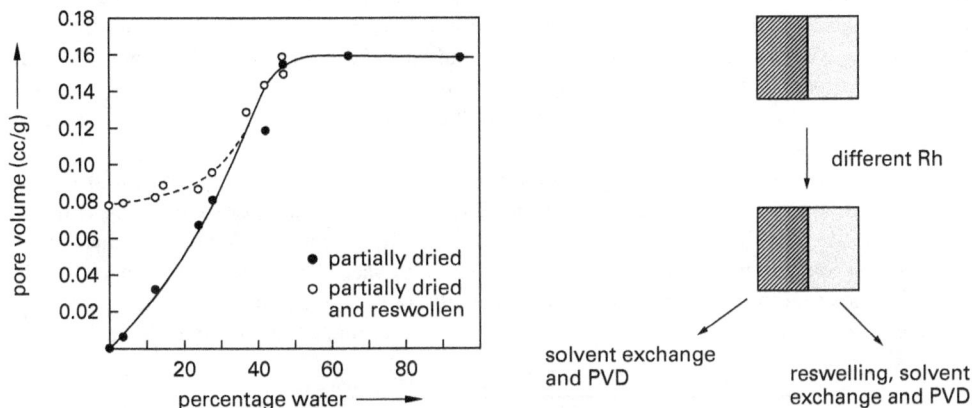

Figure 3.13. Reversible and irreversible pore closure of wood fibre walls upon drying. To the right in the figure a schematic description of the experimental design is shown. Samples were prepared to different water contents and then split in two parts. One was solvent exchanged and dried whereas one was reslushed, dewatered and then subjected to solvent exchange and drying. Both samples were then subjected to pore volume distribution (PVD) measurements through nitrogen adsorption. Adopted from Stone and Scallan (1966).

Stone and Scallan (1966) that the pores on the outer surface of the fibres were irreversibly closed since the water first is removed from the surface of the fibres whereas pores in the interior of the fibre could be reopened upon reslushing. This finding is very important since the external part of the fibres is the part exposed to other fibres during sheet consolidation in papermaking and closed surfaces, i.e. closed pores, will result in a poor interaction between the fibres. It should be noted though that the pore volume on the ordinate is rather low compared with the volumes shown in *Figure 3.11* and actually corresponds to the water found only in macropores, i.e. micropores (see *Figure 3.11*) as determined with other techniques, are closed during the solvent exchange and drying. Regardless of this the results shown in *Figure 3.13* shows that the pore closure can be quantified with this technique and that the pore closure is not homogeneous throughout the fibre wall.

3.2 Swelling of Fibres

3.2.1 Background

In the earlier section, the focus was on the adsorption of water vapour on fibres from low P/P_0 to $P/P_0 = 1$ and some of the discussion were also devoted to understanding the change in fibre structure upon drying of fibres from the completely wet state. The review of the thermodynamics of water adsorption showed that the adsorption was enthalpically driven and that the entropy of adsorption was negative almost over the entire moisture range. The negative enthalpy is linked to the formation of hydrogen bonds between water and the constituents of the fibre wall. As long as there is a favourable interaction between the water molecules and the fibres the water will adsorb to the fibres. In other words, as long as the enthalpy decrease is larger than the loss in entropy upon water adsorption water will adsorb to the fibres. When the fibres are immersed in water they will be totally surrounded by water and then additional terms, describing the fibre-water interaction, has to be included to give a full description of the swelling of the fibre wall. The present section will be devoted to describing the swelling of fibres when they are immersed in water.

A detailed description of swelling of the fibres is very important since the imbibitions of water into the fibres will lead to a change of the fibre wall. It will turn from a stiff non-conformable composite of cellulose, hemicellulose and lignin with a surface roughness of the order of micrometers to a rather soft gel with a much smoother surface. This change of properties of the fibres is crucial for the preparation of paper since the fibres have to conform to each other in order to create a large area of molecular contact between the fibres and thus to enable a large area

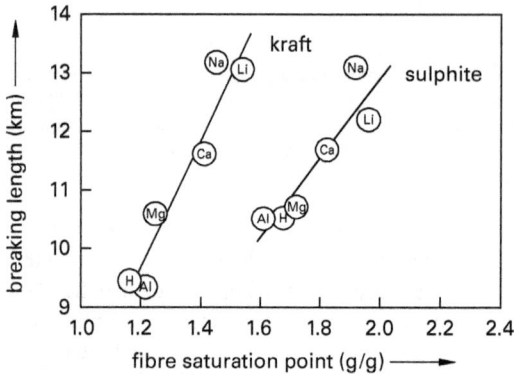

Figure 3.14. Relationship between the water holding capacity of the fibre wall (measured as the Fibre Saturation Point (FSP) i.e. amount of water in the fibre wall that is not accessible to a high molecular mass polymer) and the tensile strength of the paper formed from these fibres. The figure illustrates the almost linear relationship between FSP of the fibres and strength of the paper, the higher the swelling of the fibres, the higher the tensile strength. In the figure it also shown that addition of different types of electrolytes Al^{3+} to H^+ will affect the water holding capacity of the fibres and the paper strength. Adopted from Scallan and Grignon (1979).

of interaction between molecules on adjacent fibres. Therefore a rather large effort has been undertaken to both measure and describe the swelling of the fibres from a theoretical point of view and a summary of this will be given in the present section. The strong connection between swelling and tensile strength of papers from kraft and sulphite fibres have been shown by Scallan and Grignon (1979) and a result from their investigation shown in *Figure 3.14* can serve as a representative example of the importance of the swelling of fibres .

3.2.2 Theoretical Considerations

In addition to the driving force for water adsorption to wood fibres, as described with the integral Gibbs free energy according to equations (3.4), (3.7) and (3.10), additional terms have to be added to describe the gel swelling of the fibre wall. A large contribution to the swelling of the fibres is given by the presence of charged groups, mostly carboxyl groups on the hemicellulose, lignin and cellulose within the fibre wall. Since these groups are fixed to the fibres they will give rise to a concentration gradient of small mobile ions between the interior of the fibre wall and the external solution. This will in turn create an osmotic pressure within the fibre that will result in an expansion of the fibre wall. The expansion is stopped by the structural integrity created by the cellulose fibrils and their three dimensional organisation in the different cell wall layers. In mathematical terms this can be described by equation (3.12)

$$\Delta G_{swell} = \Delta G_{el} + \Delta G_{net} + \Delta G \tag{3.12}$$

where
ΔG_{swell} = integral Gibbs free energy for the swelling of the fibre wall
ΔG_{el} = integral Gibbs free energy from the osmotic pressure created by charges in the fibre wall
ΔG_{net} = integral Gibbs free energy describing the restraining action of the fibrillar structure of the fibre wall
ΔG = integral Gibbs free energy from the interaction between water molecules and the chemical components constituting the fibre wall; given by equation (3.4)

At equilibrium, the terms to the right hand side will balance each other and the sum will be zero. The term describing the influence of the osmotic pressure is, as discussed earlier, linked to the concentration gradient of small ions across the fibre wall and this situation can schematically be described by *Figure 3.15*.

The magnitude of the osmotic pressure will be critically determined by the concentration of charged groups in the gel, $c(1-\alpha) + c\alpha$, and the concentration of OH^-, H^+ and the salt ions present in the system (in the figure represented by Na^+ and Cl^-). With a methodology introduced by Farrar and Neal (1952), Grignon and Scallan (1980) developed a technique where the concentration difference across the virtual membrane of the fibre wall could be estimated By knowing the dissociation constant of the acidic group in the fibre wall, k_a, and assigning a value of the concentrations of Cl^- and H^+ in the fibre wall, the concentration difference can be estimated by utilising of the following equations, (3.13–3.15).

$$k_{\mathrm{a}} = \frac{y \cdot c\alpha}{c(1-\alpha)} \qquad (3.13)$$

$$\lambda = \frac{y}{x} = \frac{[H^+]_{\mathrm{g}}}{[H^+]_{\mathrm{s}}} = \frac{[Na^+]_{\mathrm{g}}}{[Na^+]_{\mathrm{s}}} = \frac{[OH^-]_{\mathrm{s}}}{[OH^-]_{\mathrm{g}}} = \frac{[Cl^-]_{\mathrm{s}}}{[Cl^-]_{\mathrm{g}}} \qquad (3.14)$$

$$\lambda = \sqrt{1 + \frac{c\alpha}{k_{\mathrm{w}}/y + n}} \qquad (3.15)$$

where the same abbreviations have been used as in *Figure 3.15 b*. This system of equations is used in such a way that a salt concentration in the fibre wall is set together with a pH , $-\log y$, of the fibre wall, the number of charges of the fibre wall, c, (from experiments) and the dissociation constant of the charged groups in the fibre wall, k_{a}, the distribution constant, λ, can be determined, equation (3.15). Having determined the distribution constant it is then possible to determine the concentration of the other ions in the solution and in the gel phase.

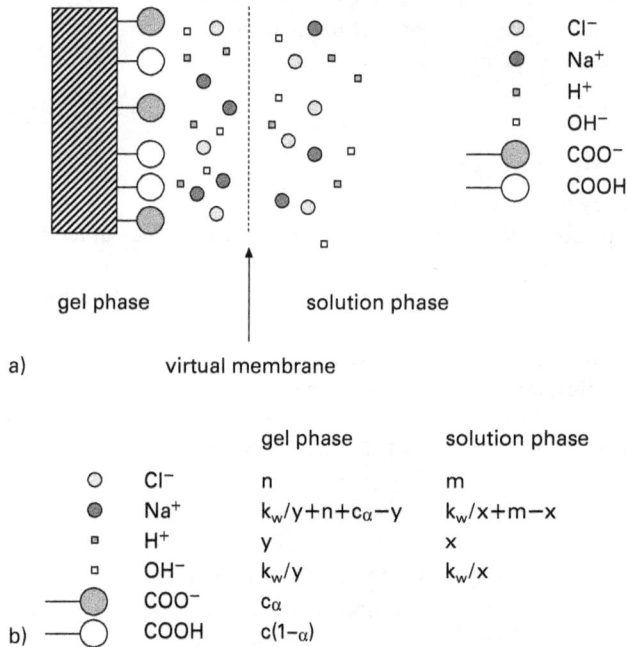

		gel phase	solution phase
○	Cl$^-$	n	m
●	Na$^+$	$k_{\mathrm{w}}/y + n + c\alpha - y$	$k_{\mathrm{w}}/x + m - x$
▫	H$^+$	y	x
▫	OH$^-$	k_{w}/y	k_{w}/x
—●	COO$^-$	$c\alpha$	
—○	COOH	$c(1-\alpha)$	

b)

Figure 3.15. a) Schematic representation of the distribution of charged groups (COOH-groups used as examples) and typical mobile ions that can move freely between the interior of the fibre wall and the external solution. In the figure, a virtual membrane has been included just to stress the similarity to the creation of osmotic pressure across a simple membrane. b) Concentrations of different species in the gel phase and in the solution phase. k_{w} is the autoprotolytic constant of water. Freely adopted after Grignon and Scallan (1980).

This treatment is very illustrative but it suffers from two serious simplifications.

a) Concentrations are used instead of activities in equation (3.14) and this will be a large simplification for highly charged fibres. For most native fibres this assumption is however reasonable.

b) A constant dissociation constant, k_a, is used regardless of the degree of dissociation. This simplification is more serious and to get a better estimate of the real dissociation constant, k_a should be multiplied with $e^{-F\Psi/RT}$ where F is faradays constant, R is the gas constant and T is the absolute temperature and finally Ψ is the electrostatic potential around the charged groups.

However, by applying the described methodology Grignon and Scallan 1980) showed that there is a large difference in pH in solution and in the gel phase for a typical wood fibre. A summary of their calculations are shown in *Figure 3.16* where a concentration of charged groups of 0.1 (corresponding to a charge of 100 µeq./g and a water in the fibre wall corresponding to 1 cm^3/g) have been used. Two different salt concentrations, n, 0 and 0.01 M have been used for these calculations.

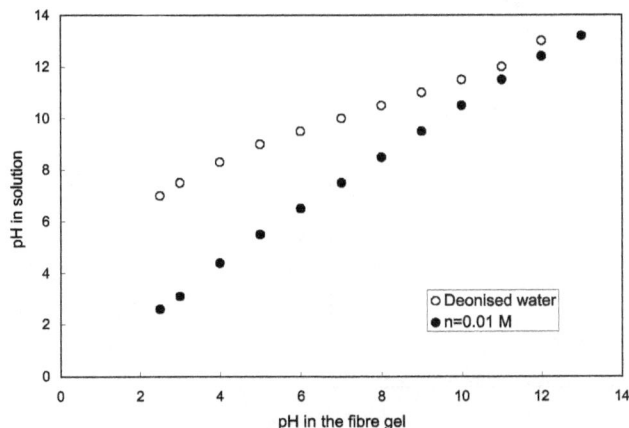

Figure 3.16. Difference in concentration of H$^+$, shown as pH in the figure, in the gel phase and in solution induced by the non-mobile charges in the fibre wall. The concentration difference has been calculated according to the methodology developed by Grignon and Scallan (1980). The data used for the calculations are $c = 0.1$, $n = 0$ and 0.01 respectively and $k_a = 0.0001$. Adopted form the data in Grignon and Scallan (1980).

As can be seen there is a large difference in pH in the gel phase and in the solution for the case with deionised water whereas the difference in 0.01 M Cl$^-$ in the fibre wall is very limited. This means that it is not useful to measure the pH in the solution around the fibres at low salt concentrations if the ambition is to estimate the degree of dissociation of the charged groups in the fibre wall. Furthermore, when performing titrations of the charged groups in the fibre wall it is necessary to have a supporting electrolyte in order to suppress this effect.

With this methodology it has been shown that the non-mobile charges will create an uneven distribution of mobile ions across the fibre wall and Grignon and Scallan (1980) in accordance with Farrar and Neale (1952) also concluded that the total concentration difference of ions between the interior of the fibre wall and the exterior solution is directly proportional to the created osmotic pressure. In terms of free energy this osmotic pressure is hence entropic in nature

58

and the entity E in equations (3.16) and (3.17) will be proportional to the osmotic pressure in the fibre wall.

$$E = [H^+]_g + [OH^-]_g + [Na^+]_g + [Cl^-]_g - [H^+]_s - [OH^-]_s - [Na^+]_s - [Cl^-]_s \quad (3.16)$$

and with equations (13–16) this can be transformed to

$$E = \left(\frac{\lambda - 1}{\lambda + 1)} \right) c\alpha \quad (3.17)$$

For a certain type of fibre with a certain restraining action of the network, corresponding to ΔG_{ne}, this entity will naturally also be proportional to the degree of swelling of the fibre wall.

By calculating the difference in ion concentrations for different salt concentrations it is hence possible to calculate the osmotic pressure and hence swelling of the fibre wall and an example of this is shown in *Figure 3.17*.

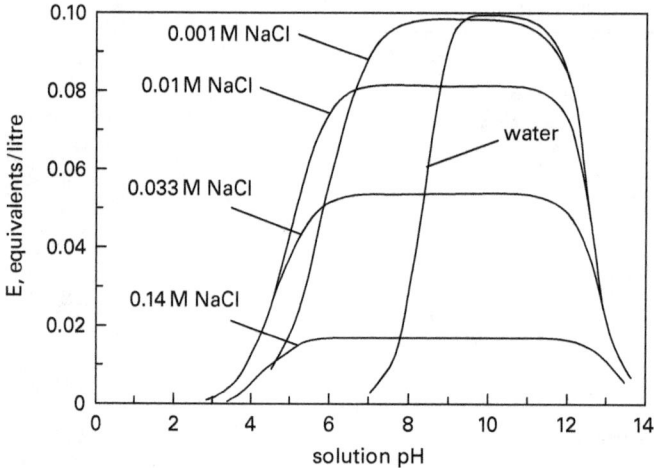

Figure 3.17. Osmotic pressure according to equation (3.17) for a typical wood fibre at different salt concentrations and different pH. Since the restraining action of the fibre network probably is the same for the fibre regardless of pH and salt concentration this diagram is also a representation of the degree of swelling of the fibre wall. Adopted from Grignon and Scallan (1980).

As can be seen in *Figure 3.17*, the pH has to be raised to rather high levels before the osmotic pressure is increased. The reason to this is that the pH inside the fibre wall will be much lower as shown in *Figure 3.16*, due to the presence of the charged groups (carboxyl groups in this case since the dissociation constant was set to 0.0001). As the salt concentration is increased the difference in pH will diminish and the swelling will start at a lower pH but the difference in salt concentration still creates almost the same pressure within the fibre wall. The difference in salt concentration between the interior of the fibre wall and the solution will decrease as the amount of added salt is further increased and this can be noted as a decreased osmotic pressure at even further increased salt addition.

With this rather straightforward treatment of the swelling forces in the fibre wall and despite the simplifications mentioned earlier this treatment can really aid in the understanding of the swelling of wood fibres in different media.

3.2.3 Examples of Swelling of Fibres Under Different Conditions

Numerous investigations have been devoted to study the swelling properties of the fibre wall following the introduction of the polyelectrolyte gel concept as a model for describing the swelling of wood fibres. Generally it has been found that the concept works, i.e. the swelling shows changes with pH, fibre charge and salt concentrations in accordance with the predictions in for example Figure 3.16. There is, however, still very few quantitative data available describing the thermodynamic properties of the swelling in terms of equation (3.12) and most data are qualitative in nature. This is naturally a drawback since the general comparison between different fibres and different fibre treatments is difficult and not linked back to fundamental properties of the fibre wall.

The qualitative data that are available show, however, that the fibre wall behaves as a polyelectrolyte gel but that the swelling is very restricted by the restraining action of the fibrillar structure of the different fibre wall layers. This is for example demonstrated in *Figure 3.18* where the deswelling of wood fibres as a function of concentrations of mobile ions with different valency is shown. As can be seen in this figure the water holding capacity of the fibre wall is decreased as the salt concentration is increased and the deswelling also starts at lower concentrations for the ions with higher valency. The larger sensitivity towards ions of higher valency is expected from polyelelectrolyte theory where the ionic strength is the parameter entering most equations describing the swelling of the gel.

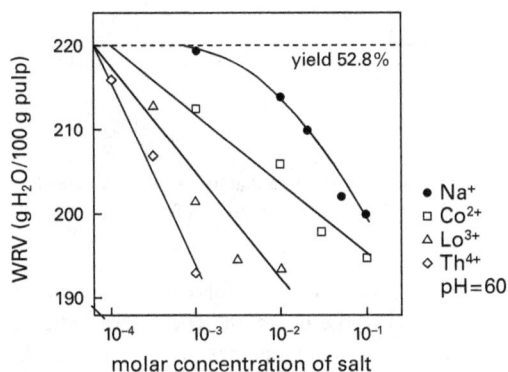

Figure 3.18. Swelling of wood fibres, measured as water retention value (WRV), from an unbleached sulphate pulp (yield 52.8 %) as a function of salt concentration. The counterion to all different cations was NO_3^-. Adopted from Lindström and Carlsson (1982).

It is also shown in the figure that the deswelling is of the order of 10 % of the total volume indicating that fibrillar structure restricts the swelling to a large extent. If, however, the charge of the fibre wall should be significantly increased a limit will be reached where the swelling forces will overcome the reastraining forces and eventually the fibre wall will start to expand. An illustrative example of this is shown in *Figure 3.19* where the swelling of fibres that had been carboxymetylated to different degrees, i.e. to different charges, was investigated. Three different experimental series are shown in the figure. First the influence of charge on the swelling of never dried fibres was tested and then the influence of how the fibres were dried was investigated. It was found that if the fibres were dried with the carboxyl groups in their H-form the formation of hydrogen bonds within the fibre wall was so extensive that the swelling forces from the charged groups was not sufficient to overcome the restraining forces of the fibre wall. If on the other hand the fibres were dried in their sodium form fewer hydrogen bonds were formed and the swelling forces were in this case sufficient to expand the fibre wall. For the never dried fibres the swelling is increased but not to the same extent as for the fibres that had been dried in their sodium form.

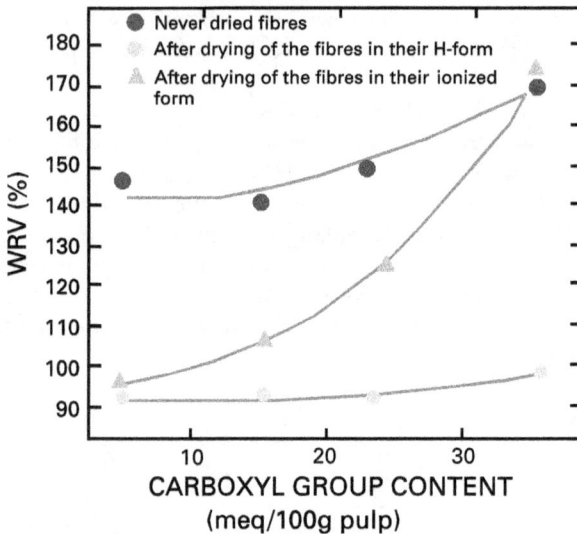

Figure 3.19. Swelling, measured as WRV, for a bleached chemical softwood pulp that had been carboxymethylated to different degrees, i.e. to different charges, and as function of the drying conditions of the pulp. Redrawn from Lindström and Carlsson (1982).

This is an illustrative example of the applicability of equation (3.12) but it is also a good example of the difficulty of identifying the different terms in the equation. The figure indicates that during drying the terms constituting ΔG_{net} is changed and a detailed description of these different terms are therefore necessary in order to be able to describe and predict the behaviour of the fibre wall.

Ignoring the contribution to the swelling by the ΔG component of equation (3.12) Scallan and Tigerström (1992) applied a simple model where the swelling pressure induced by the charges simply was expressed according to equation (3.18)

$$P_{\text{swell}} = RT\left(\frac{n_c}{V}\right) \tag{3.18}$$

where

P_{swell} = swelling pressure induced by the charges in the fibre wallN/m²)
n_c = moles of charges in the fibre wall (moles/gram)
V = liquid volume of the fibre wall (m³/gram)

This means that by determining the amount of charges of the fibre wall it will be possible to get an estimate of the swelling pressure induced by these charges. For a full description of the swelling pressure of the fibre wall much more advanced models are needed but due to the lack of such models equation (3.18) can serve as a simple tool to get a rough estimate of this pressure. Extending this model to describe also the elastic modulus of the fibre wall in the cross-direction of the fibres, E_{fcd}, Scallan and Tigerström (1992) used the volume changes detected when converting the charged groups of the fibre wall from their undissociated form, i.e. $\alpha = 0$, to the fully dissociated form, i.e $\alpha = 1$, in combination with the swelling pressure according to a Hooks law type of equation (3.19).

$$E_{fcd} = \frac{RT\left(\dfrac{n_c}{V}\right)}{\left(\dfrac{V - V_0}{V_0 + V_c}\right)}$$

(3.19)

where

E_{fcd} = Elastic modulus of the fibre wall in the cross direction of the fibres
V = Volume of the fibre wall at $\alpha = 1$ (m³/gram)
V_0 = Volume of the fibre wall at $\alpha = 0$ (m³/gram)
V_c = Volume of the cellwall material (m³/gram)

So simply by determining the volume of the fibre wall at different pH and knowing the charges of the fibre wall it was possible to determine how different fibre treatments would influence the elastic properties of the fibre wall in a quantitative way even if the model is simplistic. An example of this treatment is shown in *Figure 3.20* where it can be seen how the stiffness of the fibre wall is influenced by the yield of the fibres and the drying of the fibres.

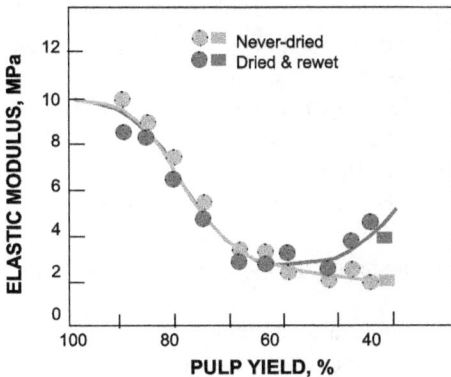

Figure 3.20. Elastic modulus of the fibre wall for chemical softwood pulps as a function of the yield of the pulp and as an effect of the drying of the fibres. The elastic modulus was calculated according to equation (3.20). Redrawn from Scallan and Tigerström (1992).

As can be seen in this figure the fibre wall becomes softer, i.e. the E-modulus decreases, as the yield of the pulp is decreased. This is due to the removal of lignin that together with the fibrillar structure of the cell wall layers contributes to the stiffness of the fibre wall. Below a yield of 60 % it is also obvious that the drying of the fibres will stiffen the fibre wall. This is hence a quantification of the change in swelling upon drying that was demonstrated in *Figure 3.19* and it is most probably caused by hydrogen bonding within the fibre wall upon drying.

3.2.4 Recent Findings and Future Developments

It is striking to note that the relatively modest differences in WRV shown in *Figures 3.14* and 3.18 will have such a large influence on the properties of papers formed from these fibres. In combination with the fact that the relative decrease in swelling upon salt addition is low, *Figure 3.18*, it is most probable that something else than the swelling and deswelling of the entire fibre wall is controlling the ability of the fibre to conform towards each other and/or other materials and form strong joint with these components. Recent investigations have also demonstrated that it is the outermost layers of the fibre wall that will control the ability of the fibres to form strong joints with other fibres (Barzyk, Page and Ragauskas 1997, Laine et al. 2000, Wågberg et al. 2002). Considering the structure of the fibre wall this might not be so surprising since the swelling forces per m^2 in the outer layers of the fibre will be of the same order of magnitude as in the rest of the fibre wall but the restraining actions will be lower due to the organisation of the fibre wall. In the future it will hence be vital to characterise the outer layers of the fibre wall and to find theoretical models for a description of this part of the fibres. Recent investigations with nano-indentation of wood fibres with the aid of Atomic Force Microscopy (AFM) (Nilsson, Wågberg and Gray 2001) show promising results.

3.3 References and Recommended Reading

3.3.1 References

Argue, G.H., and Maass, O. (1935) Measurement of the heats of wetting of cellulose and wood pulp. *Canadian J. Res.* 12: 564

Barzyk, D., Page, D.H., and Ragauskas, A. (1997) Acidic group topochemistry and fibre-to-fibre specific bond strength. *J. Pulp Paper Sci.* 23,2: 59

Duchesne, I. (2001) *PhD Thesis.* Agricultural University of Uppsala

Farrar, J., and Neal, S.M. (1952) The distribution of ions between cellulose and solutions of electrolyte. *J. Colloid Sci.* 7: 186

Grignon, J., and Scallan, A.M. (1980) Effect of pH and neutral salts upon the swelling of cellulose gels. *J. Appl. Pol. Sci.* 25: 2829

Hollenbeck, R.G., Peck, G.P., and Kildsig, D.O. (1978) Application of immersional calorimetry to investigation of solid-liquid interactions: Microcrystalline cellulose-water system. *J. Pharmaceutical Sci.* 67,11: 1599

Kouris, M., Ruck, H., and Mason, S.G. (1958) *Can. J. Chem.* 36: 931

Laine, J., Lindström, T., and Glad-Nordmark, G. (2000) Studies on topochemical modification of cellulosic materials. *Nordic Pulp Paper Res. J.* 15,5: 520

Lindström, T., and Carlsson, G. (1982a) The effect of the chemical environment on fibre swelling. *Sven. Papperstidn.* 85, 3: R14.

Lindström, T., and Carlsson, G (1982b) The effect of carboxyl groups and their ionic form during drying on the hornification of cellulose fibres. *Sven. Papperstidn.* 85,15: R146

Morrison, J.L., and Dzieciuch, M.A. (1959) The thermodynamic properties of the system cellulose-water vapor. *Can. J. Chem.* 37,11: 1379

Nilsson, B., Wågberg, L., and Gray, D. (2001) Conformabiltiy of wet pulp fibres at short length scales. In: C.F. Baker (Ed.) *The science of papermaking. Proceedings from Fundamental Res. Conf. in Oxford, September 17th-22nd 2001*, p. 211

Scallan, A.M., and Grignon, J. (1979) The effect of cations on pulp and paper properties. *Sven. Papperstidn.* 82,2: 40

Scallan, A.M., and Tigerström, A. (1992) Swelling and elasticity of the cell walls of pulp fibres. *J. Pulp Paper Sci.* 18,5: 188

Stone, J.E., and Scallan, A.M. (1965) Study of cell wall structure by nitrogen adsorption. *Pulp Paper Mag. Can.* 66,8: 407

Stone, J.E., and Scallan, A.M. (1966) Influence of drying on the pore structures of the cell wall. In: F. Bolam (Ed.) *Consolidation of the paper web.* Trans.3rd Fund. Res. Symp., Cambridge, Techn. Sect. BPBMF, p.145

Stone, J.E., and Scallan, A.M. (1967) The effect of component removal upon the porous structure of the cell wall of wood. II Swelling in water and fibre saturation point. *Tappi* 50,1: 496

Tydeman, P.A., Wembridge, D.R., and Page, D.H. (1966) Transverse shrinkage of individual fibres by micro-radiography. In: F. Bolam (Ed.) *Consolidation of the paper web.* Trans.3rd Fund. Res. Symp., Cambridge, Techn. Sect. BPBMF, p.119

Wågberg, L., Forsberg, S., Johansson, A., and Juntti, P. (2002) Engineering of fibre properties by application of the polyelectrolyte multilayer concept. Part 1.:Modification of paper strength. *J.Pulp Paper Sci.* 28,7: 222

3.3.2 Recommended Reading

Water Sorption in Fibres and the Effect on Pore Structure

Eklund, D., and Lindström, T. (1991) *Paper Chemistry-An introduction.* Grankulla, Finland: DT Paper Science Publications

Haselton, W.R. (1955) Gas adsorption by wood, pulp and paper. II The application of gas adsorption techniques to the study of the area and structure of pulps and the unbonded area of paper. *Tappi* 38,12: 716

Scallan, A.M. (1977) The accommodation of water within pulp fibres. In: *Fibre-Water Interactions in Papermaking, Trans. 6th Fundamental Res. Symp. Oxford Sept 1977.* BPBIF, p.9

Stone, J.E., and Scallan, A.M. (1969) "The effect of component upon the porous structure of the cell wall of wood. III A comparison between sulphite and kraft process.". In: *Proc. Techn Sect. CPPA.*, T288–T293

Weatherwax, R.C. (1977) Collapse of cell-wall pores during drying of cellulose. *J. Colloid Interf. Sci.* 62,3: 432

Swelling of Wood Fibres and Polyelectrolyte Gels

Arvanitidou, E., Klier, J. and Aronson, C.L. (1992) In: R.S. Harland and R.K. Prud´homme (Eds.) *Polyelectrolyte Gels - Properties, Preparation and Applications. ACS Symposium Series 480*. Washington, D.C.: American Chemical Soc., p.190

Katchalsky, A. (1954) In: J.A.V. Butler and J.T. Randall (Eds.) *Progess in Biophysics and Biophysical Chemistry, Volume 4*. New York: Academic Press Inc., and London: Pergamon Press Ltd., p.1

Lindström, T. (1986) The concept and measurement of fibre swelling. In: *Paper-Structure and Properties*. Marcel Dekker, Inc., p.75

Lindström, T. (1986) The porous lamellar structure of the fibre wall. In: *Paper-Structure and Properties*. Marcel Dekker, Inc., p.75

4 Chemistry of the Fibre Surface

Lars Wågberg
Department of Fibre and Polymer Technology, KTH

4.1 Interactions between Fibres in the Wet State and between Dry Paper and other Substances

The surface chemistry of the fibres and its importance for the surface chemistry of the paper has been given a lot of attention during recent years. This interest has been focused mainly on the interaction between paper and other materials, see for example Berg 1993[a], but the surface chemistry is also important for the interaction between the fibres in the wet state during the consolidation of the paper. During this process it has earlier been shown that the state of swelling of the fibre wall and the apparent modulus of the fibre wall will determine the deformation of the fibres in the fibre/fibre contacts (see for example Lindström 1986, and Scallan and Tigerström 1992). Before entering into a more detailed overview of the work conducted to determine both the surface energy of the fibres and fibre swelling it is necessary to review the detailed interaction between the fibres during consolidation of the wet fibre web in order to clarify why the chemistry of the fibres surface is so important. A short description will also be given on how the surface energy of the paper will determine the interaction between the dry paper and other solids or liquids.

Two fibres coming into close contact with each other can as a first approximation be represented by the interaction between two crossed cylinders as illustrated in *Figure 4.1*. When water is removed, capillaries are formed in the fibre/fibre contact and capillary forces pull the fibres together.

The forces acting between the fibres, apart from the capillary forces, are the wellknown electrostatic forces, van der Waals forces and polar type of interactions often referred to as acid/base interactions. For cellulose fibres, the electrostatic interactions are dominated by carboxylic groups or sulphonic acid groups (see for example Sjöström 1989). The origin of these groups will be discussed in more detail below. Naturally this view of the fibres as crossed cylinders is a rough approximation and is used here to demonstrate the influence of different factors on the interaction between the fibres. In a real situation, the fibres are deformed as the capillary forces

D = distance between fibres
A = affected area

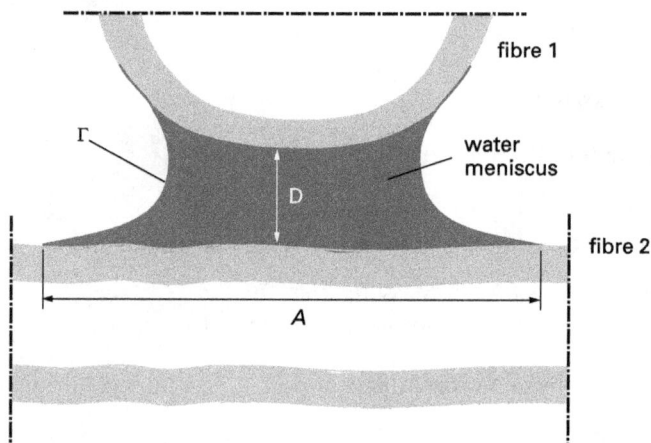

Figure 4.1. Schematic representation of the interaction between two fibres during water removal. The radius of the fibres is R, the distance between the fibres D and the affected area in the contact zone is A.

pull the fibres together and in the dry state the fibres have a shape which more resembles two ribbons brought close to each other. For crossed cylinders the total potential energy (W_{cyl}) between the cylinders can be represented by the equation:

The forces acting between the fibres, apart from the capillary forces, are the wellknown electrostatic forces, van der Waals forces and polar type of interactions often referred to as acid/base interactions. For cellulose fibres, the electrostatic interactions are dominated by carboxylic groups or sulphonic acid groups (see for example Sjöström 1989). The origin of these groups will be discussed in more detail below. Naturally this view of the fibres as crossed cylinders is a rough approximation and is used here to demonstrate the influence of different factors on the interaction between the fibres. In a real situation, the fibres are deformed as the capillary forces pull the fibres together and in the dry state the fibres have a shape which more resembles two ribbons brought close to each other. For crossed cylinders the total potential energy (W_{cyl}) between the cylinders can be represented by the equation:

$$W_{cyl} = \frac{128 \pi R n_0 k T Y_0}{\kappa^2} \exp(-\kappa \cdot D) - \frac{A \cdot R}{6D} \qquad (4.1)$$

where
R = Radius of the cylinders (i.e. Fibres)
κ = Debye length
A = Hamaker constant of the cellulose/water system
D = Distance between the cylinders
$$Y_0 = \frac{\exp(z e \psi_0 / 2\,kT) - 1}{\exp(z e \psi_0 / 2\,kT) + 1}$$
ψ_0 = Surface potential

$n_0 =$ Number of ions per unit volume in solution
$z =$ Valency of the counter ion

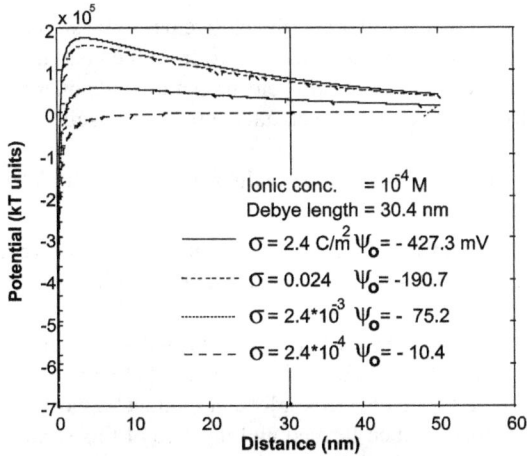

Figure 4.2a. Interaction energy between two crossed cylinders, i.e. fibres, calculated from eq. 1. A salt concentration of 10^{-4} M NaCl was used for these calculations and different degrees of neutralisation of the surface charges were chosen to demonstrate how this influences the interaction between the cylinders. The vertical line shows the Debye length.

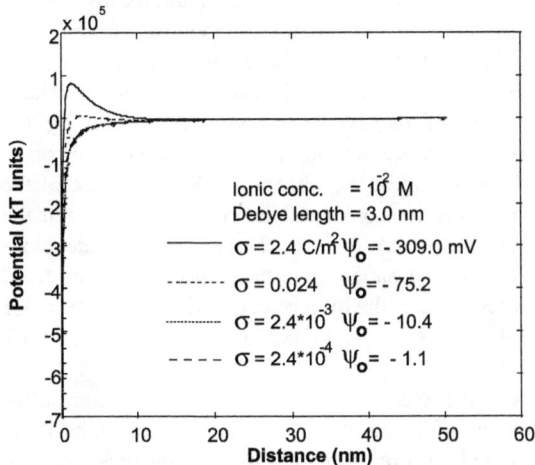

Figure 4.2b. Interaction energy between two crossed cylinders, i.e. fibres, calculated from eq.1. A salt concentration of 10^{-2} M NaCl was used for these calculations and different degrees of neutralisation of the surface charges were chosen to demonstrate how this influences the interaction between the cylinders.

By inserting appropriate values for the surface charge and Hamaker constant for the cellulose (in the present calculations a surface charge of 25 µeq/g, a specific surface area of the external surface of the fibres of 1 m²/g, a Hamaker constant of cellulose of $6.7 \cdot 10^{-20}$ J, Winter (1987), and a Hamaker constant of water of $3.82 \cdot 10^{-20}$ J) the results presented in *Figure 4.2a* and *Figure 4.2b* were obtained.

It is evident in these figures that:

a) the electrostatic repulsion is very large in deionised water and the charge has to be significantly reduced to allow for the fibres to come close enough together to form the bond necessary for paper strength to develop. The forces created by this electrostatic repulsion are naturally counteracted by the capillary forces existing in the contact zone, as is discussed below.

b) at higher ionic strengths this repulsion is naturally reduced, but values of about $1 \cdot 10^5$ kT are still found at 0.01 M NaCl.

In *Figure 4.2a and Figure 4.2b*, it is also seen that very large surface potentials are achieved. In reality counterion binding leads to a decreased surface potential, but this has not been taken into account in these very simple calculations.

At the same time, the capillary pressure, caused by the water meniscus formed in the contact zone between the fibres, draws the fibres together. To overcome the repulsion between the fibres, a sufficiently small capillary radius has to be formed before molecular contact between the fibres can occur. A contact zone between the fibres is then formed and the size of this contact zone is very dependent on the elastic modulus of the fibre wall, which in turn is very dependant on the degree of swelling of the fibre wall. The true nature of the surface of the cellulosic fibres in the contact zone determines to what extent there will be molecular contact or not between the fibres. It is strongly believed that the flexibility of the outer surface layers of the fibres, as discussed by Pelton (1993) is very important for the degree of bonding between the fibres.

This discussion shows that the electrostatic charge, the Hamaker constant and the swelling of the cellulosic fibre wall are very important for how the interaction between the fibres takes place. The swelling forces within the fibre wall are not however sufficient to create a flexible fibre with a large bonding ability to other fibres. In order to achieve this, it is necessary to delaminate the fibre wall and thereby remove some of the swelling restraints created by the interlamellar bonds within the fibre wall. This delimitation is brought about by the beating operation which is thus indirectly essential for the swelling of the fibre wall. A knowledge of the size of these entities, i.e. electrostatic charge, Hamaker constants and fibre swelling, and the factors controlling their magnitude is hence very important in order to understand the development of bonding between the fibres and the creation of paper strength. It should also be stated that in this simplified initial calculations in order to show the importance of different factors for the interaction between wet wood fibres the influence of swelling on the efficient Hamaker constant of cellulose has not been taken into onsideration.

Another important factor, which has to be considered, is how the pressing operation influences these interactions. To have a complete view of the interaction, the force-balance between the fibres should be with the force complemented with the force applied during pressing. The pressing force will naturally compress the fibres, i.e. flatten the "cylinders", and help the capillary forces to overcome the electrostatic repulsion. On the other hand, the pressing naturally removes more water and the size of the contact zone (i.e. smaller capillaries are formed directly) between two fibres will hence become smaller when more water is removed. All this shows that the model of the fibres as two interacting cylinders must be used with caution.

According to Fowkes (1962 and 1964) and as described in detail by van Oss (1994) and Berg (1993[b]) there are additional polar interactions between substrates, i.e. fibres, in aqueous media and currently there is a large scientific debate on how these interactions should be determined and also summarised theoretically. Without entering this debate, it can be stated that there will be additional forces between the fibres in the wet state depending on the acid/base properties of the fibres, and these will naturally add to the forces already mentioned. Today the change of the acid/base interaction with distance between the fibres is not known and these forces have therefore not been included in the calculations.

When the dry paper is contacted with other types of substances, the interaction will be controlled by the van der Waals interactions and the acid/base interactions. This was described for example by Berg (1993[a]) and the total work of adhesion between two different materials can be described by the equation:

$$W_{total} = W^{LW} + W^{AB} \tag{4.2}$$

where
$W^{LW} =$ Lifshitz-van der Waals interactions
$W^{AB} =$ Acid-Base Interactions

The Lifshitz-van der Waals interactions can in turn be described by the following relationship.

$$W^{LW} = 2 \cdot (\gamma_s^{LW} \cdot \gamma_l^{LW})^{1/2} \tag{4.3}$$

where
$\gamma_s^{LW} =$ Lifshitz-van der Waals contribution to the surface energy of the solid
$\gamma_l^{LW} =$ Lifshitz-van der Waals contribution to the surface energy of the liquid

and

$$W^{AB} = 2 \cdot ((\gamma_s^+ \cdot \gamma_l^-)^{1/2} + (\gamma_s^- \cdot \gamma_l^+)^{1/2}) \tag{4.4}$$

where
$\gamma_s^+, \gamma_l^+ =$ Acid contribution to the surface energy of the solid and liquid respectively
$\gamma_s^-, \gamma_l^- =$ Base contribution to the surface energy of the solid and liquid respectively.

There are several ways of calculating the acid/base interaction between the surfaces and the van Oss approach (van Oss 1995) was chosen here to illustrate how the composition of the fibre and paper surface influences the interaction with other materials. From equation (4.2)–(4.4) it is clear that

a) the van der Waal interactions are always important for the total interaction between different materials.
b) polar interactions are also important for many materials the type of polarity of the materials is important. If both materials have electron donor properties there will be no polar interaction, despite the fact that both materials have polar properties.

Berg, (1993[a, b]) also gave several examples of how differences in the acid-base properties of different materials really influence the interaction with liquids and solids.

It is obvious that a knowledge of the chemical composition of the fibre surface is important in order to optimise the interaction of the fibres in the wet state, to enhance fibre/fibre bond formation and to optimise the surface chemistry of papers for use in different practical applications. The following sections will therefore be devoted to a summary of what is known about

a) the surface chemical properties of the constituents of the fibres
b) the surface chemical properties of different fibres (pulps) and papers.

4.2 Surface Energy of the Chemical Constituents of the Fibre Wall, Fibre and Paper

Before starting the detailed description of the available data on the surface chemistry of different materials it may be appropriate to mention that the results are naturally very dependent on the techniques used to collect the data. The most easily available technique for surface energy determinations is contact angle measurements with different liquids. This technique makes it possible to determine both the LW-part of the surface energy and the acid/base part. However, there are severe drawbacks and no satisfactory method has been defined for determining contact angles on porous substrates such as paper. Another popular method for the determination of surface energy and the different components of the surface energy is Inverse Gas Chromatography (IGC) where the retention times of different probe molecules are measured. The evaluation techniques for the data collected by this method often give results different from those given by the contact angle methods and the reason for this is still not exactly known. As a further complication, there are different evaluation principles for the data collected with both contact angles and IGC. These problems are discussed in the different sections below, but due to the large variation in the techniques used, the data in the tables must be treated with caution.

4.2.1 Surface Energy Determined by Contact Angle Measurements

The origin of the use of contact angle measurements is the Young equation which gives a relation between the surface energy of the solid, the surface energy of the liquid, the interaction between the solid and the liquid and the contact angle of the liquid on the solid.

$$\gamma_s = \gamma_l \cos\theta + \gamma_{s,l} \tag{4.5}$$

where

γ_s = Surface energy of the solid in the environment used in the experiments
$\gamma_{s,l}$ = Surface energy of the interface between the solid and the liquid
γ_l = surface energy of the liquid in the environment used in the experiments
It is also commonly known that $\gamma_{s,l}$ can be characterised by the relationship:

$$\gamma_{s,l} = \gamma_s + \gamma_l - I^d - I^p \tag{4.6}$$

where

I^d = Contribution of the dispersive forces to the surface energy of the interface
I^p = Contribution of the polar forces to the surface energy of the interface

It must here be emphasized that the dispersive contribution in equation (4.6) is *not* the same as the Lifshitz- van der Waals term mentioned in e.q. (4.2). In the treatment given in equation (4.6), the dispersion forces include only the London dispersion forces and the dipole-dipole, induced dipole- dipole interactions and hydrogen bonding interactions are all included in the polar term. The treatment in equation (4.4) is based on the work by Chaudhury (1984) who showed that the Lifshitz- van der Waals component is simply a sum of the London dispersion forces, the induction forces (Debye) and the dipole forces (Keesom). This means that it is not possible exactly to compare the results from evaluations by equation (4.6) and equation (4.4) even though they should give the same trends when characterising different surfaces.

It has been shown for example by Wu (1982), that the term I^d in equation (4.6) can be written as

$$I^d = 2 \cdot (\gamma_s^d \cdot \gamma_l^d)^{1/2} \tag{4.7}$$

By inserting equations (4.7) and (4.6) into equation (4.5) the following useful relationship is obtained:

$$\gamma_l(1 + \cos \theta) = 2 \cdot (\gamma_l^d \cdot \gamma_s^d)^{1/2} + I^p \tag{4.8}$$

By using different liquids with known properties, it is possible to determine γ_s^d and I^p and, knowing how to resolve the I^p term mathematically, it is also possible to determine the polar properties of the solid surface. The derivation of equation (4.8) has been included since a fair amount of work was conducted before the acid/base concept was introduced. Some of the earlier contact angle data has been transformed into a dispersive component of the surface energy and a polar interaction with the liquid in question, since the mathematical form of the polar interactions was not known. However, later work has made use of equation (4.4) and liquids characterised from the Lifshitz- van der Waals and acid-base concept and therefore these data are also used in the summary below.

Sandell and Luner (1969) prepared films from cellulose and hemicellulose, Lee and Luner prepared films from lignin (1972) and Winter (1987) prepared films from different cellulose materials and the results of their work are given in *Table 4.1*. In this table, the critical surface tension of the hemicellulose films has also been included. This entity is achieved by measuring cos θ of a liquid on the sample for different apolar liquids, extrapolating to cos θ= 1 and reading the surface tension of the liquid at this position. This value can then be taken as a rough estimate of the dispersive part of the surface energy of the solid. It should be stated that this treatment holds true for nonpolar solids but for polar solids, it must be treated with much caution (Wu 1982).

Table 4.1 shows that the dispersive part of the interaction between cellulose, hemicellulose and lignin and different liquids is always important and that the greatest difference between the cellulose and the lignin can be found in the polar interactions with water. This difference means that more polar liquids will wet the cellulose much more easily than they will wet the lignin. To illustrate this more the I^p term in table 2 was recalculated into a γ_s^p by using a geometric average as in equation (9), also for the polar part of the surface energy. This is naturally not correct but it can be used to illustrate the difference between cellulose and lignin.

Table 4.1. Summary of the dispersive component of the surface energy of films from lignin, cellulose and hemicellulose. In the table, the polar interaction with water has been included as well as the critical surface energy of some materials. The data was used by applying equations (6)–(8). No full chemical characterisation of the materials was given by the authors. The lignin film was prepared by Lee and Luner (1972) from a commercially available kraft lignin.

Material	γ_s^d (mNm/m^2)	I^p (mNm/m^2)	γ_c (mNm/m^2)
Lignin (hardwood)	41	50	36
Cellulose (Avicel)[a]	45	78.6	36
Cellulose (Rayon)[b]	25.1	89	–
Cellulose (MCC)[c]	37.2	78	–
Hardwood xylan [d]	–	–	34
Softwood xylan [e]	–	–	35

[a]Sandell and Luner (1969), from commercial microcrystalline (MCC) cellulose
[b]Winter (1987), from a non-extracted commercial rayon material
[c]Winter (1987), from specially prepared microcrystalline (MCC) cellulose
[d]Sandell and Luner (1969) extracted with DMSO
[e]Sandell and Luner (1969) extracted with water

$$I^p = 2 \cdot (\gamma_l^p \cdot \gamma_s^p)^{\frac{1}{2}} \tag{4.9}$$

With this treatment the data given in *Table 4.2* were obtained.

Table 4.2. Summary of the surface energy data for lignin and cellulose from the work of Felix (1993)*, Westerlind (1988)** and Lee and Luner (1972)[1]. The data in row 1 and 3 have been calculated by applying a harmonic mean equation according to Wu (1982) instead of the geometric mean used for the rest of the data in the table. The data from the work by Felix and Westerlind were collected from measurements with single fibre wetting whereas the rest of the data were collected from goniometric measurements.

Material	Liquids	γ_s^d (mNm/m^2)	γ_s^p (mNm/m^2)	γ_s (mNm/m^2)
Cellulose*	Water/CH$_2$I$_2$	25.5	43.2	68.7
Cellulose (cotton)**	Water/CH$_2$I$_2$	27.5	41.0	68.5
Cellulose (cotton)**	Water/CH$_2$I$_2$	31.0	47.0	78.0
Cellulose (Avicel)[1]	Water/αBr.N	42.2	26.5	68.7
Cellulose[1]	Glycerol/αBr.N	42.2	17.6	59.8
Lignin[1] (Hardwood)	Water/αBr.N	43.5	10.9	54.4
Lignin[1](Hardwood)	Glycerol/αBr.N	43.5	2.4	45.9

These results show more clearly that the polar component is considerably smaller for the hardwood lignin, which means that polar liquids will wet the cellulose more easily. It also means that, when fibres with surfaces covered with lignin are brought into contact with other polar materials there will be a much lower interaction between these materials than between fibres containing only cellulose and the material in question. Another factor, which will influence the interaction, is naturally the hemicellulose, but no comparable data are available for hemicellulose.

The results in *Table 4.2* also show that there is a large scatter in the data, probably due both to the nature of the material used and the method used for evaluation. There is hence a great need for comparative investigations of the different methods and a standardisation of the measurements, in order to be able to compare the data on a quantitative and not just a qualitative basis.

Later work conducted after the introduction of the acid-base concept includes this type of evaluation of the fibres or films. A large volume of data has been prepared by van Oss (1994) and these data also include cellulose, but the origin of this cellulose material is not defined. Work by Berg (1993[a]) and by Toussaint and Luner (1993) also include fibres and films respectively. Berg (1993[a]) used single fibre wetting with different liquids and Toussaint (1993) used cellulose films prepared from cellulose acetate and subsequently deacetylated. The results of these investigations are summarised in *Table 4.3*.

Table 4.3. Summary of the surface energy of a number of different cellulosic materials. Apart from the Lifshitz-van der Waals component of the surface energy the acid base properties of the materials, calculated according equations (3) and (4) are summarised.

Material	γ_s^{LW} (mNm/m^2)	γ_s^+ (mNm/m^2)	γ_s^- (mNm/m^2)	γ_s^{ab} (mNm/m^2)	γ_{total} (mNm/m^2)
SWBK[a]	41.8	0.2	24.5	4.4	46.2
HWBK[a]	43.2	0.4	16.3	5.1	48.3
CTMP[a]	30.8	1.0	67.8	16.5	47.3
Cellulose[b]	44	1.6	17.2	10.5	54.5
Cellulose[c]	39.1	2.0	39.7	17.8	56.9

[a]Berg (1993[a])
[b]van Oss. (1994)
[c]Toussaint and Luner (1993)

The Lifshitz-van der Waals components of these materials have approximately the same values except for the CTMP fibres. This is also expected since these fibres are covered with lignin. The result is not however in accordance with the data in *Table 4.2*, where it was shown that the dispersive component of the surface energy was about the same for lignin and cellulose. The data are not however fully comparable, since the Lifshitz van-der Waals component by definition is different from the γ_s^d as determined according to equation (4.7). Another explanation can be that the lignin materials in the two investigations are not the same, since the CTMP fibres contain hydrogen peroxide bleached lignin whereas the sample reported in *Table 4.2* was not peroxide bleached.

Another factor, which was not given by the authors, was the content of sulphonic acid groups, which has a large influence on the properties of the fibres. Considering all this, it can only be concluded that the samples cannot be compared thoroughly, since the characterisation of the samples is not sufficiently good. In order to make a satisfactory comparison, it is necessary to know

1) the concentration of extractives.
2) the types and degree of modifications of the lignin in the pulps and also whether there is a difference between softwood and hardwood lignin.
3) how the fibres have been defibrated, which parts of the cell wall are exposed.
4) even when rayon was used it is necessary to know how the rayon has been produced, since the process may leave contaminates which in turn may be very important for the surface properties of the rayon films.

The *Table 4.3* also shows that samples, which should be similar, i.e. the SWBK, HWBK and the cellulose samples still have very different properties. This might be caused by some flaws in the method but it is probably only a sign that the materials have different properties. This statement is based on the results presented by Toussaint and Luner (1993) who showed that the measured changes in the surface energy and its separate components could be used to predict the interaction between cellulose and synthetic polymers.

Another result presented in *Table 4.3* is that the fibres have a basic character, which is rather unexpected since the fibre surface, at least of the CTMP, should have a lot of acidic groups. It is not only the sulphonic and carboxylic acid groups that can contribute to the interactions as defined by the definition of the Lewis acid/base concept, but also these groups are believed to play a dominating role for this material. On the other hand, the occurrence of a basic nature of the fibres is not surprising, since the method used to determine these properties (Berg 1993[a]) often results in surfaces with basic properties, and Berg (1993[b]) also claims that this is a sign that this evaluation principal is not totally correct. Without going into too much detail on this topic, it should also be mentioned that the properties of the different liquids used to determine the contact angle have been determined under the assumption that the electron donor and electron acceptor properties of water are the same, i.e. 25.5 mNm/m^2. Recent publications have shown that the Lewis acid component for water should be 34.2 mNm/m^2 and that the Lewis base component should be 19 mNm/m^2, (Lee 1996). This results in lower values for the Lewis base component for all surfaces, but not sufficiently to change the properties to acidic properties. As for the results in *Table 4.3*, it must again be concluded that further investigations are needed where the surface properties of the fibres are systematically changed and then characterised with these methods. In view of this, the large base component of the surface energy of the CTMP is difficult to discuss without any further information, but this large value indicates that it should be possible to find certain specific properties of these fibres.

It is also interesting to note that there are no systematic characterisations of extractives available in the literature regarding Lifshitz-van der Waals components and acid base properties of the pure components. This is surprising, since these components are believed to have a profound effect on the adhesive interaction between fibres and other materials. It was shown by Aberson (1970) that the existence of low amounts of benzene extractives (<0.1 %) resulted in drastic changes in the contact angle of water on these fibres. This is shown in *Figure 4.3* where the contact angle is shown as a function of the extractives content.

Aberson (1970) also showed that this difference in contact angle between different pulps had a very large effect on the absorption properties of pulps containing these fibres. The same type of results has also been found in later investigations by Hodgson and Berg (1988). The ageing effect shown by Aberson (1970) in *Figure 4.3* has been discussed in terms of vapour phase migration of extractives, as originally shown by Swanson (1959). Without going further into the discussion of extractives it should be stated that further investigations of pure extractives components regarding both Lifshitz-van der Waals parts of the surface energy and acid/base properties of the surface energy are definitely needed to understand the complete interaction between fibres and other materials.

As a final comment on the surface energy evaluation of cellulosic fibres through contact angle measurements, it should be mentioned that the fibres are very difficult to characterise both since they have a heterogeneous surface with a mixture of lignin, cellulose, extractives and hemicellulose and since even fairly clean surfaces of bleached chemical sulphite fibres for viscose production have moisture- and temperature; dependant surface properties. This was shown in early work by Borgin (1961) where it was shown that the water contact angle could be changed from 35 °C to 55 °C by heat treatment at 140 °C and 0 % RH for 7 hours. This was interpreted as being due to a reorientation of the hydroxyl groups at elevated temperatures towards the interior of the fibres. This changes the properties of the fibres and also shows that a definition of the ground state used for evaluation of surface energy of the fibres is also very important. A word of caution is needed for this interpretation, however since the heat treatment used during the drying might have caused a migration of extractives, which are probably present in these fibres.

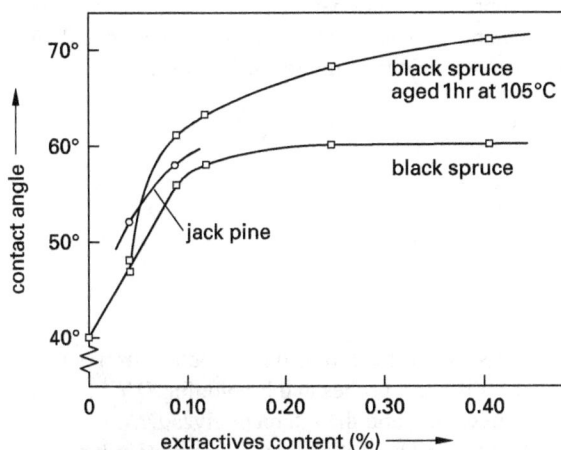

Figure 4. 3. The influence of extractives content on the contact angle of bleached sulphite fibres against water. In the figure results from both "fresh" fibres and aged fibres are shown. Reprinted from Aberson (1970).

In summary, it must be concluded that despite the large efforts made by different researchers, it is striking to note that the chemical characterisation of most materials investigated is not thorough enough to allow for unambiguous interpretations of the results. The same is also true of the different techniques and it is obvious that a standardisation of how the contact angle can be and should be used is very important for future research in this area.

4.2.2 Surface Energy Determined by Inverse Gas Chromatography

The discussion so far has been focused on the acid/base concept as defined by van Oss et.al. (1994) and the evaluations of the acid-base properties have also been focused on different concepts of this evaluation principle. Since the techniques are very different for the evaluation of the acid-base properties and since the entities achieved with the different evaluation principles are also very different, the following text is fairly detailed in order to compare the different data in a clear way.

As proposed by Fowkes (1978), the acid/base part of the work of adhesion, W^{AB} as defined in equation (4.2), is a direct function of the enthalpy of the formation of an acid/base linkage between different materials according to:

$$W^{AB} = fn\,(-\Delta H^{AB}) \tag{4.10}$$

where

n = Number of acid/base linkages per unit area
f = Conversion factor to equate the Gibbs free energy with an enthalpic quantity
ΔH^{AB} = Enthalpy of the acid/base linkage formation

One common way of determining the enthalpy of an acid/base linkage formation is that of Drago as described by Jacob and Berg (1994). However, this approach not has been applied to cellulose to any great extent since the Drago approach cannot handle probe molecules with both acidic and basic sites it. Another approach which can handle probes with both acidic and basic sites is the Gutmann (1966) approach which can be written in the following form.

$$(-\Delta H^{AB}) = (DN_S AN_L)/100 + (DN_L AN_S)/100 \tag{4.11}$$

where

DN_S = Electron-donating ability of the solid
DN_L = Electron-donating ability of the probe molecules
AN_L = Electron-accepting ability of the probe molecules
AN_S = Electron-accepting ability of the solid

This equation means that it is possible to determine the electron-donor and electron-acceptor properties of the solid by determining DH^{AB} for different probes and by plotting DH^{AB}/AN_L as a function of DN_L/AN_L. The slope of this graph gives AN_S and the intercept gives DN_S.

It was found by Dorris and Gray (1980) that the Gibbs free energy of interaction between a probe molecule and a solid could be determined by Inverse Gas Chromatography (IGC). In this procedure a GC column is packed with the material to be characterised and probe molecules are then injected on this column and the retention time of the probe molecules is then used to calculate the interaction energy between the probe molecules and the column material. Schematically this is shown in *Figure 4.4*.

The free energy of interaction can then be calculated from the following relationships:

$$W_A = -\Delta G_{ads}/a_{mol} \tag{4.12}$$

where

W_A = Work of adhesion between a probe molecule and a solid
ΔG_{ads} = Gibbs free energy of adsorption
a_{mol} = Molar area of the probe molecule on the solid

$$-\Delta G_{ads} = RT\ln V_N + C \tag{4.13}$$

where
   R = Gas constant
   T = Absolute temperature
   V_N = Retention volume in the IGC equipment
   C = Constant

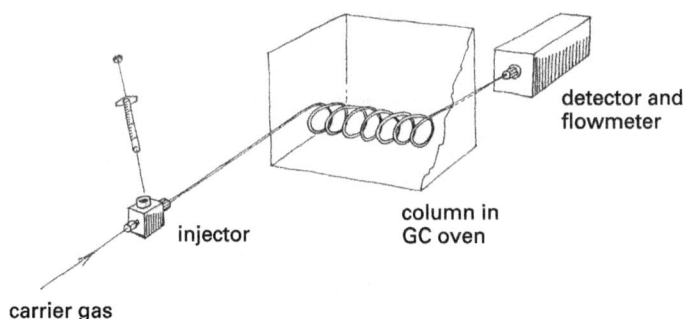

Figure 4.4. Schematic representation of the Inverse Gas Chromatography (IGC) set-up used to determine the free energy of interaction between probe molecules and the column material, i.e. wood based fibres.

Dorris and Gray (1980) also showed that the Lifshitz-van der Waals part of the surface energy could be determined by combining equations (4.3), (4.12) and (4.13) into the equation:

$$RT \ln V_N = 2a_{mol} \left(\gamma_s^{LW} \gamma_l^{LW} \right)^{\frac{1}{2}} + C_3 \qquad (4.14)$$

where
   γ_l^{LW} = Lifshitz-van der Waals component of the surface energy of the probe molecule

By using probe molecules having only London dispersive interactions with the solid and by plotting $RT \ln V_N$ as a function of $a_{mol}(\gamma_l^{LW})^{1/2}$, it is possible to determine γ_s^{LW} from the slope of this graph. In order to determine the acid/base interaction between a probe molecule and the solid, a probe molecule with a known molecular area on the solid is injected in the IGC equipment and its distance from the reference line according to equation (4.14) is then a measure of the acid-base contribution to the Gibbs free energy of adsorption. In mathematical terms this can be written as:

$$-\Delta G^{AB} = RT \ln \left(V_N^{Probe} / V_N^{Alkane} \right) C_3 \qquad (4.15)$$

From these equations, it is possible to determine both the Lifshitz-van der Waals component of the surface energy of the solid and the acid-base contribution to the energy of interaction between a probe molecule and the solid by using IGC. By performing calorimetric measurements with different probe molecules and by applying the approach of Gutmann (1966) it is also possible to separate this acid-base interaction into an electron donor and an electron-acceptor property of the solid.

This is very tedious work and only a few publications available where the electron; acceptor and electron; donor properties of cellulosic fibres have been determined with the described technique. More publications are available where the Lifshitz-van der Waals components of the surface energy of the solid have been determined and where the acid-base interaction between the fibres and a probe molecule has also been determined. In *Table 4.4*, data from some fairly recent publications have been summarised and γ_s^{LW}, ΔG^{AB} (with different probe molecules), K_a and K_d (constants describing the electron-acceptor and electron-donor properties of the solid respectively) values have been collected.

Table 4.4. Summary of surface energy determinations of different cellulosic fibres, and Micro Crystalline Cellulose (MCC), with the aid of IGC. Both data for γ_s^{LW}, constants describing the electron acceptor (K_a) and electron donor (K_d) properties of the materials and $-\Delta G^{AB}$ for interaction with different probes are included in the table. There is no full chemical characterisation of the different pulps used.

Material	γ_s^{LW} (mNm/m²)	K_a	K_d	$-\Delta G^{AB}$(mNm/m²) Probe (acid/base)
60% Beechwood 40% Birchwood (1*)	44.0	0.31	0.24	Several
CTMP (2*)		5.7	-5	Several
SWBK (2*)		4.0	0	Several
MCC (2*)		4.0	0	Several
Whatman filter paper (4*)	48.4	0.41	0.26	Several
Softwood ClO₂ Bleached(3*)	47.0			5.0/27.6 chloroform/ p-dioxane
Softwood H₂O₂ Bleached (3*)	43.5			3.9/29.5 chloroform/p-dioxane
Softwood O₃ Bleached (3*)	42.1			3.7/28.9 chloroform/p-dioxane
Hardwood ClO₂ Bleached (3*)	46.5			5.8/25.8 chloroform/p-dioxane

1* = Felix (1993), 2* = Jacob (1994), 3* = Lundqvist (1996), 4* = Lee (1989)

It is very difficult to obtain a clear picture of the acid-base properties of the fibres. If the data of Felix (1993) and Lee (1989) are compared with the data of Jacob (1994), it is found that the absolute values differ and so does also the trend of the data. The data of Felix (1993) and Lee (1989) are in fairly good agreement and they show that paper is both acidic and basic with dominance for the acidic properties. This is also found in the work of Lundqvist (1996) where the interaction energies with the basic probes is much larger than the interaction with the acidic probe, which again indicates that materials from bleached chemical pulps with a dominating content of cellulose have a predominantly acidic character. In contrast, Jacobs data (1994) show that both fibres containing mainly cellulose and MCC have a monofunctional acidic nature and that lignin-containing fibres (CTMP) give anomalous results with negative basic properties (!). Since Jacob (1994) also points out that the interpretation of her data, which is made at a surface coverage of 2 % and not at infinite dilution, as for the rest of the data in *Table 4.4,* is not straightforward these data must be treated with some caution. The common picture emerges that fibres are both acidic and basic in nature but that they have a predominantly acidic character and the constants given in the *Table 4.4* are then also in fairly good agreement. It is also clear that the different investigations give fairly comparable results with regard to the Lifshitz-van der

Waals components of the surface energy of the fibres containing mainly cellulose. This kind of fibre always contains hemicellulose and the data in *Table 4.4* are therefore typical for fibres containing both cellulose and hemicellulose.

When these data are compared with the data in *Table 4.3*, it is clear that there is a fairly good agreement between the different methods regarding the Lifshitz-van der Waals components, with the minor difference that the contact angle approach gives values slightly lower than the IGC method. This indicates that the different methods are measuring the same entity and that this entity is probably the Lifshitz-van der Waals component of the surface energy. A similar comparison has also been made by Lundquist (1996) who compared different laboratory-prepared fine paper sheets. The sheets consisted of 50 % bleached softwood pulp, 50 % bleached hardwood pulp, and when filler was added it was added at an amount of 20.5 % and, regardless of pH, kaolin was used as filler. Furthermore, the sheets were either sized with rosin dispersion size (and alum and the sheets were prepared at a pH of 4.5) or with alkylketene dimer size. The results of this comparison are shown in *Figure 4.5*.

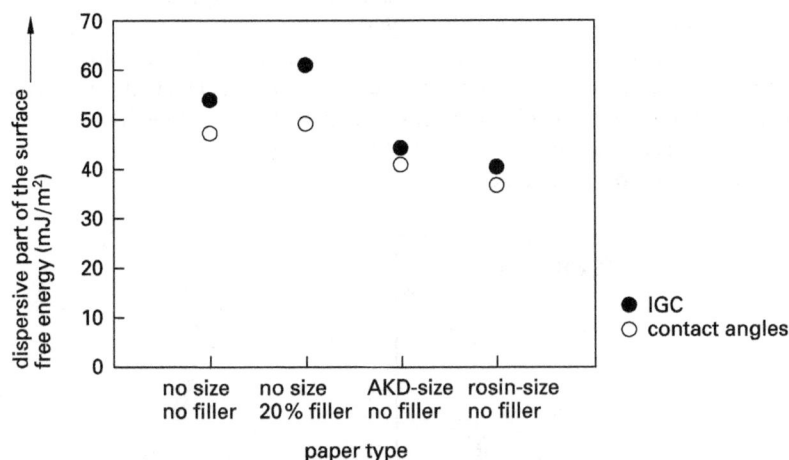

Figure 4.5. The Lifschitz-van der Waals part of the surface energy for different laboratory-prepared fine paper sheets obtained by IGC and contact angle measurements. Reprinted from Lundquist (1996).

As can be seen in *Figure 4.5*, the contact angle data are generally lower than the IGC data but the trend for the different sheets is the same and it is also clear that the difference in the case of the more hydrophobic sheets, i.e. sheets containing sizing agents, is lower than the difference for the more hydrophilic sheets. This same trend was found by Felix (1993) where contact angle measurements (single fibre wetting in a Cahn balance) were compared with IGC measurements. These measurements showed that the dispersive part of the surface energy was 25.5 mNm/m^2 as determined with contact angle measurements and 44.0 mNm/m^2 as determined with IGC. Felix ascribes this discrepancy to the fact that the IGC measures the surface at zero surface coverage whereas the contact angle measurements measures the surface properties "through" a multilayer of the contact liquid on the adsorbate. This statement is however contradicted by the results in *Table 4.2*, where it is shown that it is mainly single fibre wetting which gives very low values for the dispersive part of the surface energy of the fibres. The reason for this is not known. However, from these comparisons, it may be concluded that there is a good agreement between the contact angle method, from goniometric measurements, and the IGC method when the dispersive part of the surface energy is considered and that the difference is larger the higher the surface energy of the solid.

As has already been mentioned, the discrepancy is larger when the acid-base properties of the fibres, as determined by the two methods, are compared. In the IGC method the determination yields virtually acidic fibres whereas the contact angle method yields almost totally basic fibres. This discrepancy has not been resolved at present but a recent publication, (Lee 1996), has shown that the division of the polar properties of water in equal acidic and basic parts is probably not correct and that a more realistic value would be a Lewis acid component value of 34.2 mNm/m^2 and a Lewis base value for water of 19.0 mNm/m^2. As was discussed in connection with *Table 4.3*, this will result in a lower Lewis base component for the tested surfaces but it will not be enough to change the basic properties to acidic properties.

Since there are, today, unexplained differences between the methods, systematic investigations are needed to clarify the reasons. Until this is done the acid-base properties as determined with the two methods must be treated with considerable caution. In this respect, a very interesting investigation has recently been published, (Lundqvist 1996), where fibres have been carboxymethylated to different degrees and then measured with the IGC technique. Unfortunately, contact angle measurements were not published, but the IGC data clearly show that the fibres become more and more acidic with an increasing degree of carboxymethylation and as shown in *Figure 4.6*, there is an almost linear relationship between D.S of the fibres and the interaction energy with a basic probe, p-dioxane.

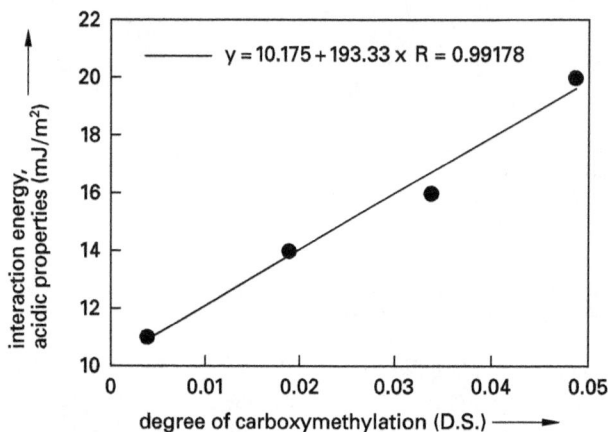

Figure 4.6. The interaction energy between carboxymethylated bleached softwood cellulosic fibres and a basic probe, p-dioxane, as a function of degree of substitution of the fibres (Lundqvist 1996).

This shows that the IGC method can be used to determine correctly a change in the acid-base properties of the fibres. Lundqvist (1996) also showed that a modification of the fibres with a basic component, i.e. a commercial diethylaminoethyl cellulose was used, resulted in a cellulosic surface with a predominantly basic nature. This gives further support to the belief that the IGC method can be used not only to determine the Lifshitz-van der Waals component of the surface but also by using well characterised probe molecules, to detect correctly changes in the acid-base properties of the surface.

4.3 Chemical Characterisation of the Fibre Surface

The chemical composition of the fibre surface can be significantly different from the bulk composition of the fibres. Due to the lack of suitable techniques, it has however been very difficult to study the surfaces of the fibres separately from the bulk of the fibres. The development of the ESCA (Electron Spectroscopy for Chemical Analysis)-technique opened up new opportunities and several workers have used this technique to study fibre surfaces over the last 20 years. With this technique a sample is exposed to x-ray radiation and the energy from this radiation releases electrons from different energy levels of the materials in the substrate. By analysing the energy of the released electrons it is possible to identify from which energy level in the atoms the electrons are released. All elements show different patterns with this methodology and it is therefore possible to identify the chemical composition on the surface of the substrate with this technique. A schematic description of this technique is shown in *Figure 4.7*.

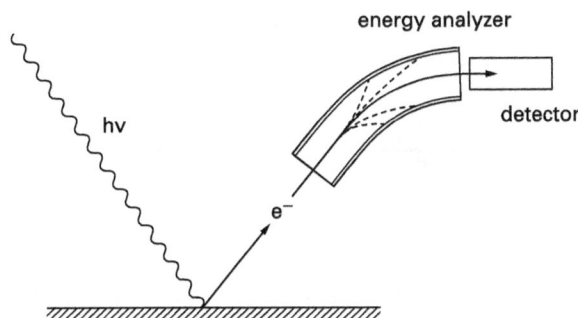

Figure 4.7. Schematic representation of the working principles behind the ESCA measurements.

The following section gives a short review of how this technique has been applied and a summary of the results achieved in these investigations.

Apparently Dorris and Gray (1978) were the first to study fibres with the ESCA technique. They used the C(1s) and the O(1s) peaks to characterise the chemical composition of the surfaces of the fibres and they claimed that the escape depth was 1–5 nm and that this was a typical measure for the depth of the analysis in their investigation. They worked on the assumption that the carbon atoms could be divided in different classes depending on the chemical surroundings of the atoms according to the following:

1) carbon atoms bonded only to carbon and/or hydrogen (C–C)
2) carbon atoms bonded only to a single oxygen other than a carbonyl oxygen (C–O)
3) carbon atoms bonded to two non-carbonyl oxygen's or to a single carbonyl oxygen (O–C–O, C=O)
4) carbon atoms bonded to a carbonyl oxygen and to a non carbonyl oxygen (O=C–O)

In the early paper they worked only with the ratio between O(1s) oxygen atoms and C(1s) atoms, N_O/N_C. It was first in a later paper, (Dorris and Gray 1978^c) that they showed how the C(1s) peak could be deconvoluted into peaks corresponding to the different groups and, in this later work, they also showed that the results with only the relative intensities used in their first work were a good approximation. By comparing calculated values for lignin and cellulose with measured values, they achieved the results given in *Table 4.5*.

Table 4.5. Calculated oxygen-to-carbon ratios for different materials (from Dorris and Gray 1978[a]). The different values for kraft pulps and Whatman filter papers correspond to different samples and should represent the scatter of properties for different materials.

Material	N_O/N_C
Cellulose (Calculated)	0.83
Spruce millwood lignin (calculated)	0.33
Spruce dioxane lignin (calculated)	0.36
Whatman filter paper	0.80, 0.83, 0.79
Bleached kraft pulp	0.72, 1.14, 0.62
Bleached sulphite pulp	0.73
Dioxane lignin on filter paper	0.36
Dioxane lignin on glass	0.31

The authors concluded that the difference between the calculated values and the measured values could be ascribed to the existence on the surface of organic compounds other than cellulose and lignin.

In a later work, the same authors, (Dorris and Gray 1978[b]) presented a way to calculate the weight fraction of cellulose and lignin on the surface of the fibres from the theoretical composition of cellulose (C_6O_5) and lignin ($C_{9.92}O_{3.32}$) according to the equations:

$$W_L = 1.132 S_L / (1 + 0.132 S_L) \tag{4.16}$$

where

W_L = Weight fraction of the lignin on the surface
S_L = Segment mole fraction of lignin on the surface (according to eq. 4.17 below)

$$S_L = (5 - 6(N_O/N_C)) / (1.68 + 3.92(N_O/N_C)) \tag{4.17}$$

where N_O and N_C are defined as above.

With these definitions, the authors characterised many different mechanical pulps and, by measuring both unextracted and Benzene/acetone/water-extracted pulps, it was possible to determine both the influence of the extractives and the weight fraction of lignin on the surface of the fibres, assuming that the extracted fibres contained only cellulose and lignin. The results given in *Table 4.6* were presented by Dorris and Gray (1978[c]).

Table 4.6. Summary of the data for mechanical pulps characterised by Dorris and Gray (1978[c]). The table shows both the original data for the pulps as well as the values for the extracted sheets and the corresponding weight fraction of lignin on the surface according to eq. (16) and (17).

Material	N_O/N_c	Solvent-Extracted (Benzene/acetone/ water)	Lignin Weight fraction (on the surface)
SGW (Mill deckers)	0.47	0.63	0.32
-"- (fines free)	0.52		
-"- (aged)	0.46	0.62	0.34
SGW (Mill grinders) (fines free)	0.49	0.58	0.41
-"- (Aged)	0.50		
RMP (Mill deckers) (fines free)	0.48		
RMP (Mill grinders) (fines free)	0.46	0.65	0.28
TMP (pilot plant)	0.44	0.56	0.45

It is clear that the amount of lignin on the surface of the fibres corresponds approximately to the average bulk concentration of lignin of the fibres, i.e. all except TMP have values somewhat higher than the bulk concentration of lignin. *Table 4.6* also shows that the TMP fibres seems to have a higher content of lignin on the surface, which may also be expected considering the process conditions used to produce these pulps. It is also clear that there is a large influence of the extractives despite the fairly low total amount of extractives in the sheets. This finding was supported by experiments with a single-stage high temperature TMP where the N_O/N_C ratio decreased from 0.47 to 0.39 when the energy was increased in the refining, indicating that the surfaces were covered with material rich in carbon. The authors were not able to determine the exact source of this material but the extractives were indicated as being mainly responsible for this decrease. This statement is also supported by later work by Katz and Gray (1980). It may thus be concluded that ESCA proved to be a very useful tool to characterise the outermost surface of the fibres and, using simple mathematical formulae and careful extractions of the pulps, it was possible to determine the weight fraction of cellulose and lignin on the surface and also the influence of the extractives on the N_O/N_C ratios detected.

Several workers continued to apply ESCA to different pulp grades and a good summary, see *Table 4.7*, of this work was given by Carlsson (1996) who summarised both own work and similar work conducted by others. This summary also follows very closely the summary given by Laine (1994).

Table 4.7. The N_O/N_C ratio for different materials and a summary of the distribution of how the carbon atoms are linked to oxygen (in % of total). From Carlsson (1996) but the summary follows very closely the summary given by Laine (1994). The bleached chemical pulp was extracted with DiChloroMethane (DCM).

Material	N_O/N_C	C–C (%)	C–O (%)	O–C–O (%)	O–C=O (%)
Cellulose (theoretical)	0.83	0	83	17	0
Extr. Bleached kraft	0.80	6	75	18	1
Galactoglucomannan(theoretical)	0.81	3	78	16	3
Arabinoglucuronoxylane(theoretical)	0.81	0	78	19	3
Xylan (Laine 1994)	0.83	5	67	24	4
Lignin (theoretical)	0.33	49	49	2	0
Kraft lignin (Laine 1994)	0.32	52	38	7	3
Kraft lignin(Mjöberg 1981)	0.40	–	–	-	-
Oleic acid (theoretical)	0.11	94	0	0	6
Extractable mtrl.	0.12	93	5	0	2

This table shows that the extracted bleached pulps are very close to the theoretical value for cellulose but some C–C linkages are nevertheless found in this pulp indicating the existence of remaining lignin and/or extractives. However, according to Laine (1994), the C–C linkages found in the extracted pulp are probably due to a strongly adsorbed extractive, which it is not possible to remove with the DCM treatment.

It is also clear in the table that the lignin data of Laine (1994) fit the theoretical data, according to Freudenberg, very well, whereas the data of Mjöberg (1981) showed a considerably higher value. Both investigations used lignin isolated from spent liquors from kraft cooks and the reason for the difference between the investigations seems to be linked to the existence of small amounts of carbohydrates in the samples tested by Mjöberg, according to Carlsson (1996). The results also show the importance of the extractives for the N_O/N_C ratio on the fibres, since the value of the ratio for the extractives is so much lower than for example for cellulose.

Ström and Carlsson (1992) introduced a simple way of calculating the surface coverage of the fibres of extractives and lignin using the N_O/N_C ratios for the original pulps, extracted pulps and pure carbohydrates according to the equations

$$\Phi_{extr} = \frac{N_o/N_{c(\text{after extraction})} - N_o/N_{c(\textit{before extraction})}}{N_o/N_{c(\text{after extraction})} - N_o/N_{(\text{extractives})}} \quad (4.18)$$

$$\Phi_{lignin} = \frac{N_o/N_{c\,(\text{after extraction})} - N_o/N_{c\,(\textit{carbohydrates})}}{N_o/N_{c\,(\text{lignin})} - N_o/N_{c\,(\textit{carbohydrates})}} \quad (4.19)$$

These equations assume that the depth of analysis is lower than the depth of the covering layer. This is naturally a fairly rough assumption and, as was pointed out by Carlsson (1996), the thickness of extractive layers may be smaller than the depth of analysis in the ESCA equipment and this means that the surface coverage of extractives as determined with equation (4.18) is underestimated. Carlsson (1996) also claims that the depth of analysis in the ESCA experiments was 5–10 nm, which is fairly thick in relation to a monolayer coverage of for example extractives. The influence of the depth of analysis was also discussed by Laine (1994) who found, using ESCA analysis with different angles of emission, that the thickness of the surface layer of lignin was very thin. With these assumptions in mind, the summary presented in *Table 4.8* can be made regarding the surface coverage of extractives and lignin for different pulps, Carlsson (1996) and Carlsson (1995):

Table 4.8. Summary of composition of a number of different pulps (total amounts of DCM and lignin) (evaluated from the kappa number of the pulps and appropriate conversion factors) and the surface coverage of extractives and of lignin (on extracted pulps). From Carlsson (1995) and Carlsson (1996).

Material	DCM extractives (%)	Bulk Lignin (%) by weight	$\Phi_{extractives}$(surface composition)	Lignin/carbohydr. (surface composition)
Highyield kraft	0.2	12	13	66/34
-"-, aged	0.2	12	11	66/34
Unbleached kraft	0.09	3.3	9	30/70
-"-, aged	0.09	3.3	11	30/70
Bleached kraft	0.04	< 0.1	2	14/86
-"-, aged	0.04	< 0.1	0–5	14/86
Unbleached sulphite	0.6	4.3	14	28/72
-"-, aged	0.6	4.3	28	28/72

Table 4.8 shows that the ESCA technique is a very powerful tool to characterise different pulps and the influence of ageing and extraction on the surface properties of these pulps with regard to the interaction between fibres and fibres and other materials. It is also clear in the table that it is possible to determine the migration of extractives when sulphite pulps, i.e. pulps with a high extractives content, are stored. The table shows that the surface coverage increases from 14 to 28 % when the pulps are aged. Considering that the concentration of DCM extractives is only 0.6 %, it is astonishing to find a surface coverage of 28 %.

It is also obvious in *Table 4.8* that the surface concentration of lignin is considerably higher than the bulk concentration of the lignin. As was mentioned earlier, this was also found by Laine (1994) using ESCA with different angles of emission, which means different depths of analysis of the fibres.

In a later work, Laine et.al. (1996) investigated how different bleaching conditions affected the surface composition of fibres from an unbleached chemical pulp with an initial (unbleached) kappa value of 25.9. In general the author found that after the cooking there is a higher content of lignin on the fibre surfaces and that this slows down the bleaching response of different bleaching chemicals. The investigation also showed that the fraction of surface lignin which is removed by O_2 and H_2O_2 bleaching is lower than the average removal of the lignin, whereas O_3 gives the same fractional removal of lignin regardless of the position of the lignin in the fibre wall. The influence of ClO_2 depends on the number of bleaching stages before the addition of the ClO_2. This large investigation is summarised in *Figure 4.8* which shows the fraction of surface lignin, calculated according to eq. (4.19), as a function of the total lignin content. The sequence of bleaching and not only the chemicals as such has a large influence on the results.

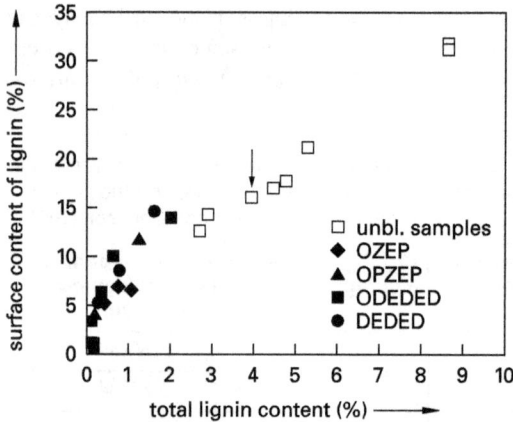

Figure 4.8. The surface coverage of lignin as a function of the total amount of lignin in softwood kraft pulps. O = oxygen, P = peroxide, Z = ozone, D = chlorine dioxide and E = alkaline extraction. The unbleached sample, marked with an arrow, is the starting material for the bleaching and results from the other unbleached samples are from Laine (1994). From Laine (1996[b]).

In this study, the surface coverage of the extractives was also investigated and it was found that the surface coverage varied between 2 and 9 % for the different bleaching conditions, whereas the total amount of extractives varied between 0.06 and 0.15 %. This is probably due to a reprecipitation of the extractives on the fibre surfaces. The bleaching studies also showed that the surface fraction of extractives decreases with O_2 and slightly with H_2O_2 whereas O_3 and ClO_2 bleaching had only a minor effect on the extractives on the fibre surface. In this respect it is believed that the neutral components of the extractives play a major role, but it must be kept in mind that bleaching operations represent two alkaline and two acidic treatments and that this, as such, has a large impact on the achieved results.

By combining enzyme treatment of unbleached chemical pulps, Buchert et.al. (1996) were able to draw some interesting conclusions regarding the relative location of the lignin, carbohydrates and extractives on the fibre surface. A xylanase treatment of a birch pulp, with a kappa number of 18.2, resulted in the removal of 29 % of the xylan originally present in the pulp and a simultaneous removal of 10 % of the lignin originally present in the pulp, but no change in the surface coverage of lignin, which indicates the existence of lignin/carbohydrate complexes. Treatment of the same pulp with a mannanase did not result in the same lignin removal. An analysis of the extractives of these pulps showed that the overall weight fraction of the extractives was 0.5 % but that the surface coverage was 35 %. The xylanase treatment reduces the surface concentration of the extractives from 35 to 20 % and this also indicates a link between the extractives and the xylan. A similar treatment of a pine kraft pulp with an initial kappa number of 25.9 did not give the same result. When treating these fibres with xylanase, 21 % the xylan originally present was removed and, at the same time, the lignin coverage of the surface increased from 17 to 28 % and only 2 % of the lignin was removed with the xylan. This indicates that the xylan is located outside the lignin on the surface of the fibres and also that the lignin is precipitated first before the xylan is reprecipitated on top of the lignin. In turn, these results also show, according to the authors, that there is no linkage between the lignin and the xylan as was found for the birch pulp. A treatment of the fibres with mannanase resulted in a decrease in the galactoglucomannan content by about 24 % of the material originally present. All these results show the very fruitful effect of combining selective removal of certain compounds from the fibres with an analysis of the difference with specific techniques such as ESCA. The investigation will probably lead to similar work in the future, even though some of the conclusions drawn are rather far reaching. Instead of cleavage of linkages between lignin and carbohydrates, pure washing can be used to explain the results. Nevertheless the combination of the techniques is very interesting.

Similar studies have been conducted by Heijnesson et.al. (1995) where the surfaces of different unbleached pulps were mechanically peeled with a novel procedure, and both the remaining fibres and the material peeled from the fibres were analysed. The results were essentially the same as those already mentioned, which means that the surfaces are richer in lignin-rich materials which can be peeled off with the described technique. A possible future application would be to combine the mechanical peeling methodology with the ESCA technique.

Another very interesting technique for chemical characterisation of the fibre surface is the Atomic Force Microscope (AFM) technique, where a very fine tip interacts with the fibre surface and either the force between the surface and the tip is detected or a topological map of the surface is drawn, Hanley (1992, 1994 and 1996). Fibres represent a very difficult task for this very sensitive technique with their extremely rough surfaces but, as was demonstrated by Hanley, both fibrillation, (Hanley et.al. 1992) and fibril angle (Hanley 1994) can be detected by the technique. Pereira et.al. (1995) also used the AFM technique to characterise the residual lignin in kraft pulps.

In order to achieve a more specific mapping of the fibre surface, a better characterised tip is needed in the AFM and, in a recent work by Akari (1996), it has been demonstrated that by coating the tip with carboxyl groups it is possible to obtain an image of single polyethyleneimine molecules on polystyrene latices. With this technique, it should also be possible to investigate the distribution of anionic charges on fibres and by modifying the tip with different probe molecules it should be possible to study other types of interactions with high lateral resolution.

As a summary of this description of the use of ESCA to characterise the fibre surface, it may be mentioned that the surface composition of the fibres can be significantly different from the bulk composition. In the case of the lignin the differences are not very large, i.e. smaller than one order of magnitude, but for extractives the difference in composition between the bulk and the surface is even larger than an order of magnitude. It should also be emphasised that the results achieved with well characterised pulps and modern ESCA equipment are very interesting and that the combination of ESCA analysis and e.g. the effects of treatment with well characterised enzymes looks very promising for the future chemical characterisation of the fibre surface. Many of these conclusions are based on the assumptions in eq. (18) and (19), and model experiments are definitely needed to show the validity and limitations of these assumptions. The AFM techniques with chemically modified tips or latices are also very interesting techniques for the surface characterisation of fibres.

88

4.4 Literature

Aberson, G.M. (1970) In: *Tappi STAP.* No.8:, p. 282

Akari, S., Schrepp, W., and Horn, D. (1996) *Langmuir* 12, 4: 857

Alince, B. (1991) *Tappi J.* 11: 200

Ampulski, R.S. (1989) *Nordic Pulp Paper Res. J.* 4, 1: 38

Ampulski, R.S., and Högfeldt, E. (1989) *Nordic Pulp Paper Res. J.* 4, 1: 42

Annergren, G. (1994) *Unpublished results.*

Annergren, G., Rydholm, S., and Vardheim, S. (1963) *Svensk Papperstidn.* 66(1): 1

Annergren, G.E., and Rydholm, S.A. (1959) *Svensk Papperstidn.* 62(20): 737

Arvanitidou, E., Klier, J., and Aronson, C.L. (1992) In: R.S. Harland and R.K. Prud´homme (Eds.) *Polyelectrolyte Gels - Properties, Preparation and Applications. ACS Symposium Series 480.* Washington, D.C.: American Chemical Soc., pp. 190

Atack, D. (1977) Advances in beating and refining. In: *Fibre-Water Interactions in Paper-Making. Transactions of the Symposium held at Oxford: September 1977, volume 1.*, p. 261

Atalla, R.H. (1995) Cellulose and the Hemicelluloses: Patterns for Cell Wall Architecture and the Assembly of Lignin. In: *8th International Symposium on Wood and Pulping Chemistry, Helsinki 1995, Proceedings*, p. 77

Attwood, T.K., Nelmes, B.J., and Sellen, D.B. (1988) *Biopolymers* 27: 201

Berg, J.C. (1993a) *Nordic Pulp Paper Res. J.* 8,1: 75

Berg, J.C. (1993b) In: J. Berg (Ed.) *Wettability. Surfactant science series v.49.* New York, Basel and Hongkong: Marcel Dekker Inc., p. 75

Berthold, J. (1996) In: *PhD Thesis.* Stockholm, Sweden: The Royal Institute of Technology, Dept. Pulp and Paper Chemistry and Techn.

Buchert, J., Carlsson, G., Viikari, L., and Ström, G. (1996) *Holzforschung* 50: 69

Buchert, J., Teleman, A., Harjunpaa, V., Tenkanen, M., Viikari,L., and Vuorinen, T. (1995) *Tappi J.* 78 (11): 125

Buchert, J., Tenkanen, M., Ek, M., Teleman, A., Viikari, L., and Vuorinen, T. (1996) Effects of Pulping and Bleaching on Pulp Carbohydrates and Technical Properties. In: *1996 International Pulp Bleaching Conference, Washington D.C:, Proceedings, Book 1.*, p. 39

Budd, J., and Herrington, T.M. (1989) *Colloids Surfaces* 36: 273

Carlsson, G. (1996) *PhD Thesis.* Stockholm, Sweden: Royal Institute of Technology

Carlsson, G., Kolseth, P., and Lindström, T. (1983) *Wood Sci. Technol.* 17: 69

Carlsson, G., Ström, G., and Annergren, G. (1995) *Nordic Pulp Pap. Res. J.* 10, 1: 17

Chaudhury, M.K. (1984) *Short-range and long-range forces in colloidal and macroscopic systems (PhD Thesis).* Buffalo, USA: SUNY

Clark, J. d'A. (1978) *Pulp Technology and Treatment for Paper.* San Francisco: Miller Freeman Publications, Inc., p. 147

Dorris, G.M., and Gray, D.G. (1978a) *Cell. Chem. Techn.* 12: 9

Dorris, G.M., and Gray, D.G. (1978b) *Cell. Chem. Techn.* 12: 721

Dorris, G.M., and Gray, D.G. (1978c) *Cell. Chem. Techn.* 12: 735

Dorris, G.M., and Gray, D.G. (1980) *J. Colloid Interface Sci.* 77, 2: 353

Engstrand, P., Sjögren, B., Ölander, K., Htun, M. (1991) Significance of carboxylic groups for the physical properties of mechanical pulp fibers. In: *6th International Symposium on Wood and Pulping Chemistry Proceedings, vol. I*. Appita, Parkville, Victoria, Australia:, p. 75

Farrar, J., and Neale, S.M. (1952) *J. Colloid Sci.* 7: 186

Felix, J. (1993) *PhD Thesis*. Gothenburg, Sweden: Chalmers University of Technology, pp. 39

Felix, J.M., and Gatenholm, P. (1993) *Nordic Pulp Pap. Res.J.* 8, 1: 200

Fengel, D. (1970) *Tappi* 53(3): 497

Fengel, D., and Wegener, G. (1989) *Wood. Cemistry, Ultrastructure, Reactions*. Berlin, New York: Walter de Gruyter, p. 66

Flory, P.J. (1953) *The Principles of Polymer Chemistry*. Ithaca and London: Cornell University Press

Fowkes, F. (1962) *J. Phys. Chem.* 66: 382

Fowkes, F. (1964) *Ind. Eng. Chem.* 56: 40

Fowkes, F., and Mostafa, M.A (1978) *Ind. Eng. Chem. Prod. Res. Dev.* 17: 3

Gellerstedt, G., and Lindfors, E.-L. (1991) On the structure and reactivity of residual lignin in kraft pulp fibers. In: *International Pulp Bleaching Conference 1991, Stockholm, Sweden, June 11–14, 1991, Proceedings, vol. 1*, p. 73

Goring, D.A.I. (1971) In: K.V. Sarkanen, and C.H. Ludwig (Eds.) *Lignings, Ocurrence, Formation Structure and Reactions*. New York, London, Sydney, Toronto: Wiley-Interscience, p. 695

Grignon, J., and Scallan, A.M. (1980) *J. Appl. Polymer Sci.* 25: 2829

Gutmann, V., Steininger, A., and Wychera, E. (1966) *Monatsh. Chem.* 97: 460

Hanley, S. (1996) *PhD Thesis*. Montreal, Canada: McGill University

Hanley, S., and Gray, D.G. (1994) *Holzforschung* 48: 29

Hanley, S., Giasson, J., Revol, J.-F., and Gray, D.G. (1992) *Polymer* 33,21: 4639

Hartler, N. (1963a) *Svensk Papperstidning* 66(11): 443

Hartler, N. (1963b) *PhD Thesis*. Stockholm, Sweden: Royal Institute of Technology

Hasa, J., and Ilavsky, M. (1975) *J. Polym. Sci.* 13: 263b

Hasa, J., Ilavsky, M., and Dusek, K. (1975) *J. Polym. Sci.* 13: 253a

Heijnesson, A-C., Simonsson, R., and Westermark, U. (1995a) *Holzforschung* 49: 75

Heijnesson, A-C., Simonsson, R., and Westermark, U. (1995b) *Holzforschung* 49: 313

Herrington, T.M., and Midmore, B.R. (1984) *J. Chem. Soc. Faraday Trans.* 1, 80: 1525

Herrington, T.M., and Petzold, J.C. (1992) *Colloids Surfaces* 64: 97

Hodgson, K.T., and Berg, J.C. (1988) *Wood Fibre Sci.* 20: 3

Jacob, P.N., and Berg, J.C. (1994) *Langmuir* 10, 9: 3086

Jayme, G., and Islam, M.A. (1973) *Papier* 27(3): 81

Johansson, M.H., and Samuelson, O. (1977) *Sven. Papperstidn.* 80: 519

Katchalsky, A. (1954) In: J.A.V. Butler, and J.T. Randall (Eds.) *Progess in Biophysics and Biophysical Chemistry, Volume 4*. New York: Academic Press Inc., and London: Pergamon Press Ltd., p. 1

Katz, S., and Gray, D.G. (1980) *Svensk Papperstidn.* 83: 226

Katz, S., Beatson, R.P., and Scallan, A.M. (1984) *Svensk Papperstidn.* 87: R48

Kerr, A.J., and Goring, D.A.I. (1975) *Cell. Chem. Techn.* 9: 563

90

Kettunen, J., Laine, J.E., Yrjala, I., and Virkola, N.E. (1982) Aspects of strength development in fibres produced by different pulping methods". *Paperi ja Puu* 4: 205

Laine, J., Buchert, J., Viikari, L., and Stenius, P. (1996a) *Holzforschung* 50, 3: 208

Laine, J., Lövgren, L., Stenius, P., and Sjöberg, S. (1994) *Colloids Surfaces* 88: 2772

Laine, J., Stenius, P., Carlsson, G., and Ström, G. (1994) *Cellulose* 1: 145

Laine, J., Stenius, P., Carlsson, G., and Ström, G. (1996) *Nordic Pulp Pap.Res. J.* 11,3: 2012

Laivins, G.V., and Scallan, A.M. (1996) *J. Pulp and Paper Sci.* 22, 5: J178

Lee, L.H. (1996) *Langmuir* 12, 6: 1681

Lee, L.H., and Luner, P. (1989) *Nordic Pulp Pap. Res. J.* 4, 2: 164

Lee, S.B., and Luner, P. (1972) *Tappi* 55, 1: 116

Lennholm, H. (1994) *Investigation of Cellulose Polymorphs by 13C-CP/MAS-NMR Spectroscopy and Chemometrics (PhD Thesis)*. Stockholm: Royal Institute of Technology, Department of Pulp and Paper Chemistry and Technology

Li, T.-Q., Henriksson, U., and Ödberg, L. (1993) *Nordic Pulp Paper Res. J.* 8,3: 326

Lindström, T. (1980) *Das Papier* 34: 561

Lindström, T. (1986) In: A. Bristow, and P. Kolseth (Eds.) *"Paper- Structure and performance"*. New York, Basel: Marcel Dekker, Inc., p. 75 and p. 99

Lindström, T., and Carlsson, G. (1979) *Proceedings EUCEPA Conf., Warzaw.*, pp.32

Lindström, T., and Carlsson, G. (1982a) *Sven. Papperstidn.* 85,3: R14

Lindström, T., and Carlsson, G. (1982b) *Sven. Papperstidn.* 85,3: R146

Lloyd, J.A., and Horne, C.W. (1993) *Nordic Pulp Paper Res. J.* 8,1: 48

Lundqvist, Å. (1996) *Lic. Thesis*. Stockholm, Sweden: Royal Inst. Techn.

Luner, P., and Sandell, M. (1969) *J. Polymer Sci. Part C* 28: 115

Mitikka, M., Teeaar, R., Tenkanen, M., Laine, J., and Vuorinen, T. (1995) In: *8th International Symposium on Wood and Pulping Chemistry, Helsinki 1995, Proceedings, 3: Poster Presentations.*, p. 231

Mjöberg, J. (1981) *Cell. Chem. Techn.* 15: 481

Nordman, J. (1987) In: *Pulp and Paper Manufacture, Volume 2. Mechanical Pulping*. Montreal, Atlanta: Joint Textbook Committee of the Paper Industry of the United States and Canada, CPPA/TAPPI, p. 272

Page, D.H. (1989) Beating of Chemical Pulps - the Action and the Effects. In: *Fundamentals of Papermaking. Transaction of the Research Symposium Held at Cambridge, September 1989, Volume 1.*, p. 1

Page, D.H., Seth, R.S., and El Hosseiny, F. (1985) Strength and Chemical Composition of Wood Pulp Fibres. In: *Transactions of the Eight Fundamental Research Symposium held at Oxford, September 1985.*, p. 77

Parham, RA (1983) In: MJ. Kocurek and C.F.B. Stevens (Eds.) *Pulp and Paper Manufacture. Vol. 1. Properties of Fibrous Raw Materials and their Preparation for Pulping. VI Ultra structure and Chemistry*. Montreal, Atlanta: Joint Textbook Committee of the Paper Industry, pp. 35–45

Pelton, R. (1993) *Nordic Pulp Paper Res. J.* 8,1: 113

Pu, Q., and Sarkanen, K. (1989) *J. Wood Chem. Techn.* 9,3: 293

Rydholm, S.A. (1965) *Pulping Processes*. New York: Interscience Publishers, p. 1115

Scallan, A.M. (1974) *Wood Sci.* 6, 3: 266

Scallan, A.M. (1978) In: *Fibre-Water Interactions in Paper-Making. Transact. Sympos. at Oxford.* London, UK: Tech. Div. Br. Paper Board Ind. Fed., pp 9

Scallan, A.M. (1989) *Tappi J.* 72, 11: 157

Scallan, A.M., and Grignon, J. (1979a) *Sven. Papperstidn.* 82.3: 4

Scallan, A.M., and Grignon, J. (1979b) *Sven. Papperstidn.* 82,3: 40

Scallan, A.M., and Tigerström, A. (1992) *J. Pulp Paper Sci.* 18, 5: J188

Scallan, A.M., Katz, S., and Argyropoulos, D.S. (1988) In: C. Schuerch (Ed.) *Cellulose and Wood - Chemistry and Technology. Proceedings of the Tenth Cellulose Conference.* John Wiley & Sons Inc., p. 1457

Sears, K.D., Alexander, W.J., Goldschmid, O., and Hamilton, J.K. (1978) *Tappi* 61(9): 105

Sjöström, E. (1981) *Wood Chemistry, Fundamentals and Applications.* New York: Academic Press

Sjöström, E. (1989) *Nordic Pulp Paper Res. J.* 4, 2: 90

Sjöström, E. (1993a) *Wood Chemistry. Fundamentals and Applications, second edition.* San Diego: Academic Press, Inc., p. 54

Sjöström, E. (1993b) *Wood Chemistry. Fundamentals and Applications, second edition.* San Diego: Academic Press, Inc., p. 133

Stone, J.E., and Scallan, A.M. (1965a) In: F. Bolam (Ed.) *Consolidation of the Paper Web. Transact. Symp. Cambridge, Vol. 1.* London, UK: Tech. Sect. Brit. Paper Board Makers Assoc., p. 1453

Stone, J.E., and Scallan, A.M. (1965b) *J.Polym. Sci.: Part C* 11: 132

Stone, J.E., and Scallan, A.M. (1965c) *Pulp Paper Mag. Canada* 66, 8: T4071

Stone, J.E., and Scallan, A.M. (1967) *Tappi* 50, 10: 496

Stone, J.E., and Scallan, A.M. (1968a) *Cellulose Chem. Techn.* 2, 3: 3432

Stone, J.E., and Scallan, A.M. (1968b) *Pulp Paper Mag. Canada* 69, 6: T2881

Ström, G., and Carlsson, G. (1992) *J. Adhesion Sci. Technol.* 6: 745

Swanson, J.W., and Cordingly, S. (1959) *Tappi* 42,1: 812

Swerin, A., and Wågberg, L. (1994) *Nordic Pulp Paper Res. J.* 9,1: 18

Teleman, A., Harjunpää, V., Tenkanen, M., Buchert, J., Hausalo, T., Drakenberg, T.,, and Vuorinen, T. (1995) *Carbohydrate Research* 272: 55

Terashima, N., and Atalla, R. (1995) In: *The 8th International Symposium on Wood and Pulping Chemistry, Helsinki.*, p. 69

Tieman, H.D. (1906) In: *U.S. Dep. Agric. For. Serv. Bull. No 70.*

Timell, T.E. (1967) *Wood Sci. Technol.* 1, 1: 45

Toussaint, A.F., and Luner, P. (1993) In: K.L. Mittal (Ed.) *"Contact Angle, Wettability and Adhesion".* VSP Publications, p. 383

Van de Ven, T.G.M. (1996) *Private Communication*

van Oss, C.J. (1994) *Interfacial forces in aqueous media.* New York, Basel and Hongkong: Marcel Dekker, Inc.

Wågberg, L., Ödberg, L., and Glad-Nordmark, G. (1989) *Nordic Pulp Paper Res. J.* 4, 2: 71

Wågberg, L., Winter, L., and Lindström, T. (1985) Papermaking Raw Materials. In: *Transactions of the Eight Fundamental Res. Symposium held at Oxford, Mechanical Eng. Publ. Ltd., London.*, p. 917

Wågberg, L., Winter, L., Ödberg, L., and Lindström, T. (1987) *Colloids Surfaces* 27: 163

Weiss, N., and Silberberg, A. (1976) In: J.D. Andrade (Ed.) *Hydrogels for Medical and Related Applications. ACS Symposium Series 31*. Washington D.C.: American Chem. Soc., p. 69

Westerlind, B., and Berg, J.C. (1988) *J. Appl. Polym. Sci.* 36: 523

Westermark, U., and Vennigerholz, F. (1995) Morphological Distribution of Acidic and Methy-lesterified Pectin in the Wood Cell Wall. In: *8th International Symposium on Wood and Pulping Chemistry, June 6-9, 1995, Helsinki, Finland; Proceedings, 1: Oral Presentations.*, p.103

Winter, L. (1987) *Lic. Thesis*. Stockholm, Dept. Paper Technology: Royal Inst. Tech.

Wu, S. (1982) *Polymer Interface and Adhesion*. New York, Basel and Hongkong: Marcel Dekker Inc.

Yin, Y.-L., Prud´homme, R.K., and Stanley, F. (1992) In: R.S. Harland and R.K. Prud´homme (Eds.) *Polyelectrolyte Gels - Properties, Preparation and Applications. ACS Smposium Series 480*. Washington, D.C.: American Chemical Soc., p. 91

Zhang, Y., Sjögren, B., Engstrand, P., and Htun, M. (1994) *J. Wood Chem. Techn.* 14, 1: 83

5 Fibre Suspensions

Bo Norman
Department of Fibre and Polymer Technology, KTH

5.1 Introduction

Elongated, elastic cellulose fibres suspended in water have a strong tendency to form connected fibre networks. This influences the flow picture for a fibre suspension, and is important in a number of process steps. Most important is to minimize the occurrence of fibre flocks with any essential network strength at the moment of the sheet forming. In this chapter, the mechanical flocculation of cellulose fibres will be treated in detail.

A fibre under free rotation in a dilute fibre suspension covers a spherical volume with a maximum diameter equal to the fibre length. The maximum concentration of freely mobile fibres with length L is then represented by the case of closely packed spheres of diameter L. The corresponding volume concentration c_v of fibres will then depend not only on the fibre length L but also on the fibre diameter d, and in reality it will be unambiguously determined by the slenderness ratio of the fibre L/d. The more slender the fibre, the lower is the volume concentration required to give the fibres full individual freedom of movement.

5.2 Sediment Concentration

Add a small amount of cellulose fibres to a beaker of water, circa 0.5 g fibres per litre water. If the beaker is left undisturbed, the fibres will gradually settle at the bottom of the beaker under the influence of gravity (density of cellulose ≈ 1.5 g/cm^3). If the weight of the added fibres is known, it is possible from the height of the sediment to calculate the weight concentration of the fibres in the sediment, the so-called sediment concentration. The sediment concentration is usually expressed either as a percentage or in the ten times smaller unit g/l.

The sediment consists of a mechanically connected fibre floc. The sediment concentration is actually the lowest concentration at which a connected floc can be formed by these fibres. The least stirring in the beaker is sufficient to cause the fibres in the sediment to whirl up, which indicates that the floc strength in the sediment is very low. In the treatment of wastewater, sediment is concentrated through careful stirring.

The level of the sediment concentration depends mainly on the mean slenderness of the fibres. For normal cellulose fibres, the slenderness lies within an interval of 50 to 300. The higher the slenderness, the lower is the sediment concentration. *Table 5.1* shows the approximate sediment concentration for some different types of pulp.

Table 5.1. Sediment concentration in g/l.

Softwood fibres	2–3
Hardwood fibres	4–5
TMP	4–6
Groundwood pulp	5–9

In order to be able to make a paper with a low occurrence of fibre flocs from a fibre suspension, the fibre concentration must not greatly exceed the sediment concentration of the pulp.

5.3 Network Generation and Flocculation

Carefully add dry, unflocculated fibres (e.g. cut polymer fibres) to a beaker with water to a concentration clearly greater than the sediment concentration, and mix gently with a rod. Place the rod in the middle of the beaker, and it will immediately fall against the edge of the beaker.

Add turbulence energy with the help of a small propeller agitator. When the stirring has been stopped, the rod can be placed in the middle of the beaker, without falling against the beaker edge. The fibre suspension thus no longer behaves as a liquid (a liquid cannot take up shear stresses without continuing deformation) but as an elastic body, a fibre network, with considerable network strength.

The explanation of this network strength is that, during the turbulent stirring process, the fibres have been deformed from their natural shapes, which they then strive to regain when the stirring ceases. When they then strike adjacent fibres, the springing back is hindered, and the fibres become locked in fixed shapes in a network. As shown in *Figure 5.1*, at least three contact points are required with the surrounding fibres for a fibre to be kept in an interlocked state.

It is traditionally suggested that fibre flocs are broken down by turbulent shear, and thus it is often suggested that turbulence generation in a headbox is a useful way of avoiding fibre flocculation in the final paper. The above-described experiment clearly demonstrates, however, that when turbulence decays, fibre flocs are reformed. Temporary floc breakdown can therefore be obtained by the introduction of turbulent energy, but re-flocculation will always take place.

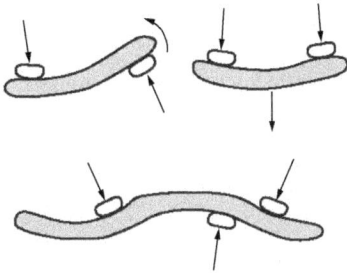

Figure 5.1. Top: Mobile fibres; Bottom: Interlocked fibre.

Meyer and Wahren analysed how the average number of contact points n with surrounding fibres depends on the slenderness of the fibres L/d and the volume concentration c. They then derived an approximate relationship according to the equation:

$$c_v = \frac{16\pi \dfrac{L}{d}}{\left(\dfrac{2}{n}\dfrac{L}{d} + \dfrac{n}{n-1}\right)^3 (n-1)} \qquad (5.1)$$

Kerekes and Soszynski have reported two experiments, which confirm the validity of the above model for network forming. First they poured a fibre suspension into a rotating beaker with a sloping axis. After a period of time, uniform, nearly spherical fibre flocs then form, so-called Jaquelin-flocs, see *Figure 5.2.*

When flocs had been formed, the fibre suspension was diluted with water, under continued rotation of the beaker. If the original fibre concentration was low, the fibre flocs were broken up by this dilution. When the original concentration was progressively increased, conditions were eventually reached where the flocs resisted the subsequent dilution. They call this critical, original concentration the threshold concentration, and the way in which it varies with the slenderness of the fibres is shown in *Figure 5.3.*

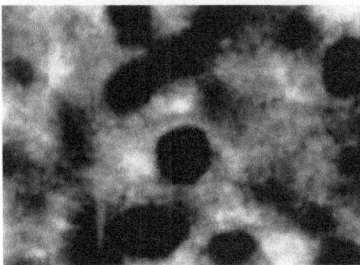

Figure 5.2. Jaquelin-flocs.

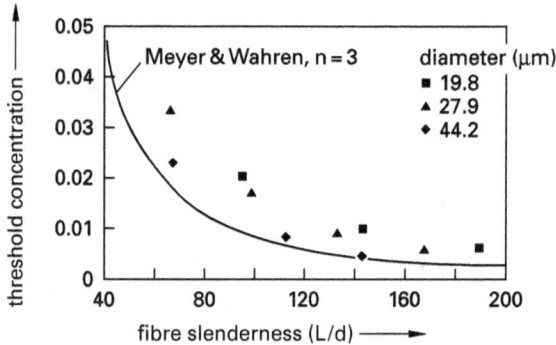

Figure 5.3. Threshold concentration as a function of fibre slenderness. The full line corresponds to equation 17-1 with $n = 3$ (Kerekes and Soszynski).

In *Figure 5.3*, a curve has also been drawn according to equation 5.1 with an average of three contact points between a fibre and the surrounding fibres. It can be seen that there is good agreement between this curve and the experimentally determined threshold concentration.

Kerekes and Soszynski then verified by another experiment that the mechanical interlocking of the fibres is decisive for the strength of the flocs. They produced Jaquelin-flocs from polymer fibres and then heated the suspension above the softening temperature of the polymer. Flocs exposed to this temperature treatment were considerably easier to break up than non-treated flocs. This can be explained by the fact that the stress relaxation in the fibres, which has been possible at the higher temperature level, has really reduced the connecting forces between the fibres in the flocs.

5.4 Crowding Factor

Kerekes has introduced the concept of crowding factor N for the number of fibres of length L and diameter d within a reference volume, at a volume concentration of c_v. The reference volume is chosen as the volume of the sphere created by a freely rotating fibre of length L. The following relationship is then valid for the crowding factor N:

$$N = \frac{2}{3} c_v \left(\frac{L}{d} \right)^2 \tag{5.2}$$

Since the density of cellulose is approx. 1.5, it would be expected that the volume concentration of the fibres is less than the weight concentration, but considering the amount of water inside the fibre wall and lumen, the volume concentration can be as high as twice the mass concentration.

The mass concentration c_m is much easier to evaluate experimentally, and the Crowding factor N can alternatively be evaluated from the equation:

$$N = \frac{\pi}{6} c_m \frac{L^2}{\omega} \qquad (5.3)$$

where L is fibre length in mm and ω is fibre coarseness in g/m

Using a small laboratory device, Kerekes studied flocculation after an initial dispersion with grid induced mixing, see *Figure 5.4*, left. The grid was initially in top position, and moved quickly downwards through the suspension. When turbulence had decayed, flocculation was evaluated from photographs taken in transmitted light, see *Figure 5.4*, right.

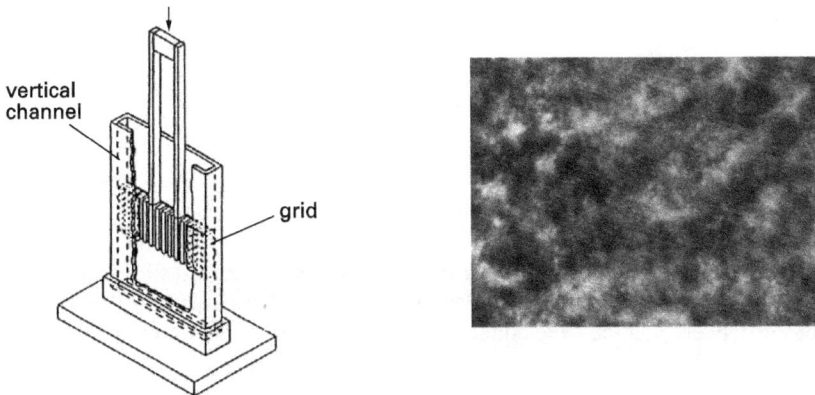

Figure 5.4. Left: Device with grid and fibre suspension in vertical, narrow channel. Right: Final state of flocculation for Douglas fir, 0.5 %, Crowding Factor $N = 75$ (Kerekes).

Applying the above method, Kerekes has derived a relationship between the Crowding factor and fibre contacts at increasing concentrations, see *Table 5.2*.

Table 5.2. Fibre contacts – Crowding factor.

Concentration	type of fibre contact	corwding factor N
Dilute	Rare collisions	$N < 1$
semi-conentrated	Frequent collisions	$1 < N < 60$
conentrated	continuous contact	$N > 60$

5.5 Floc Stretching

Floc breakdown by stretching in an elongational flow is a more efficient way of reducing floc size than turbulent shear, since reflocculation will not take place. This mechanism has been known for a long time, but was earlier not emphasised in papermaking. Nordström demonstrated that paper formation is improved by a strong acceleration flow in the headbox nozzle (see further in the chapter on Forming). This was a result of stretching the flocs to breakage

Physically, in a contracting nozzle flow the fluid at the front end of a floc will move at a higher velocity than that at the back end. The longitudinal flow friction forces then created inside a floc will stretch it and eventually break it apart, see *Figure 5.5*.

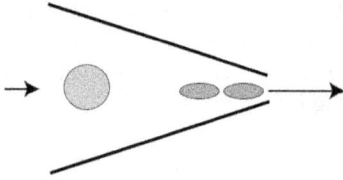

Figure 5.5. Floc stretching and breaking in a contracting nozzle flow.

5.6 Floc Strength

The assumption that the mechanical interlocking of the individual fibres creates the strength of the fibre network predicts that the network strength increases with increasing fibre concentration. With measurements in a rheometer, Wahren showed a relationship between the shear failure strength τ_B of the network and the fibre concentration c according to the formula:

$$\tau_B = \tau_o \left(c - c_s\right)^k \tag{5.4}$$

where τ_o and k are characteristic parameters for the fibres and c_s is the sediment concentration. The network strength is thus zero at the sediment concentration and then increases rapidly. At high concentrations, the network strength is approximately proportional to c_k, where the exponent k is close to two for most pulps. At concentrations close to the sediment concentration, colloidal phenomena can influence the network strength. In this way, the occurrence of fibre flocs in the finished paper can e.g. be influenced by the choice of retention chemicals.

If considerable amounts of air are present in the system, this interacts with the fibres. Small air bubbles may get stuck in the fibre network and may influence the network formation, especially at low concentrations.

5.7 Pipe Flow

In pipe flow, fibre suspensions will in principle behave in different ways depending on the flow rate, see *Figure 5.6*.

At low speeds, a plug flow occurs, where all the fibres move as a connected network. The whole speed difference is located in a boundary layer which is laminar at low speeds, and which consists of practically fibre-free water. If a T-junction is connected, almost pure water can then be withdrawn at low discharge rates. In this case, suspension sampling therefore requires very large discharge flows, if the concentration of the sample is to be representative of the average concentration in the pipe.

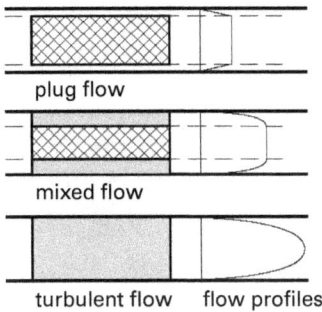

Figure 5.6. Different pipe flow mechanisms for fibre suspensions.

At higher flow rates, the boundary layer becomes turbulent, and the surface of the plug begins to break up. A flow region is entered which is usually called mixed flow, and which may eventually transform into a completely turbulent flow at very high flow rates.

When a suspension flows in a pipe, it is exposed to a retarding shear force τ through friction against the pipe wall. The frictional force gives rise to a pressure drop along the pipe, as shown in *Figure 5.7*.

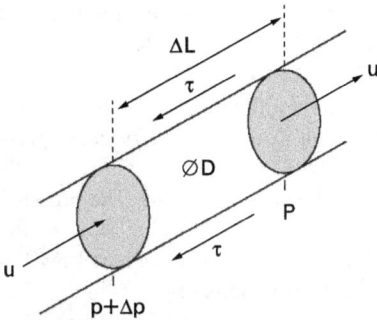

Figure 5.7. Pressure Δp during a flow along the pipe length ΔL.

At equilibrium, the frictional force must balance the force generated by the pressure drop:

$$\tau \pi D \Delta L = \Delta p \frac{\pi D^2}{4} \Rightarrow \tau = \frac{D}{4} \frac{\Delta p}{\Delta L} \tag{5.5}$$

The pressure drop Δp can thus be represented by the shear stress τ.

In *Figure 5.8*, the flow of water is compared with a fibre suspension, and the pressure drop is shown as a function of the flow rate.

To start the fibre suspension to move from standstill, a considerable initial pressure drop is required. This is due to the fact that the fibre network is pressed against the pipe wall, and that the static friction between fibres and pipe wall must be overcome.

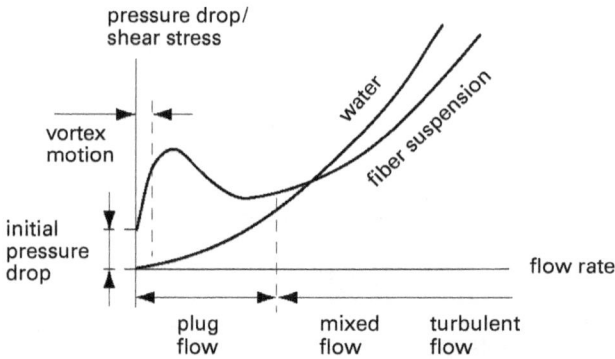

Figure 5.8. Pressure drop as a function of flow velocity in pipe flow for a fibre suspension and for water.

When fibre suspensions are pumped through complex pipe systems, the start itself can give rise to problems. At standstill, parts of the suspension can be drained (e.g., at valves), so that local high fibre concentrations occur, and thereby related high initial pressure drops. It is actually easier to start the pumping of a fibre suspension the greater the pipe diameter is. The explanation is that the frictional forces are proportional to the circumference area of the pipe (i.e. to the diameter) whilst the driving pump pressure force is proportional to the cross-sectional area of the pipe (i.e. to the square of the diameter). This is never the case for the flow of Newtonian fluids, like water.

At very low flow rates, rolls of loose fibres are formed between the pipe wall and the network covering the main part of the cross section. With increasing flow rate, the frictional pressure drop increases rapidly within the roll-generating region.

If the rate is further increased, the boundary layer begins to become turbulent. The pressure drop then passes through a maximum, drops and reaches a minimum when the flow in the boundary layer has become completely turbulent. Thereafter, the region of turbulent flow begins to stretch further into the fibre suspension, and the pressure drop increases.

As shown in *Figure 5.8*, the curves for water and fibre suspension cross each other and the fibres therefore induce a frictional decrease at high flow rates. The explanation is that the fibres have a dampening effect on the generation of turbulence. It can be pointed out that a further friction decrease can be obtained if friction-reducing polymers are added. This is due to the fact that these polymers, which have a considerably smaller geometrical extension than the fibres, dampen the generation of very small-scale turbulence.

Figure 5.9 shows experimentally determined pressure drop curves for a long-fibre pulp. Within normally used ranges, the following approximate relationship (5.6) is valid for the concentration dependence of the pressure drop Δp, also for a concentration range considerably above the maximum 3.4 per cent concentration shown in *Figure 5.9*.

$$\Delta p = k \frac{L}{D} c^2 \tag{5.6}$$

Here, k is a fibre-dependent parameter, L/D the pipe length expressed in number of diameters D, and c the fibre concentration.

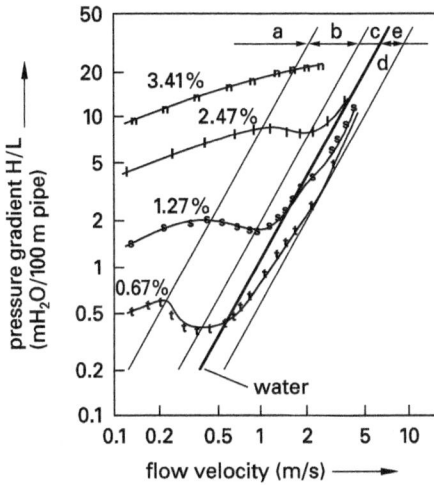

Figure 5.9. Pipe flow pressure drop at different concentrations of long fibre pulp (Möller et al). a) Roll generation range; b) Plug flow range; c) Range with increasing boundary layer turbulence; d) Mixed flow with a lower pressure drop than for water, and where the pressure drop curve approaches the water curve from below at high velocities.

The dependence on the square of the concentration is also valid for the strength of the fibre network according to equation (5.2). This confirms that the friction is intimately related to the strength of the fibre network close to the pipe wall.

It can be noted that mechanical sensors for fibre concentration in the interval of 2–5 per cent measure a frictional quantity, which is also closely related to the network strength. The sensor signal therefore becomes almost proportional to the square of the concentration. When the curves for water and fibre suspension intersect each other, the following approximate relationship between the flow velocity u and fibre concentration c is valid (5.7):

$$u = 1.22c^{1.4} \tag{5.7}$$

The constant 1.22 is not dimensionless.

5.8 High Concentration Fibre Suspensions

The pipe flow apparatus used by Möller to obtain the results in *Figure 5.9*, was not dimensioned for higher concentrations than approx. 3.5 per cent. Gullichsen has later shown that the principal appearance of the curves from the pipe flow tests is the same at much higher concentrations. Gullichsen made tests with high-concentration fibre suspensions in a cylindrical vessel, and noted the torque required to drive a central agitator at a given rotational speed for different concentration levels, see *Figure 5.10*.

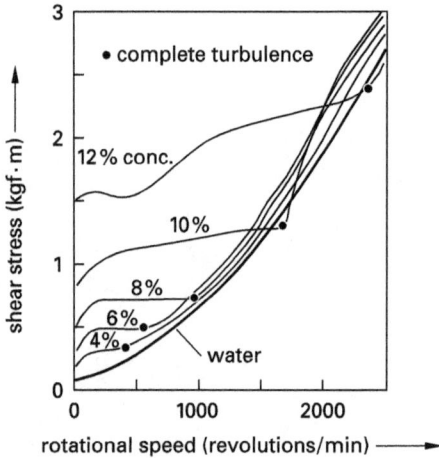

Figure 5.10. Shear stress versus rotational speed at different fibre concentration levels. The black circles represent start of wall range fluidisation (Gullichsen).

With a transparent housing of the vessel, it was possible with high-speed filming to determine when the state of the fibre suspension at the wall became "fluidised". The results show a great similarity to the pipe flow tests in *Figure 5.9*, indicating that fibre suspensions behave in principle in similar ways regardless of the concentration. At consistencies as high as 12 per cent, fibre suspensions can still be fluidised if the local energy supply is sufficiently high. It may seem surprising that even at fibre concentrations as high as 12 per cent, the friction pressure drop at high flow rates according to Gullichsen's test becomes lower for a fibre suspension than for pure water.

In the pumping of fibre suspensions with conventional centrifugal pumps (with open pump wheels), total separation is obtained at the inlet at fibre concentrations higher than about 4 per cent. Gullichsen therefore equipped the pump wheel with a dispersion propeller at the inlet, so that a fluidised fibre suspension reached the pump wheel itself. In this way, it was possible to pump fibre suspensions with consistencies of up to 10–15 per cent. For this development, he was awarded the Wallenberg prize ("The Forest Products Industry's Nobel Prize"). Increased concentration is now being used within different sections, specifically in the pulp-making processes, which may lead to both an energy saving and reduced apparatus costs. Common to these applications is that, as a result of a local high-energy supply, the fibre network is fluidized, which is a basic condition for carrying out the processes.

The terminology to designate the increased concentration differs between e g pulp- and papermaking, since the designation is based on the conventional concentration in the processes concerned. In high-concentration forming, the concentration is around 3 per cent, which in the pulping area would be considered as low concentration. In beating, on the other hand, low-concentration beating takes place at a concentration of 3–4 per cent and high-concentration beating at about 30 per cent. Beating at a concentration range of 10–15 per cent can then be designated beating at *medium-concentration*, a term also used for many of the pulp-making processes.

6 Water and Material Balances in the Paper Mill

Bo Norman
Department of Fibre and Polymer Technology, KTH

6.1 Introduction

The task of a paper machine is to produce a product with a given grammage and composition, constant in time. For this to be possible a constant flow of the correct material components must be fed to the paper machine and the retention of each single component must be constant. It is possible to maintain accurate volume flows with the help of speed controlled pumps and magnetic flow meters. If accurate equipment also existed for the online measurement of concentration, it would therefore be possible accurately to control the material flow even if the concentration of the fibre suspension fed to the paper machine varied. Today's concentration meters are not, however, sufficiently accurate for this purpose. It is therefore important that the process equipment is chosen and dimensioned with special regard to the task of keeping all concentration variations as small as possible.

A flow schedule is a "connection schedule" for the papermaking process, and defines how different equipments are connected and how water and material are led in different stages of the manufacturing process. Besides all the flows in the "manufacturing direction", there are also a large number of return flows. The main task of the white water system is to recycle water and material, to reduce the water consumption and to reduce material losses. The broke system feeds back the paper web from different stages of the manufacturing process, such as edge trim, during web failure and operation disturbances and portions of the finished product, which have been rejected for quality reasons.

Because of e.g. variations in the incoming raw material and consequent retention variations and variations in the amount of broke, there are always variations in the concentration and material composition at all positions in the white water system. To minimize these variations, a typical flow schedule includes not less than 20 circuits for control of the concentration. It is not our ambition to discuss in detail a realistic flow schedule in this book, but to give a certain insight into water consumption, material losses and related problems.

For economical and environmental reasons, it is important to minimize the water consumption and to reduce the material content of the discharged effluent. A high degree of closure means low water consumption and low material losses in the effluent, while the opposite is usually called an open system. To illuminate the concept of degree of closure, this chapter shows how this is improved when the process has been developed, and simple water and material balances are used for this purpose. A description is also given of the disc filter, which has a very important function in today's paper mills with regard to reducing material losses and keeping undesired material transfers between different paper machines under control.

In order to illustrate the methods available for the analysis of complete white water systems, it is shown how it is possible to analyse the conditions in a single circulation and how this can be extended with computer aids to more complete systems.

Finally, the manner in which the volume of the system influences the dynamic properties of the system is demonstrated, with special emphasis on the influence of different retention levels for individual components.

6.2 Total Balances

A paper mill can be illustrated as in *Figure 6.1*, where Q is the total material flow, including all components.

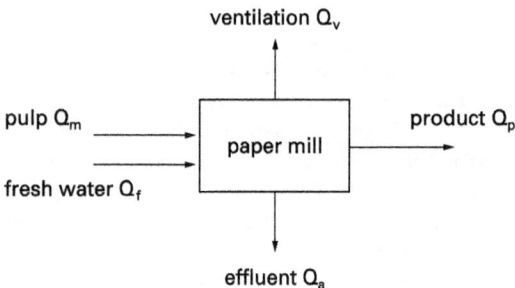

Figure 6.1. Water and material balance for a paper mill.

The total flow Q can be divided into a number of components $q(i)$, of e.g. water (1), coarse fibre (2), fine fibre (3), filler (4) and dissolved components (5). At equilibrium in the system, there is a balance for each component between incoming and outgoing component flows.

For water:

$$q_m(1) + q_f(1) = q_v(1) + q_p(1) + q_e(1) \tag{6.1}$$

In the case of the water losses $q_v(1)$ through ventilation, approximately half the water that is evaporated in the drying section is lost, but the material content of this water is low. For the other components, the following balance applies:

$$q_m(i) = q_p(i) + q_e(i) \tag{6.2}$$

where the effluent losses q_e are normally small.

If the balance is to be used for general comparisons, the specific flows required to produce one ton of paper should be calculated.

The material flow q for a given component can be calculated from the total mass flow Q and the concentration c, using the equation:

$$q = Qc \tag{6.3}$$

The effluent losses are especially important:

$$\text{effluent losses} = Q_e c_e \tag{6.4}$$

6.3 The Short and Long Circulations

To reduce the effluent losses to a minimum, the white water system in a modern paper mill is designed so that the water is reused to the greatest possible extent, and so that material, which it contains, can be added to the product, see *Figure 6.2*.

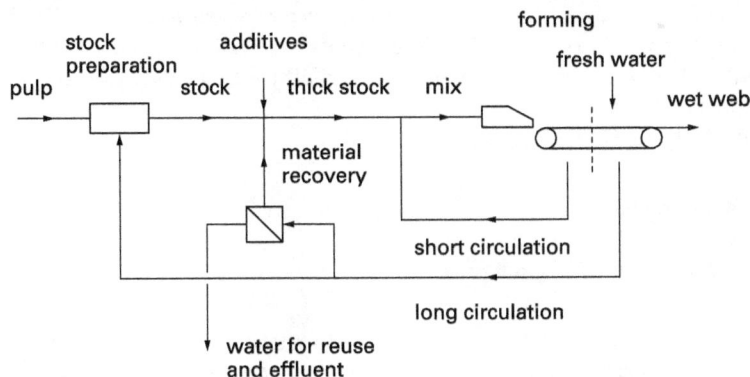

Figure 6.2. Basic principles of the white water system of a paper machine, including a short and a long circulation.

The short circulation dilutes the thick stock pulp to a mix with a suitable concentration for sheet forming, which normally means a fibre concentration of less than one per cent.

All surplus white water is recycled in the long circulation, which is drawn very schematically in the figure, since in reality it includes a large number of flow loops, including the broke handling system. The water, which leaves the system, is first cleaned through an internal recycling of material over a disc filter. A certain proportion of the water which is reused is also first be cleaned, to be useful for e g showers.

The water flow $q(1)$ in tons ($\approx$ m3) required to transport 1 ton of material (specific water flow) at a concentration of c per cent can be calculated from the equation:

$$q = \frac{100 - c}{c} \tag{6.5}$$

The specific water flow at different positions of the process, see schedule in *Figure 6.2*, has been calculated using equation (6.5) and are shown in *Table 6.1*.

Table 6.1. Specific water flow at different positions in the process.

Position	Fibre concentration (%)	Water (m³ per ton product)
Thick stock pulp	4	24
Mix (headbox)	0.5	199
Web after the wire section	20	4
Web after press section	40	1.5

6.4 Open and Closed Systems

During the first half of the twentieth century, the white water system was normally open, see *Figure 6.3*. In addition to the water, which was added with the pulp, a large amount of fresh water was also added. This was pure water, which was used for dilution of the stock, as shower water to keep headbox and the wire section clean, as sealing water to seal packing boxes in pumps, as cooling water etc. It was also typical that the effluents were not separated according to the degree of cleanliness. Everything was mixed in a common effluent. The only feedback of water, which occurred, was in the short circulation.

The following figures could apply to a typical old mill:

- effluent flow Q_e: 200 m³/ton product
- effluent concentration c_e: 1500 mg/l = 1.5 kg/ m³
- material losses $Q_e c_e$: 200 · 1.5 = 300 kg/ton product

A first, simple stage of closing the white water system was to introduce a white water tank, for the overflow from the wire pit and for the dewatering from the later parts of the wire section. The tank should be large enough to allow low flow velocities, so that some sedimentation of material was possible. From the bottom of the tank, water enriched by sedimented material was recirculated for internal use in the stock preparation plant, thus reducing the need for fresh water. The effluent water leaving as overflow was then partially cleaned, see *Figure 6.4.*

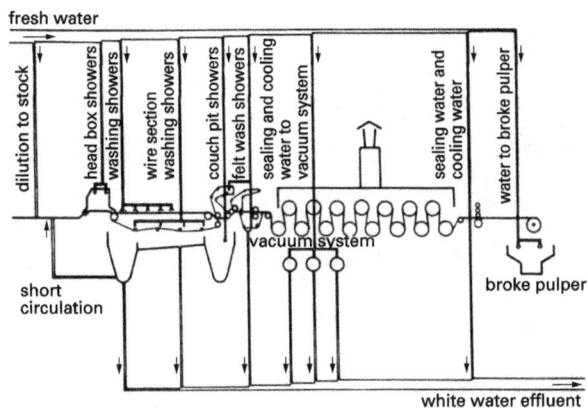

Figure 6.3. Old design of open paper machine white water system (SSVL).

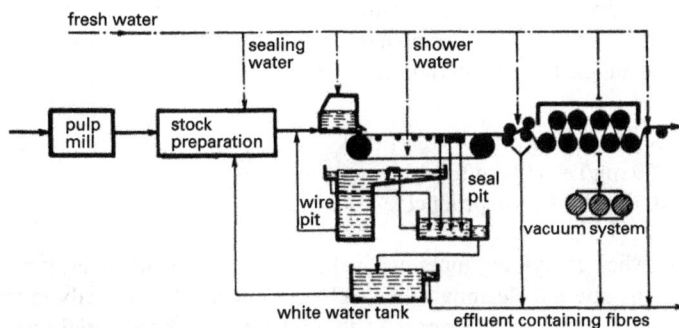

Figure 6.4. White water flow diagram, circa 1970 (SSVL).

The following values could then be representative for the effluent losses, indicating a considerable decrease in comparison with the open systems:

- effluent flow Q_e: 100 m³/ton product
- effluent concentration c_e: 350 mg/l = 0.35 kg/ m³
- material loss $Q_e c_e$: 100 · 0.35 = 35 kg/ton product

In a modern closed white water system, as shown in *Figure 6.5*, a large part of the fresh water supply is replaced with uncleaned and cleaned white water from the long circulation. To reduce the water supply to the process, it is also necessary to concentrate the incoming pulp and thus to reduce the water added with the pulp from the pulp mill. There is also a separation between fibre-free and fibre-containing effluents. For example, by avoiding contamination with fibres in the cooling water from the vacuum pumps, this water can be mixed with the fibre-free effluent.

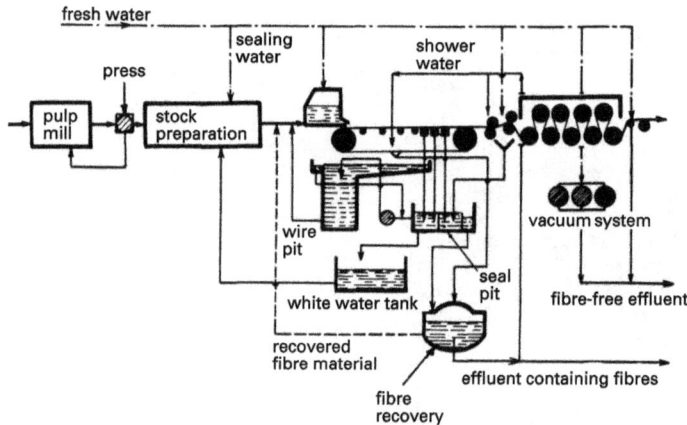

Figure 6.5. Modern design of a closed white water system. (SSVL).

It is today possible to reach effluent flows as low as 10-20 m³/ton. If the fibre-containing effluent is cleaned internally with a disc filter, a realistic value of the material content in the effluent with chemical pulps can be 50 mg/l, and it is somewhat higher for mechanical pulps.

The following figures can be attained in a modern mill:

- effluent flow Q_e: 10 m³/ton product
- effluent concentration c_e: 50 mg/l = 0.05 kg/m³
- material losses $Q_e c_e$: $10 \cdot 0.05 = 0.5$ kg/ton product

These values are valid only when everything functions well; if there are disturbances, it may be necessary to let water pass the internal cleaning. The discharge is then taken directly to the external cleaning. Temporary discharges are often responsible for the greatest material losses from a modern paper mill.

The water consumption differs considerably between different product types. Most fresh water consumptions, such as showers, sealing water etc, are fairly independent of the production rate. A considerably higher specific water consumption and specific material discharge are then obtained in the case of e g tissue paper (very low grammage) than in the manufacture of high grammage board products.

6.4.1 Disc Filter

The disc filter mentioned earlier is used to recover fibres, fines and filler material from internal flows. The principle design of a disc filter is shown in *Figure 6.6*.

The white water to be filtered is sucked through wire-cowered hollow discs, which are partly submerged in a trench. The degree of separation for fine materials would, however, be too poor if only the uncovered wire were used as filter medium. Besides the uncleaned white water, additional "sweetener" pulp is therefore added to the disc filter headbox. A suitable stock component (with low filtration resistance) is used for this purpose, and is after the filter passage returned to the stock line together with the material separated from the white water.

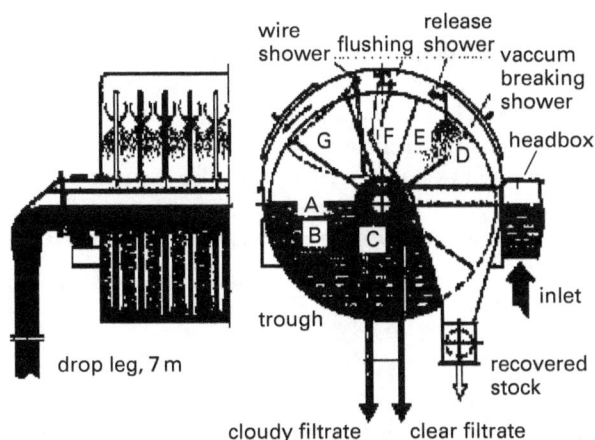

Figure 6.6. Disc filter for material recovery from white water.

The sweetener pulp is first mixed with the uncleaned white water in the headbox. A pulp cake is formed during dewatering on the wire screen, through which the white water is to be filtered according to a filtration cycle described in *Figure 6.6*. The procedure is similar to the papermaking process itself.

The filter is built around a tubular axis, the cross-section of which is divided (in the picture) into ten sectors. Each filter disc consists of ten filter sectors, which are each connected to a channel in the filter axis. A filtrate valve at the end of the pipe axis opens and breaks for vacuum to the channels (filter sectors) respectively. Vacuum is created with a drop leg or with a vacuum pump.

When the axis of the filter discs rotates, the following work cycle is obtained:

A   The filter sector enters the suspension
B   The filter sector is under vacuum and cloudy filtrate is extracted
C   The filtrate valve changes from cloudy to clear filtrate
D   The vacuum is disconnected by a vacuum breaker shower
E   The fibre cake is removed with the help of a shower and is recovered to the stock line through the funnels between the discs
F   Cleaned water is drained from the filter sector via the axis channel
G   The filter wire is cleaned with the help of an oscillating shower

The cloudy filtrate is recycled. The clear filtrate can partly be reused, partly be discharged to the external effluent treatment.

6.4.2 Separation of Paper Machine Systems

A mill with several paper machines often has, at least to a certain extent, a connection between the long circulations of the different machines. Material will thus be transferred between the different machines. This can be avoided by using a disc filter for internal process closure. A disc filter can be connected in the long circulation for a given paper machine, and the separated material can be recycled to the stock line of that machine. In this way, the machine is isolated from the other paper machines in the mill with regard to the separated material. This technology is applied particularly in the case of printing paper machines using mechanical pulp furnish, since these contain large proportions of fines with low retention. There are also examples of the utilization of two disc filters connected in series for this function, to effectively separate two machines from each other with regard to material carry-over. Examples of material transfer between machines are treated in Section 6.5.

6.5 The Retention Concept

Figure 6.7 demonstrates how material is recycled via the white water system. In a normal mill today, material flows are monitored in the short circulation, but no attempts are made to quantify the corresponding flows in the long circulation. This is due to practical reasons, since the long circulation in reality consists of large number of loops.

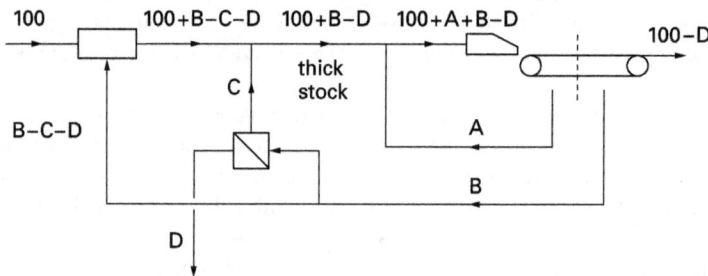

Figure 6.7. Per cent material in the white water system.
A: In the short circulation
B: In the long circulation.
C: Material recovered by the disc filter
D: To external cleaning or common stock preparation system.

A common misconception is that the thick stock pulp ($100 + B - D$) reflects the material composition of the furnish and the product. *Figure 6.7* shows clearly that this is not the case. There may actually be considerable material quantities, which are recycled, and as an example, the long circulation in newsprint manufacture can contribute with up to half the amount of fines in the thick stock, which is fed into the short circulation. It is normally impossible to obtain a relevant stock sample, which is not enriched with recirculated material. Web edge trim after the couch is often the best alternative.

In order to quantify the degree of material separation in a specific process, the retention concept is used:

Retention = The proportion of a component present in an original mixture which is found in the mixture at some stage of its treatment or in the final product.

In order to calculate a retention value, three quantities must thus be specified:

1. the material component
2. the original stage in the process
3. the later stage in the process

It is especially important to point out that the retention should be determined for each component individually, since it can differ considerably between different components. Of main interest is the retention level of low retention components like fines and filler.

6.5.1 The Retention of the Short Circulation

Figure 6.8 defines the total flow rates Q and component concentrations c in the short circulation.

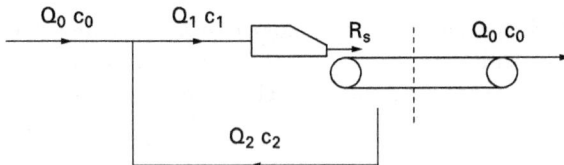

Figure 6.8. Total flows Q and component concentrations c needed to define the retention R_S in the short circulation.

The retention of the short circulation R_S can then be calculated according to equation (6.6).

$$R_S = \frac{Q_0 c_0}{Q_1 c_1} = \frac{Q_1 c_1 - Q_2 c_2}{Q_1 c_1} = 1 - \frac{Q_2 c_2}{Q_1 c_1} \tag{6.6}$$

If the retention R_S is calculated according to Equation 6.6, both the concentrations and the flow rates must be known. In practice, however, it is difficult to determine the flow rates Q, since these are normally not measured. An approximate value of the retention of the short circulation has therefore been introduced (called "Single Pass Retention" in US) by neglecting the influence of flow ratio Q_2/Q_1 in Equation (6.6). This ratio approaches unity, with an increasing degree of dilution of the thick stock.

The term Primary retention, R_P, is here proposed to designate this type of "retention", which in principle differs from the correct retention level R_S.

$$R_P = 1 - \frac{c_2}{c_1} \tag{6.7}$$

According to Equation (6.7), the primary retention R_P will underestimate the real retention R_S of the short circulation.

Since the primary retention gives only an approximate value of the retention in the short circulation, it cannot be used for comparisons between different paper machines or between before and after a rebuild of the white water system on a paper machine. The use of R_P is on the other hand a simple way to follow-up retention changes on a given paper machine. It should however not be applied to the calculation of material balances.

The primary retention takes into consideration only the concentration difference between mix and white water. It then gives a better measure of the local efficiency of the filtration process on the wire, than the correctly defined retention value would give. If only a small part of the wire section is considered, e.g. the dewatering by a single dewatering element, the use of the correct retention definition R_S would mean that the calculated value would depend mainly on the amount of dewatering. 5 % local dewatering, with the same white water concentration as that in the headbox would give $R_S = 95$ %. For the primary retention, however, we would calculate $R_P = 0$, a more informative value in this case.

6.5.2 The Retention of the Long Circulation

In a paper mill, one is normally satisfied with considering the retention of the short circulation, but with today's high degrees of system closure, there is also reason to consider the material content in the long circulation. *Figure 6.9* shows the parameters, which define the retention level R_L of the long circulation.

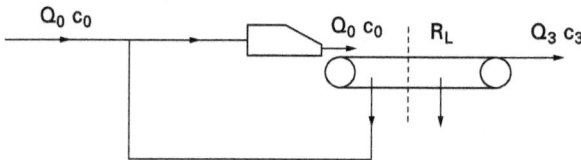

Figure 6.9. Total flows Q and component concentrations c, defining the retention R_L in the long circulation.

The figure shows that the material flow on the wire, which passes the borderline between the short, and the long circulation is equal to $Q_0 c_0$, i.e. equal to the flow, which enters with the thick stock. It therefore follows that the retention R_L in the long circulation can be calculated as the ratio between the material flows in the web after the couch and that entering with the thick stock according to Equation (6.8).

$$R_L = \frac{Q_3 c_3}{Q_0 c_0} \tag{6.8}$$

Like in the case of the retention in the short circulation, it is desirable also in the case of the long circulation to be able to calculate an approximate retention value with the help of concentrations alone, i.e. without knowledge of the flow rates involved. In this case, however, in contrast to what was the case for the primary retention, it is not possible to neglect the flow rate ratio Q_3/Q_0 since this differs considerably from unity.

If we consider a chemical pulp furnish consisting of fines and fibres only, it is possible to make a simple evaluation of fines retention in the long circulation, assuming that the fibre retention at this late stage of dewatering amounts to 100 %, which is a good approximation. The approximate long circulation retention value for fines, R_{LF}, can then be calculated using Equation 6.9:

$$R_{LF} = \frac{m_3}{1-m_3}\frac{1-m_0}{m_0}$$

(6.9)

where m_3 and m_0 are the proportions of fines components in the web after the couch and in the thick stock respectively.

In the calculation one then relies on the fines content in relation to the constant fibre content. It should be noted that if web edge trim after the couch is taken as web sample, it does not matter that the dry content is influenced by e.g. edge cut showers, since the water content (or rather concentration c) will not influence the calculations.

In the characterisation of mechanical pulps, the size distribution of the material is often determined by removing a fines fraction m with the help of a "200 mesh" wire. The rest, $(1-m)$, is then an approximate measure of the "long-fibre" fraction.

6.5.3 Wire Retention

The total wire retention R_W defined as the material flow in the wet sheet in relation to the material flow from the headbox, see *Figure 6.10*, is given by Equation (6.10):

$$R_W = \frac{Q_3 c_3}{Q_1 c_1}$$

(6.10)

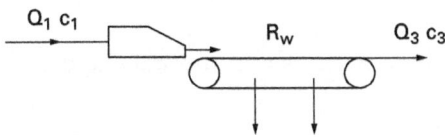

Figure 6.10. Flow rates Q and concentrations c for determination of wire retention RW.

By combining Equations (6.6) and (6.8), it can be shown that R_W is equal to the product of the retention of the short and long circulations:

$$R_W = R_S R_L$$

(6.11)

If the retention has been calculated separately in the short and long circulations, it is thus easy to calculate the wire retention. If R_S and R_L are changed so that the wire retention remains unchanged, however, the conditions as a whole are less favourable with a low value of the retention in the long circulation. This is due to the fact that a larger part of the white water system of the paper machine is influenced in such a case, which means that the time effects are larger (see Section 6.6).

An approximate value of the wire retention level of fines R_{WF} can be calculated as the product of the primary fines retention R_{PF} and the approximate long circulation fines retention, R_{LF} from Equations (6.7) and (6.9)

$$R_{WF} \approx R_{PF} R_{LF} = \left(1 - \frac{c_2}{c_1}\right) \frac{m_3}{1 - m_3} \frac{1 - m_0}{m_0} \tag{6.12}$$

Since the first is underestimated whilst the latter is overestimated, the product will be more correct than the single factors. It is worth repeating that the overall retention value for the complete furnish is of less interest than that of the low retention level components.

6.6 Simplified Balance Calculation

A simplified evaluation of fines material flow in a short circulation will be demonstrated below.

Assume that one unit of fines material is added with the thick stock, and that the wire retention for this component is 20 per cent. Assume further that, of the flow dewatered through the wire, 75 per cent is led to the short circulation whilst the remaining 25 per cent is recycled through the long circulation, see *Figure 6.11*.

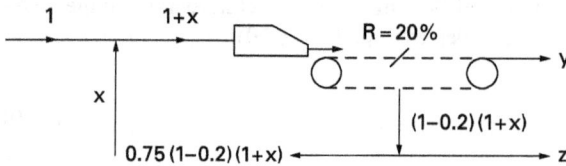

Figure 6.11. Fines balance in a short circulation.

Let the recycled material flow in the short circulation be x. The flow carried to the headbox is then

$(1 + x)$

and the proportion passing through the wire is

$(1 - 0.2)(1 + x)$

If 75 per cent is led to the short circulation, this flow amounts to

$0.75(1 - 0.2)(1 + x)$

According to the first assumption, this flow is then equal to x, i.e.

$X = 0.75(1 - 0.2)(1 + x)$, which gives $x = 1.5$

Figure 6.12 shows a summary of the fines flow. Note that, in this special case, as much fines is recycled via the long circulation as was originally added to the stock.

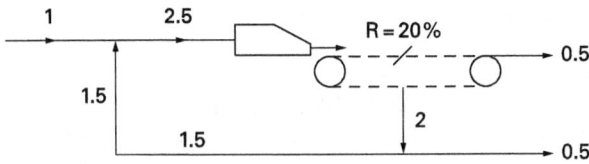

Figure 6.12. Fines flow in a short circulation.

Figure 6.13 shows an example of the transport of material, e g fines, between two parallel paper machines. It could represent newsprint manufacturing with a common TMP plant. Both long circulations are returned to the common TMP plant, which means mixing of the long circulations. One machine has a retention level of 50 per cent whilst the other has 25 per cent. Of the originally added fines material, a larger proportion will reach the product from the machine with the higher retention level. The calculation is carried out in a similar way as for the short circulation above.

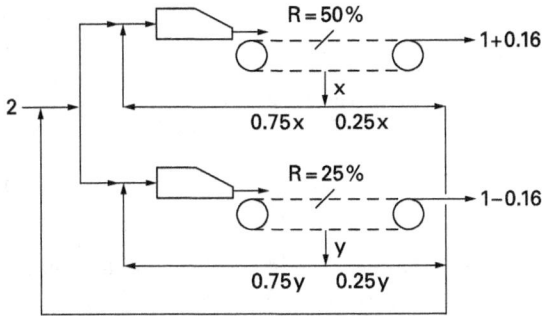

Figure 6.13. Effect on product composition of mixing between the long circulation flows for two machines with different retention levels.

Carry out the calculation and show that circa 16 per cent extra fines will end up in the product from the machine with the higher retention level, while the corresponding reduction is valid for the other machine.

This would not be acceptable regarding product quality, if the two machines are expected to produce similar products. In reality two separate TMP lines would be used. Alternatively, each long circulation could be cleaned using a disc filter, and the separated materials returned to respective machine.

6.7 Dynamics of the White Water System

In a mill, the conditions are in reality not static but are instead highly dynamic, where several days' prehistory can influence the conditions at a certain time. To reduce the amplitudes of different variations, mixing tanks with suitable residing times are used. To reduce the variations in incoming stock, three successive tanks with preceding concentration controls are traditionally applied. However, each additional tank volume will slow down the dynamics of the whole system.

Previously, we have analysed balances for systems run at stationary conditions. The following text describes a simple dynamic case: Start-up of a water-filled short circulation, see *Figure 6.14*. All flow rates Q are constant, and only the material concentrations c vary with time. Besides the steady state equations also applicable in the static case, the dynamic conditions in the wire pit must now be added, and its volume V is if deciding importance. Efficient mixing is assumed to take place at all times, so that the outlet concentration is equal to the average concentration in the pit. The following material conserving equation then applies:

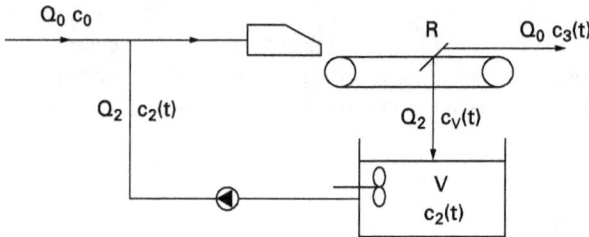

Figure 6.14. Parameters involved in the calculation of dynamics in the short circulation.

$$V\frac{d\left[c_2(t)\right]}{dt} = Q_2 c_v(t) - Q_2 c_2(t) \tag{6.13}$$

Starting with a water-filled system, the boundary condition $c_2(0) = 0$ applies. For the concentration c_3 in the sheet on the wire, the following relationship can then be derived:

$$\frac{c_3(t)}{c_0} = 1 - (1-R)e^{-R\frac{t}{V/Q_2}} \tag{6.14}$$

The normalized time $t/(V/Q_2)$ is a measure of the number of times the liquid volume in the wire pit has been replaced.

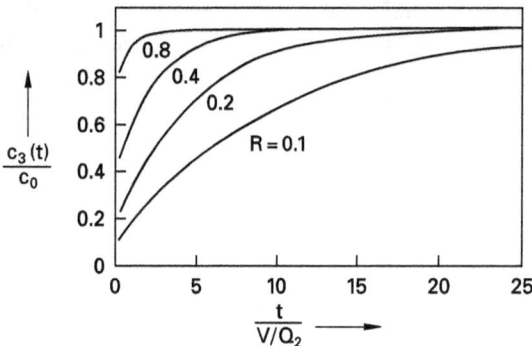

Figure 6.15. Dynamics for start up of a water filled short circulation. Normalised concentration c_3/c_0 on the wire as a function of dimensionless time $t/(V/Q_2)$ at different retention levels R.

In *Figure 6.15*, the equation has been drawn for values of the retention R of between 10 and 80 per cent. The figure shows clearly that the approach to steady state is slower the lower the retention value. The reason is that, before steady state is attained, there must be time for a correspondingly high concentration of components with low retention to be built up in the white water system. This means that e.g. fines and filler, which have a relatively low retention level, reach steady state considerably more slowly than the fibre fraction.

Note that the lowest retention level possible applies to dissolved components. If these follow the water flow, the minimum retention, with the designations in *Figure 6.14*, amounts to:

$$R_{min} = \frac{Q_0}{Q_0 + Q_2} \tag{6.15}$$

which inserted into Equation 6.14 will give the following relationship

$$\left[\frac{c_3(t)}{c_0}\right]_{Dissolved\ material} = 1 - \frac{Q_2/Q_0}{1 + Q_2/Q_0} e^{-\frac{1}{1+Q_2/Q_0} \frac{t}{V/Q_2}} \tag{6.16}$$

Since the retention level influences the time to equilibrium, there is no simple way to change the web grammage, while maintaining a constant product composition. To e.g. increase the grammage, the thick stock flow is increased. When the thick stock flow is increased, the fibre content (high retention) in the web changes rapidly, while the fines fraction (low retention) is slowly built up to its equilibrium position in the web. During the intermediate time, too low a fines content will therefore be present in the product. When the thick stock dosage is reduced, the fines content in the product will instead initially become too high. A better understanding of the dynamic conditions in the white water system of a paper machine will become increasingly important, as the demands for quality uniformity increase.

6.8 Computer Calculation of Static and Dynamic Material Balances

In the 1970's, the computer program GEMS (General Energy and Mass Balance System) was developed for the numerical solution of this type of equation system. The total flow scheme was then first divided into different sub-processes, e.g. mixing, dewatering, dilution etc. The sub-processes were then combined together to describe the system of interest, see *Figure 6.16*. The equation system was then solved (applying iterative algorithms), using a mainframe computer.

To simulate the dynamic behaviour of the system, the SIMNON program was applied. Starting from the static balance, relevant volumes and transport times must be included.

It was normally not the computer part of the work, which was the most extensive in studies of a real system, but instead the task of collecting a sufficient amount of input data (retention values, flows etc) to characterise the system. Later, commercial simulation programs with graphical interfaces have been developed, which can be run on personal computers, Flow Mac and Ideas are based on the Extend program, and several other programs exist, like Wingems, PAPMOD and Balas. Some of these programs are also able to carry out dynamic calculations in white water systems.

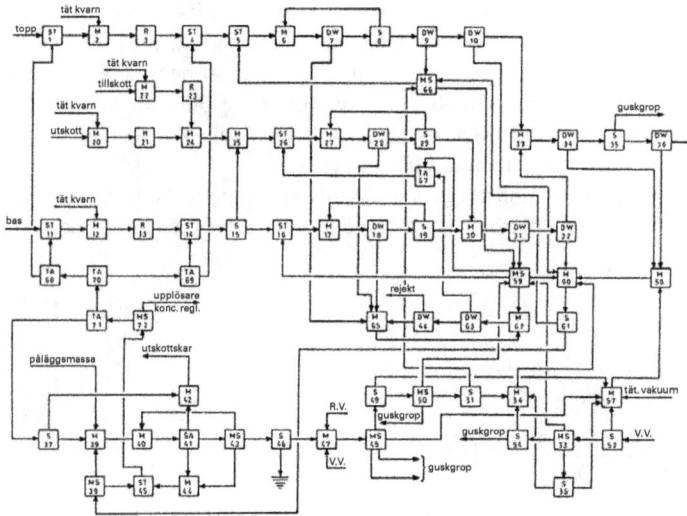

Figure 6.16. Block diagram of the GEMS type for computer calculation of system balances on a machine for the manufacture of three-ply linerboard.

6.9 The FEX Pilot Machine

Figure 6.17 shows the white water system in the pilot paper machine FEX at STFI-PF.

Pulp is prepared batch wise, including refining. The stock is stored in five large chests, and total storage capacity is ca 500 m³. Two stock lines with individual compositions can be fed for three-ply forming. A large number of chemicals addition positions are provided.

In the two short circulations, trash separator and deaeration is provided. A disc filter is installed, and the excess white water is filtered before leaving the system. After passing through the machine, paper samples can be rolled at the press section, for separate drying in a ten cylinder off-line dryer. However, most of the paper is mixed with cleaned white water, pumped back to a screw press and dumped in a container for recycling to some paper mill.

In *Figure 6.18* the FEX System is modelled using the FEXTEND Program (based on Extend).

A single-ply application is modelled starting where the thick stock is added and ending where the web leaves the press section. A main advantage of this graphical interface based program – besides being possible to run on a desk top computer – is the possibility to introduce individual drawings of the different components in the model, such as wire section, wire pit, deaerator, disc filter etc.

Figure 6.19 shows the result of a simulation for fibres, fines and solved materials respectively, regarding the dynamic behaviour during start-up of the water filled FEX System. According to the simulation, 60 minutes are required before 90 per cent of the steady-state value has been attained for the solved components, while 20 minutes are enough for fines and only 10 minutes for the fibre fraction. Experimentally determined results agree approximately with the simulations.

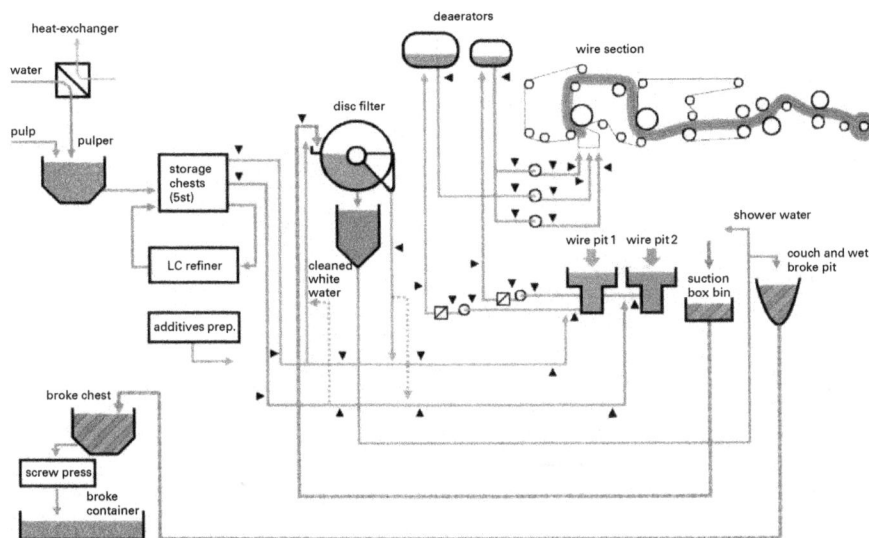

Figure 6.17. White water system for the pilot plant FEX at STFI. Black triangles indicate points for chemicals addition.

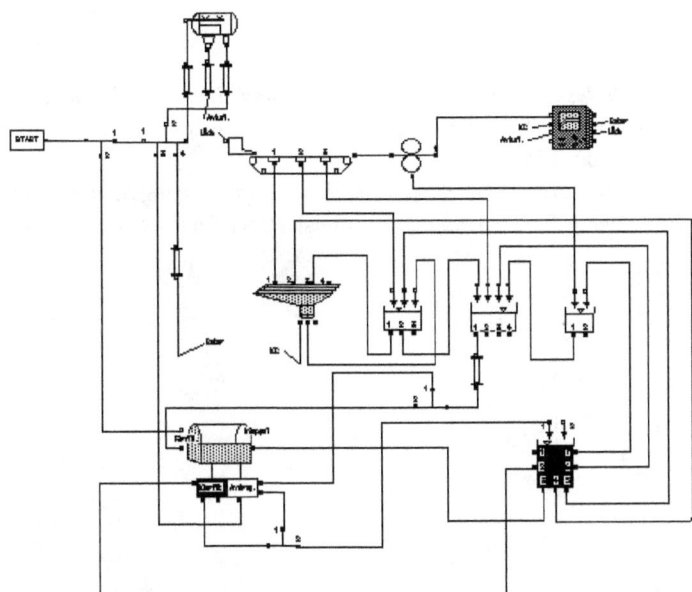

Figure 6.18. Simulation model for the white water system of the FEX machine. Piping enlargements (five places) represent transport times (FEXTEND).

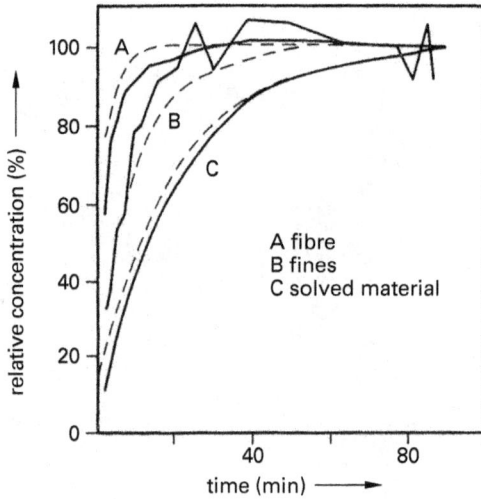

Figure 6.19. Time to equilibrium for fibre, fines and solved material after start-up of a water filled system of the FEX machine. Fully drawn curves: experimental; broken curves: simulations.

6.10 Literature

SSVL (1974) *The Forest Products Industry's environmental care project. Technical Summary, Chapter 8 Paper-making.* SSVL (The Swedish Forest Products Industries' Foundation for Water and Air Care Research), pp. 154–194

Paulapuro, H. (2000) *Stock and water systems of the paper machine. Papermaking Science and Technology. Book 8: Papermaking Part 1, Chapter 5.* Finnish Paper Engineers' Association, p. 123

7 Industrial Beating/Refining

Göran Annergren, Nils Hagen
SCA

7.1 Aim of Beating/Refining – Effect on the Fibres

Beating/refining is a mechanical treatment of a pulp suspension, with the aim to achieve pulp properties suitable for papermaking. The term beating is used mainly for treatments in laboratory equipment and in old-fashioned mill beaters. The term refining refers to a similar treatment but is more appropriate for modern mill equipment and will be used exclusively for mill processes in the following. It is mainly chemical and semichemical pulps and recycled fibres, which are treated in separate refining stages whereas the properties of for instance mechanical pulps are primarily developed already in connection with the mechanical treatment in the pulping stage/stages.

The refining makes the fibres flexible and ready to form fibre to fibre bonds in paper. A primary goal is to produce strength, usually in the form of high tensile strength, or in some cases high tensile energy absorption (achieved by high consistency refining). Refining helps to straighten the free fibre segments in paper to a certain extent and to improve the formation of paper. For the present situation with both softwood and hardwood pulps available for the furnishes the normal target of the refining is to improve bonding ability of the fibres whereas fibre cutting is reserved for excessively long fibres, see *Figure 7.1.*

Refining requires a considerable amount of energy, increases the drainage resistance of the refined pulp and decreases the bulk of the paper, the latter also being negative when bending stiffness is important for the product. In fact, refining influences all paper properties to some extent. As both positive and negative effects are obtained the refining should not be brought further in commercial operation than is absolutely necessary for satisfying the paper quality specifications.

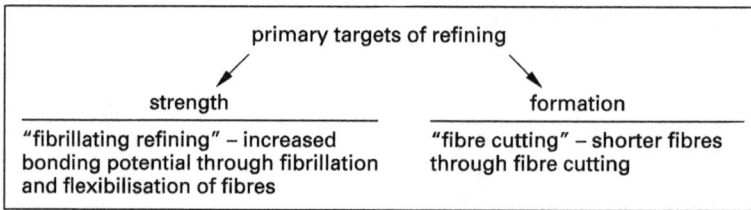

Figure 7.1. Two different primary targets in refining of papermaking fibres, to improve bonding ability of fibres ("fibrillating refining") and to reduce fibre length ("fibre cutting").

Figure 7.2 shows the characteristic features of a refining zone. (For refining plate pattern, see *Figure 7.8.*) The pulp suspension, normally at low consistency (2–6 % and most typically 3–5 %), is pumped through a refining zone. Refining takes place when the fibres are trapped between a rotating and a stationary bar or between counter-rotating bars, whereas the fibre network is reorganised in the grooves between the bars so that it again becomes accessible to refining forces in the next passage between bars.

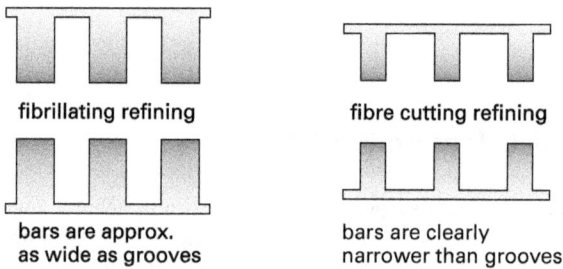

Figure 7.2. Schematic picture of refiner sets for fibrillating refining and fibre cutting respectively. Courtesy of Metso Paper.

The stresses that fibres experience under these conditions are responsible for the refining effects. As illustrated in *Figure 7.3*, fibres are influenced by tensile, compression, and shear forces, which give a complicated pattern of refining effects. Furthermore, the degree of refining varies considerably between fibres in commercial operation. In some cases both practically unrefined and strongly refined fibres can be found in the suspension after refining.

The heterogeneity of refining produces a pattern of different fibre effects, as illustrated in *Figure 7.4*. Depending on differences in refining conditions, the proportions between the different fibre effects may vary considerably. Generally internal and external fibrillation is essentially positive refining effects for strength development, although it also causes enclosure of water and increased drainage problems. Fibre cutting and fibre deformation through axial compression are more or less negative in this case. On the contrary when using excessively long fibers, fiber cutting might be the preferred effect to achieve improved formation. It is therefore quite obvious that refining is a very complex process, which puts serious demands on refiner design and refining strategy.

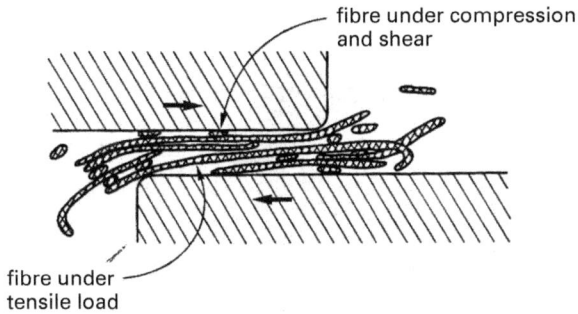

Figure 7.3. Impression of fibres trapped between counterrotating bars (Page 1989). Courtesy of Fundamental Research Committee.

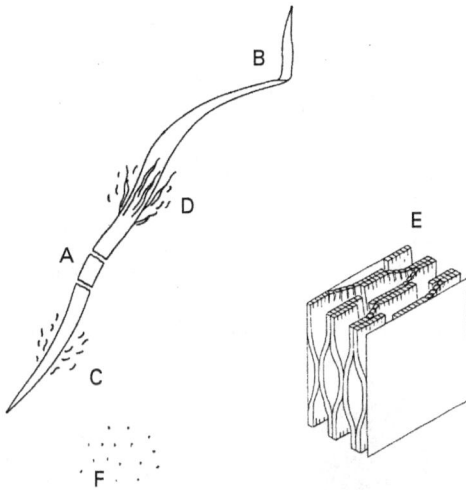

Figure 7.4. Schematic picture of the effect of refining on a fibre.
A. Fibre cutting
B. Fibre deformation (development of curl, kinks and fibre microcompressions, as well as fibre straightening
C. Creation of fines
D. External fibrillation
E. Internal fibrillation
F. Dissolution/migration of fibre wall material
G. Other changes, primarily on the molecular level

7.2 Refining Equipment – Process Conditions

The refining equipment can be either conical or disc type refiners as schematically shown in *Figure 7.5a* and *Figure 7.5b*. *Figure 7.6* shows an example of a conical refiner and *Figure 7.7* of a disc refiner.

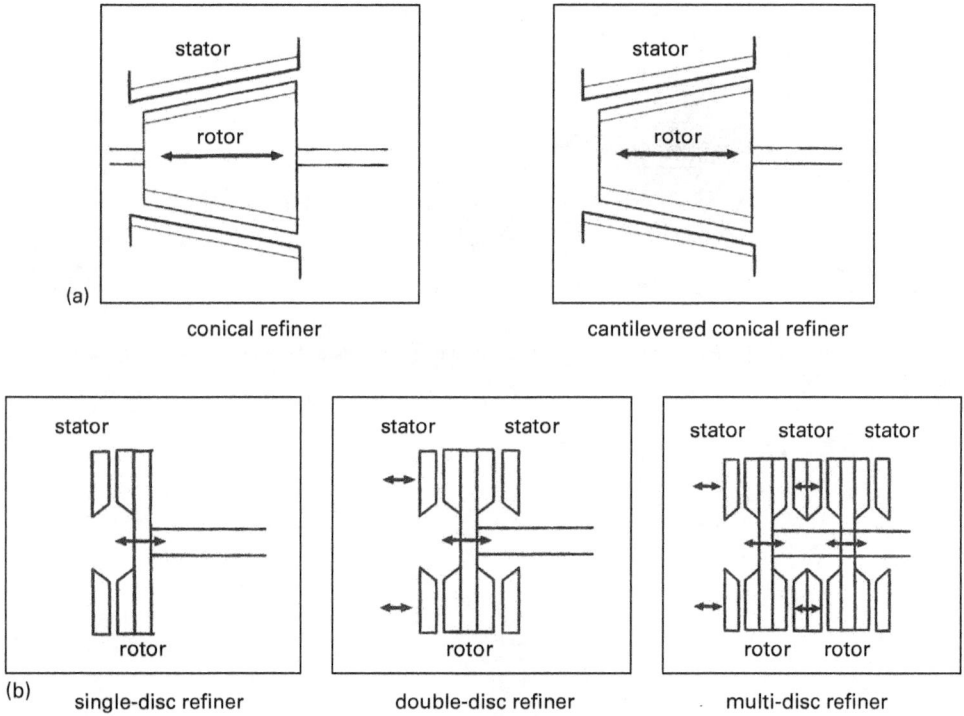

Figure 7.5. Schematic description of conical and disc type refiners. Courtesy of Metso Paper.

Figure 7.6. Picture of a Metso cantilevered conical refiner. Courtesy of Metso Paper.

Figure 7.7. Picture of a Beloit double disc refiner. Courtesy of Metso Paper.

The conical refiners today normally have a rather low cone angle (≤30°) and some also have a cantilevered design for easy access to fillings (plates). These refiners are used in low consistency refining. Disc refiners can be divided into single disc, double disc and multi disc refiners. Single disc are today mostly used in high consistency refining whereas double disc is normally used for low consistency refining. Multi disc refiners are intended for rather low intensity refining with a fine plate pattern and therefor suitable for special applications where a gentle fibre treatment is needed such as in post refining of mechanical pulps.

The refining action is normally described in terms of a theory called the specific edge load theory. This rather simple description of the complex refining process characterises the refining action with the help of only two parameters, namely the amount of energy consumed expressed as the net energy consumed per dry weight of pulp and the refining intensity expressed as the net energy consumed per length of bars crossing each other.

In mathematical terms

$$E_{net} = \frac{P - P_0}{q \cdot c}$$

$$SEL = \frac{P - P_0}{L}$$

$$L = Z_r \cdot Z_S \cdot l \cdot n$$

where
E_{net} = specific refining energy, [kWh/t] (bone dry ton)
P = total refining power, [kW]
P_0 = idling power (no load power), [kW]
q = flow, [m³/h]
c = fibre consistency, [t/m³]
SEL = specific edge load, [J/m]
L = cutting edge length per time, [km/s]

Z_r = number of rotor bars
Z_s = number of stator (or counter rotating) bars
l = common contact length of opposite bars [km]
n = rotational speed, [1/s].

The SEL theory fails to consider several important factors, such as refining consistency, bar and groove dimensions and gap clearance. More sophisticated theories have been presented but these theories have not been accepted to the same extent as the simpler SEL theory, which has proven to be industrially useful.

The refining plates with the bar and groove pattern has to be changed regularly as they are worn down. For smaller refiners the cone or disc refiner plate can be cast in one piece, but for the bigger machines the plates are manufactured in segments. A plate segment is shown in *Figure 7.8*. Observe that the direction of bars and grooves deviate from the radial direction, and furthermore that the deviations normally are opposite for the rotor and the stator (or the counter rotating plate). This is to certify that bar areas always will be in a crossing position and thus carry the load applied. An alteration between bar-bar and bar-groove load carrying will cause unwanted vibrations.

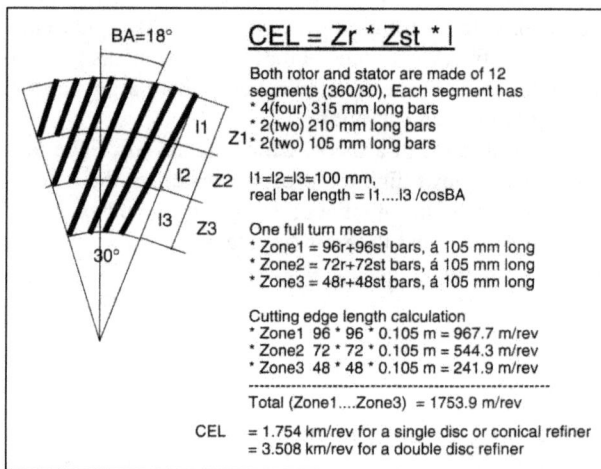

CEL = Zr * Zst * l

Both rotor and stator are made of 12 segments (360/30), Each segment has
* 4(four) 315 mm long bars
* 2(two) 210 mm long bars
* 2(two) 105 mm long bars

l1=l2=l3=100 mm,
real bar length = l1....l3 /cosBA

One full turn means
* Zone1 = 96r+96st bars, á 105 mm long
* Zone2 = 72r+72st bars, á 105 mm long
* Zone3 = 48r+48st bars, á 105 mm long

Cutting edge length calculation
* Zone1 96 * 96 * 0.105 m = 967.7 m/rev
* Zone2 72 * 72 * 0.105 m = 544.3 m/rev
* Zone3 48 * 48 * 0.105 m = 241.9 m/rev

Total (Zone1....Zone3) = 1753.9 m/rev

CEL = 1.754 km/rev for a single disc or conical refiner
     = 3.508 km/rev for a double disc refiner

Figure 7.8. Refiner plate segment. Cutting edge length calculation. Courtesy of Metso Paper.

Refiner plate manufacturing is a precision job as the normal distance between bars in low consistency refining is in the range of one or two tenth of a millimetre. This is to be compared with the width of a swollen fibre which is in the range of 20–40 μm.

For a specific pulp lower SEL promotes internal and external fibrillation whereas higher SEL gives a harsher treatment leading to more fibre damage in terms of fibre cutting and fines production. In terms of paper properties the harsher treatment at higher SEL results in lower attainable maximum strength levels though the initial increase in strength normally is comparable or faster than what is achieved at lower SEL, cf. *Figure 7.9*.

In practice, the SEL normally spans from around 0.5 to 6 J/m. For multi discs the figures are lower and for specific refining purposes such as cutting of fibres they can be higher than 6 J/m.

As sulphite and hardwood sulphate pulps are more sensitive to mechanical treatment and require less refining energy, these pulps are refined at rather low SEL (0.5–1.5) whereas softwood sulphate pulps normally are refined at higher SEL (1.5–6.0).

tensile index

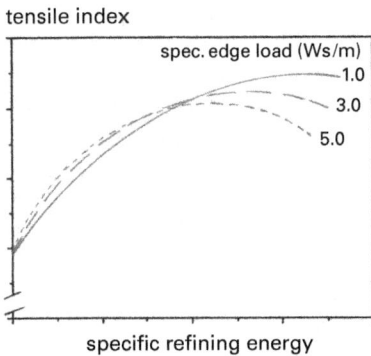

specific refining energy

Figure 7.9. Shows principally the effect of edge load in refining of chemical pulps expressed in terms of tensile index as a function of specific refining energy. Observe that the initial strength development normally is comparable or faster for higher SEL's compared to lower SEL's. On the other hand the maximum achievable strength is normally lower and is reached faster at higher SEL's. Some more easily refined hardwood pulps can deviate from this pattern.

With a given refiner supplied with a specific set of refining plates the degrees of freedom of operation is limited. According to the specific edge load theory the refining result is characterised by two parameters, namely the amount of specific energy consumed and the intensity factor. As can be seen from the mathematical expressions an increase in load through increased power will at the same time change the specific edge load.

As refiners today are limited to only one rotational speed, the only way to change the specific energy consumption at a given SEL is to regulate the flow, but this is not practical in a continuous process with limited buffer capacities. In practise the refining is therefor controlled through changes in power even though the specific edge load will vary somewhat.

Taking into account that every refiner has a range of operation in terms of power and throughput (flow) it is obvious that a careful selection of refining strategy and equipment should be made for every investment in refiner capacity. Observe that the throughput is essential not only for production capacity as such, but also for qualitative reasons. An increased throughput leads to a shorter dwell time for fibres in the refining zone and thus less probability to be treated. Thus, the amount of fibres flowing right through the grooves increases at higher throughput, resulting in a more heterogeneous refining.

7.3 Refining of Different Pulp Types

The final papermaking properties of *mechanical pulps* are developed already in connection with the fiber liberation in the mechanical pulping process. But to reach extremely low shive content post refining at low consistency in the paper mill is sometimes applied. A gentle treatment has also shown to give a limited positive effect on the strength potential.

For mixtures of chemical and mechanical pulps it is a normal practice to fine-tune the properties of the fibre furnish in a final refiner just before the paper machine.

Contrary to what is the case for pulps with low yields, the mechanical pulps retain most of their previously attained papermaking properties during *pulp drying,* to a great extent due to the wet stiffness provided by the native lignin, which effectively counteracts the contracting capillary forces in drying.

On the other hand, drying of chemical pulps does not only eliminate effects of refining on the papermaking properties, but makes it even more difficult to develop papermaking properties similar to those of never-dried pulps in a refining operation. Theoretically it would be possible to reproduce the properties of refined, never-dried pulp through increased refining, although at an increased drainage resistance and gentle beating in the laboratory, such as in a PFI beater, comes close to this theoretical standard. But a normal experience in industrial refining is inferior results due to a harsher and more heterogeneous treatment of the dried fibres.

Semichemical pulping requires that, for defibration, the dissolution in the initial cooking stage be supplemented by a mechanical treatment in a medium or high consistency refiner stage. In integrated pulp and paper mills, which is the normal concept for this pulp type, it is followed by a separate refining stage, usually at low consistency, for development of the papermaking properties. There is however no clear-cut borderline between defibration and development of papermaking properties in such a system.

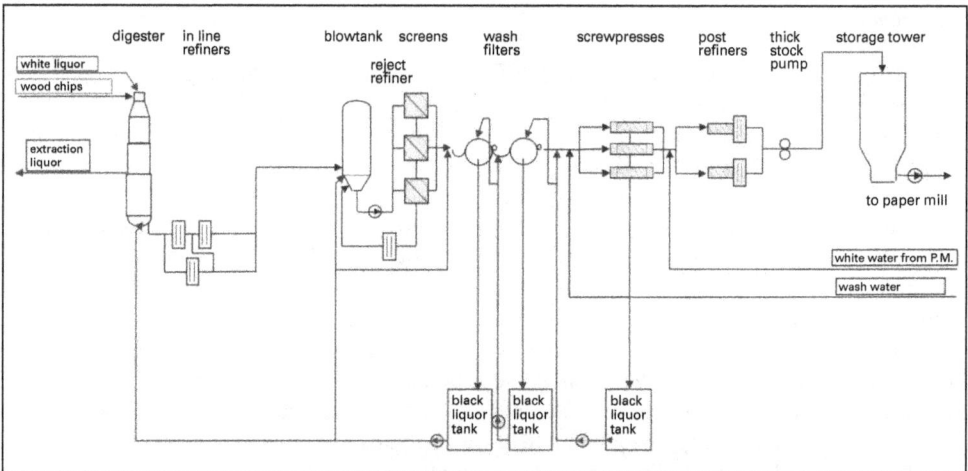

Figure 7.10. An example of refiner systems for defibration and development of desired papermaking properties of high yield kraft pulp for linerboard. Courtesy of SCA Packaging Munksund.

The production of *high yield softwood kraft pulp* for kraft linerboard, kappa number usually around 90, may be a typical example of such a semichemical system with continuous cooking *Figure 7.10.* A so-called in line refiner defibrates the cooked chips to such an extent that good washing can be obtained without drainage problems. After brown stock washing the rather coarse pulp is run through a deshiving medium consistency refiner in the paper mill, which also starts to develop papermaking properties. A low consistency refiner in the paper mill finishes the development of papermaking properties, at the same time as the remaining shives are defibrated. The optimal results are obtained when both defibration and development of papermaking properties are divided between the different refining stages after washing.

In unbleached kraft pulp, lignin is the main source of refining resistance. This is particularly important for high yield kraft pulp for which the refining resistance limits the aim towards a high kappa number and a correspondingly high yield. Gentle beating in the laboratory can indeed produce high tensile strength also in the case of unusually high kappa numbers (high lignin contents) but the less gentle mill refining systems are not capable of managing pulps with a high refining resistance. The higher the resistance, the higher the refining intensity (specific edge load) needed, which makes it even more difficult to match a high tensile strength target, since the maximum attainable tensile strength decreases with increased refining intensity. This sets an upper limit to the target kappa number as mentioned above.

Refining of an *unbleached chemical softwood kraft pulp* is carried out either at low consistency or in turn at *high consistency* (28–35 %) and low consistency, *Figure 7.11*. The latter case is specifically used in production of sack paper, where high strain and high tensile energy absorption values are needed. The refining at high consistency results in curled fibres with some micro-compressed zones, which gives a "fibre straightening" and "fibre extension" potential in the paper network, that is to say increased strain. These high-consistency refining effects are mainly due to mechanical force induced structural changes in the lignin matrix of the fibre wall. The low consistency refining stage following the high consistency refining is necessary to achieve the desired strength values, though low consistency refining also counteracts the high consistency refining effects by straightening the fibres.

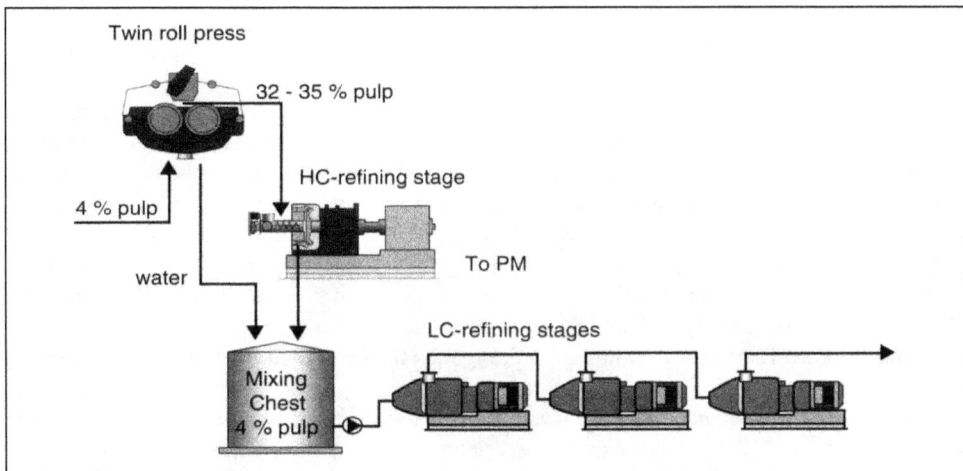

Figure 7.11. HC – LC refining of unbleached chemical kraft pulp for sack paper.
Courtesy of Metso Paper.

Even if the lignin is not the same kind of obstacle for reaching high tensile strength, as for high yield pulp, the content of lignin and its condition are important issues for the refining.   In general, the refining resistance decreases with the lignin content and is practically no longer influenced by lignin for softwood kraft pulps at kappa number 20 (lignin content about 3 %) and lower.

Swelling of the lignin in unbleached pulp promotes the refining. Hence, both pH (for its influence on the content of charged groups in lignin) and ionic strength are important refining variables in the case of unbleached kraft.

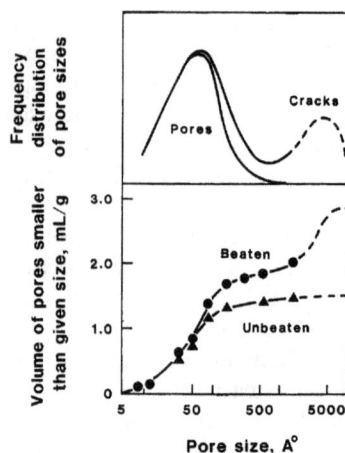

Figure 7.12. Solute exclusion data of Stone et al. Note that the data for the beaten pulp show the beginning of the second distribution, corresponding to the cracks.
(Å = ångström). (Page, Cambridge 1989.) Courtesy of Fundamental Research Committee.

Swelling as such does not give significant refining effects but it can make refining easier. To make the fibres sufficiently flexible, a *de-lamination* of the fibre wall is needed. The difference between swelling and de-lamination in this context is illustrated by *Figure 7.12*. Indeed, swelling gives a very open structure to the lignin/hemicellulose matrix between the cellulose fibrils in chemical pulp, but the matrix is obviously strong enough to help preserve the fairly high stiffness of the unrefined fibre wall. Refining on the other hand produces so-called cracks and de-lamination in the fibre wall, which radically reduces the stiffness of the fibre wall.

Another important refining effect is external fibrillation. It is a result of changes in the outer layers of the fibre, at first in the primary wall (P) and then also the outer layer of the secondary wall (S1). Extensive removal of P and S1 has a profound effect on the fibre stability but requires an unusually powerful refining. In normal refining the effect on P and S1 is smaller but not negligible.

Refining of *bleached chemical pulp* has a somewhat different fibre background. The matrix between cellulose fibrils now consists almost exclusively of hemicellulose, which has a strong affinity to water and swells more or less independently of possible charges on the hemicellulose (the charges being mainly carboxylate groups from uronic acid on xylan, to a great extent hexenuronic acid groups). Neither pH, nor ionic strength seems to have a significant effect on the refining result in the case of a bleached softwood kraft pulp. It is likely that these factors, which are known to promote swelling of the charged components, have a positive effect but that the effect seems to be so small that, with the normal scatter of the process, it cannot be proven.

Bleached softwood kraft pulp contains normally fairly little hemicellulose, also when cooked to a fairly high kappa number. Cooking to low kappa numbers leads to hemicellulose contents, which are critically low for developing a high tensile strength by the refining process. At low hemicellulose content the fibres loose elasticity, a development which is strengthened by an increased degradation of cellulose.

Bleached birch kraft pulp is quite easy to refine but may be sensitive to a high intensity in the refining process. In general, it has a high content of more or less charged hemicellulose, at the same time as the hemicellulose content varies considerably. This variation is clearly reflected in variations in refining behaviour (the more hemicellulose the more rapid strength development). Even so, it is suitable to refine only lightly, particularly in the case of never-dried pulp. It may, however, be necessary to extend the refining somewhat to straighten curled fibres in some cases.

Commercial *bleached eucalypt kraft pulps* (dried pulps) have small, rather stiff fibres with fairly little hemicellulose. They require considerably more refining for the development of tensile strength than the corresponding birch pulps, but may then develop a surprisingly good strength. Variation of the hemicellulose content is more critical to the strength development because of the low general level of hemicellulose and the short fibres, which require good bonding between the fibres to produce strong paper.

At a given capacity and specific energy consumption the shorter hardwood fibres gives a smaller clearance between the bars in the refining zone than softwood fibres. It is likely to be an effect of a less rigid wet fibre network with lower strength and thus a somewhat different rheology (a lower apparent viscosity of the pulp suspension). Fibre stiffness should also be of importance in this connection. The clearance is known to decrease during repeated circulation of a pulp through a refiner.

Sulphite pulps are easier to refine than kraft pulps. Sulphonation of the lignin makes the lignin swell and thereby considerably decreases its adverse effects in refining. This is particularly obvious when semichemical softwood bisulphite pulp at 70 % yield is compared with semichemical softwood kraft pulp at 56 % yield, *Table 7.1.* The changes in lignin structure is no doubt of great importance but there is a considerable difference in beatability also for bleached pulps, which indicates that part of the better beatability is to be explained by differences on the carbohydrate side. It is then noteworthy that a comparison between different bleached pulps show that the softwood sulphite pulps have a lower hemicellulose content than the softwood kraft pulps, even though the pulp yield in the case of softwood sulphite pulps is higher (due to a very high cellulose yield in the sulphite cook, *Table 7. 2.*

Table 7.1. Comparison between semichemical bisulphite and semichemical kraft pulp at a paper density of 750 kg/m³.

Pulp (newer dried)	Lignin content, %	Beater (PFI) revolutions	Tensile index	Tear index
Semichemical spruce bisulphate, yield 70 %	20	3300	93	6.3
Semichemical spruce kraft, yield 58 %	13	11000	92	9.5

Table 7.2. Comparison between bleached sulphite and bleached kraft pulp at a paper density of 750 kg/m^3.

Pulp (dried)	Hemicuellulose content, %	Beater (PFI) revolutions	Tensile index	Tear index
Bleached spruce acid sulphite, yield 47.5 %	15	1900	62	8.7
Bleached spruce kraft, yield 45 %	17.5	4000	86	10.5

Table 7.3 gives a summary of pulp types and energy consumption in refining of the pulps for some different paper products.

Table 7.3. Some examples of refining energy for pulps used for some different paper products.

Product	Basis weight	Pulp type	Energy kWh/t
Sack paper	70–90	Sulphate, unbleached	300–400
Kraftliner	100–300	Sulphate, unbleached	
-top layer			250–300
-bottom layer			150–200
Kraftliner (heavy duty)	400–440	Sulphate, unbleached	150–250
Fluting	115–140	Semi chemical	300–500
Carton board	200–400	Sulphate, unbleached	150–200
Printing paper	60–100	Softwood sulphate, bleached	150–200
		Hardwood sulphate, bleached	50–100
Graphic paper (News, SC, LWC)	40–80	Sulphate, bleached (or semi-bleached)	150–200
Greaseproof	30–60	Sulphite	800–1000
Bank note paper	80	Cotton	2000

7.4 Refining Systems

An important issue, which has been studied extensively, is whether different kind of pulps, such as softwood and hardwood pulps, should *be refined separately or in mixture*. In some cases it would be practical to work with mixtures so as to minimise the extent of the equipment. In other cases there is a wish to have an individual choice of refining degree for the different pulps, which then adds a degree of freedom to reach lowest possible production cost for a given paper grade with a given set of paper properties.

Softwood pulps and hardwood pulps form different kinds of fibre networks with different rheological properties. This can be illustrated by the fact that hardwood pulps gives a much smaller clearance between the bars in the refining zone than softwoods at certain given process conditions. Mixtures of hardwood and softwood pulps gives intermediate clearances which indicate that the components in mixed refining is subjected to a treatment that differs somewhat from the treatment they get when refined separately. In practise though the difference in refining result generally seems to be rather small.

Another strategic issue is whether refining should be performed *with the refiners in parallel or in series.* As long as the refiners can be operated within their recommended working range the difference in refining result between the two systems seems to be small. Parallel refining gives a longer retention time in each refiner whereas serial refining, by means of repeated passages through refining zones, may give some advantage on refining homogeneity and on process control, although this advantage seems to be marginal.

Regardless of the refining strategy the refiner system must allow for a refiner to be closed down for maintenance or change of plates without jeopardising the overall refining result. This aspect which influences both capacity, system layout and control functions, must be taken into account when designing refining systems.

7.5 Refining Process Control

A primary control strategy of refining has been to aim at constant specific energy consumption. It implies feed-forward control of pulp consistency, flow of pulp suspension, and net supply of electric power (total power - no load power). A good control demands a high accuracy of all the primary measurement signals for the control, which may be difficult to fulfil. The weakest link seems to be the pulp consistency measurement. A supplementary feed-back control loop would therefor be desirable. *Figure 7.13* shows a typical refining line including traditional instrumentation, with three refiners in series.

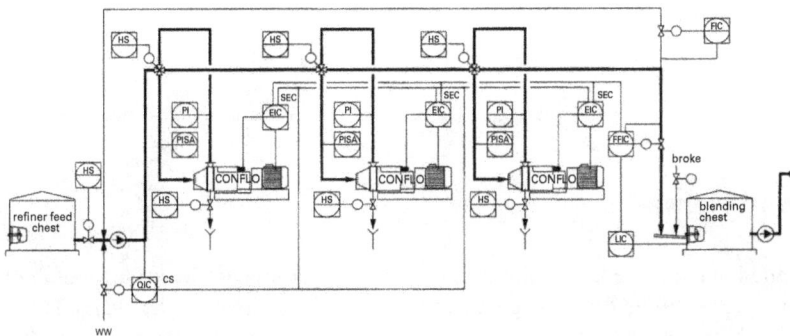

Figure 7.13. Typical refining line including traditional instrumentation, with three refiners in series. Courtesy of Metso Paper.

Feedback control may utilise some kind of drainage tester on the fibre suspension after refining, preferably on-line measurement as close to the refiners as possible so as to minimise the time lag. In closed paper mill systems the measurement can be markedly disturbed by variations in fines content of the white water used for dilution. Other measurements suitable for feedback are temperature rise of the fibre suspension, couch vacuum on the paper machine, and different properties of the finished paper. The latter measurement category represents the ultimate targets, but the value for feedback control decreases with the distance from the refiners and the time lag between the refiner process and the measurements. With a long time lag the frequency and the amplification of the control measures must be considerably subdued.

On-line measurement of fiber length is another proposal of sensor for feedback. In normal, not particularly rough refining, the resolution of the measurement may not be strong enough to permit other than a weak amplification of the fiber length signal in the control, i. e. it should serve only for minor corrections of the drainage target if any.

A major disturbance is variations in pulp beatability. This requires a more sophisticated control strategy. Unfortunately there are no sufficiently reliable measurement options available today, which leaves only two alternatives. The first priority is of course to produce pulps or pulp mixtures with constant beatability, something, which so far has not been given sufficient priority. The second alternative seems to be a rather sophisticated drainage resistance control, with compensation for different measurable conditions and pulp properties, and with due consideration to the time lag of certain data. A main problem with the control is however the reliability of the quality measurements. A possible control strategy for this is shown in *Figure 7.14.*

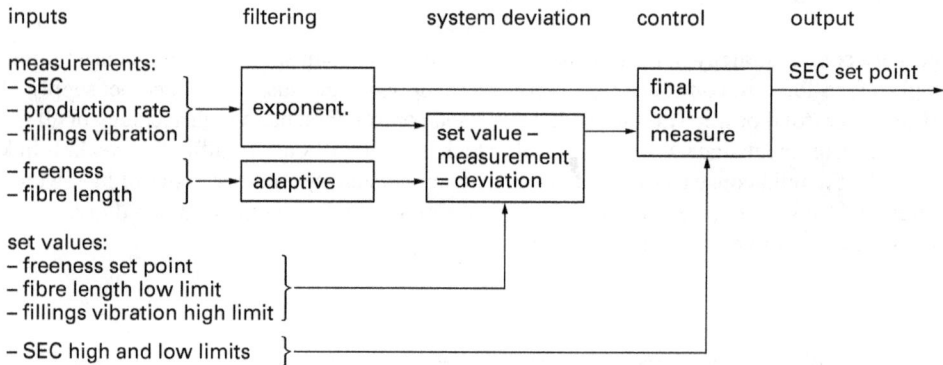

Figure 7.14. Possible refining control for difficult cases. (SEC=specific energy consumption). Courtesy of SPCI.

7.6 Concluding Remarks

As has been illustrated in the preceding, refining is a very complex process, as regards both process conditions and effects on the fibres. A general picture has been given of the available fundamental knowledge but it must be stressed that the process knowledge required in practical applications must still to a great extent be empirical. This results in a considerable variation in mill practices.

The process control has difficulty in coping with varying beatability of the incoming fibre material. A high priority should therefore be given to a consistent beatability of the pulp supply, higher than is usually the case. It is important to have a good control of fibre flow through the refiner and net energy input. Feedback of refining results such as freeness, temperature rise, couch vacuum, and paper properties is required to take care of special disturbances but can only be used with low amplification in relation to the time lag of the signals.

Since the refining conditions can greatly influence on how well the potential of the pulps is utilised and how the refining results agree with the property profile of the paper to be produced, it is important to carefully choose suitable equipment and conditions for the refining process. It is particularly important to choose such a design capacity of the refiners that they can operate reasonably close to the optimal conditions. Overloaded refiners will produce an inferior fibre strength development.

7.7 Literature

Brecht, W., and Siewert, W.H. (1958) *Das Papier* 20(1): 4

Ebeling, K. (1980) *A critical review of current theories for the refining of chemical pulps. International Symposium on Fundamental Concepts of Refining.* Appleton, USA: The Institute of Paper Chemistry

Genco, J.M. (1999) *Fundamental Processes in Stock Preparation and Refining. TAPPI Pulping Conference, Proceedings.* Orlando, USA, p. 57

Lumiainen, J. (2000) Refining of chemical pulp. In: Hannu Paulapuro (Ed.) *Papermaking Part 1, Stock Preparation and Wet End. Papermaking Science and Technology, Book 8.* Finnish Paper Engineers´ Association and TAPPI, pp. 87–122

Page, D.H. (1989) *The beating of chemical pulps – the action and the effect. 9th Fundamental Research Symposium Notes, Cambridge, UK,* p. 1

8 The Short Circulation

Bo Norman
Department of Fibre and Polymer Technology, KTH

8.1 Principles of the Short Circulation

In the short circulation, the thick stock is

- diluted with white water, to a suitable forming concentration
- machine screened
- centrifugally cleaned
- deaerated
- pumped to the headbox

The traditional design is rather conservative regarding the possibilities to deliver a highly even fibre concentration to the headbox, and a new design principle allowing a more even production is suggested.

The fibre raw material, which is delivered from a pulp mill to the paper mill, is designated "pulp", and it is prepared to a "stock"" through proper refining and product and process chemicals addition. Through successive dilutions of the stock with white water, taking into consideration on-line measured concentration variations, and including mixing in intermediate chests, variations in concentration of the stock are reduced. It is finally passing the machine chest, from which the "thick stock" is fed to the short circulation.

The thick stock has a fibre content of 3–4 per cent and must in the short circulation be diluted with white water from the wire section to a "mix" with a concentration of 0.1–1.5 per cent, depending on the product to be manufactured. The higher the strength desired in the final product,

138

the greater is the dilution required. Retention aid components can be added in the short circulation.

The principle for the design of today's short circulation is shown in *Figure 8.1*. The thick stock is first diluted with white water from the wire pit. The pulp is thereafter machine-screened, and the reject is re-screened, sometimes twice. Accepts and rejects are recycled, while the final fibre reject level is low enough to leave the system. This is followed by centrifugal cleaning, with up to six stages before the fibre reject level is low enough. The mix is then deaerated and the overflow is recycled to the short circulation. The mix is finally fed to the headbox, from which a certain overflow is recycled to the short circulation.

Several circulation circuits are thus present inside the short circulation, and since the concentrations in these are not controlled, concentration variations in the mix fed to the headbox can be introduced via these circuits.

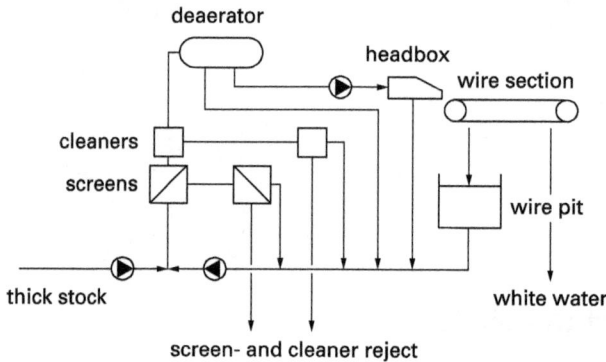

Figure 8.1. Schematic diagram of a conventional short circulation.

An example of a short circulation design in an industrial application is shown in *Figure 8.2*.

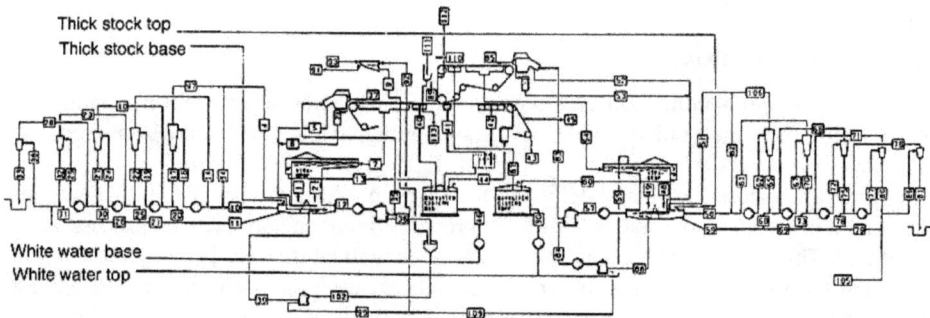

Figure 8.2. Short circulation loops for a two-ply linerboard machine

8.2 Dilution of Thick Stock Pulp

Great attention must be given to the mixing of the thick stock and the white water, since poor mixing gives rise to concentration variations and thereby grammage variations. A method for achieving good mixing is to add the thick stock in a white water pipeline at a higher flow rate in a central pipe surrounded by a concentric channel for the white water. This arrangement functions in the same way as a jet pump, and the turbulence level in the mixing zone becomes higher the more energy is added with the excess speed of the thick stock jet.

The process of mixing a thick stock jet with pure water has been studied (Tegengren, MSc thesis 1983) in a plexiglas pipe, in which the process was filmed. It was slightly surprising that the most important parameter was found to be not the absolute speed difference between the thick stock and water, but the ratio of these two speeds. A speed ratio of eight gave good mixing, as shown in *Figure 8.3*.

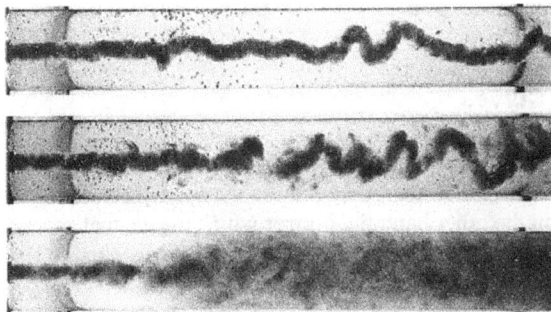

Figure8.3. Stock blending of thick stock and white water from left to right) at different velocity ratios thick stock/white water flow. (flowing Top: 1.5 Middle: 5.3 Bottom: 10.8).

In *Figure 8.4* the design of thick stock addition to white water on the FEX pilot paper machine at STFI is shown. The coarse pipes entering from above carry the white water from the two circulation loops. 90 degree bends turn the flows to a horizontal direction. Thick stock is added at the ends of the bends, coaxially with the horizontal white water flows. A considerable mixing length is allowed before the pump, feeding the trash separators and finally the deaerators located on the roof of the building.

Figure 8.4. Thick stock mixing into white water flow on the FEX pilot paper machine.

8.3 Screening

A screen placed in the short circulation is called a machine screen, and its main task is to separate impurities, which could damage the paper machine, and hard fibre flocs, stickies, etc which could lead to web breaks in the paper machine and quality defects in the finished paper. Normal shives separation must have been carried out at an earlier stage. Since the screen is placed after the addition of thick stock, all variations in the fibre reject flow from the machine screen system will result in grammage variations in the finished product. It is therefore important that the reject flow is low, which means that there must be at least one further screening of the reject and recycling of the accept. The most favourable placing of the screen would instead be immediately before the machine tub. Since the machine screen feeds the headbox, it is important that pressure pulsations are avoided, especially in headboxes, which are particularly sensitive to such variations. These require special attention to the choice of screens and pumps. Even so, a separate damper is necessary before the headbox in some cases to reduce the pressure pulsations.

8.3.1 Screening Terms and Screen Curves

By screening, it is possible to separate undesired components from a fibre suspension. The screening process is subject to two types of error, on the one hand the a-error which is that all undesired particles are not separated and on the other hand the b-error which means that acceptable fibres accompany the separated components. To reduce these two errors to acceptable levels, it is usually necessary to use several screens, connected in a screening system.

In the following, the function of a screen is first described and thereafter, the structure of a screening system is briefly treated. In screening, an *inject flow* is divided into an *accept* and a *reject* flow, see *Figure 8.5*.

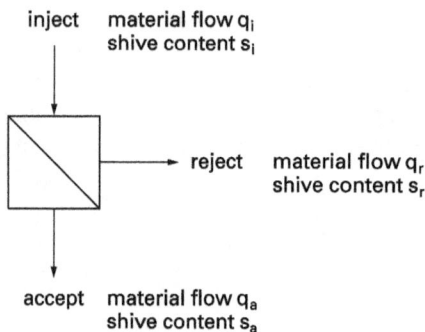

inject   material flow q_i
shive content s_i

reject   material flow q_r
shive content s_r

accept   material flow q_a
shive content s_a

Figure 8.5. Definition of screen parameters.

To achieve a positive screening effect, the contamination content must be greater in the reject flow than in the accept flow, at the same time as the fibre losses with the reject must be small. The separating function of a screen is described by the screen curve which shows the degree of shives separation S as a function of the degree of fibre reject R. The degree of shives separation S, i.e. the probability that a shive shall end up in the reject, can be described by the ratio between shives flow in the reject divided by the shives flow in the inject:

$$S = \frac{q_r \, s_r}{q_i \, s_i} \tag{8.1}$$

The degree of fibre reject R, i.e. the proportion of incoming fibres, which end up in the reject, can be written as:

$$R = \frac{q_r}{q_i} \tag{8.2}$$

The shape of a screen curve can be approximately described by the Q-factor, where Q defines the relationship between S and R according to the equation

$$Q = 1 - \frac{s_a}{s_\gamma} \tag{8.3}$$

Figure 8.6 shows the appearance of the screen curve for different values of the parameter Q. $Q = 0$ corresponds to a straight line with an inclination of 45 degrees. This represents pure flow division, where there is no screening effect. With increasing value of Q, the screening effect is improved. $Q = 1$ corresponds to the perfect screen, in which the shives separation is complete at all reject levels. The Q-factor for a working screen can be evaluated very simply using the equation:

$$S = \frac{R}{1 - Q(1 - R)} \tag{8.4}$$

This means that Q can be determined simply by analysis of the shives contents in the accept and in the reject, and that it is not necessary to determine a complete shives balance over the screen. The advantage is that it is then also possible to compare the function of two screens, which happen to be working at different reject levels, without needing to know these levels.

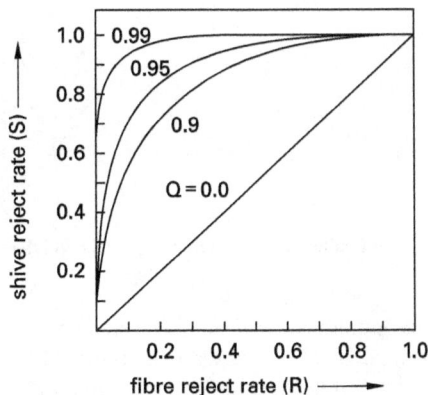

Figure 8.6. The appearance of the screen curve for different values of the Q-factor.

It should be pointed out that the screen curve defined by a certain Q-value calculated according to equation 3 naturally agrees exactly with the real screen curve only in one single point, namely at the reject level which was valid on the sampling occasion. Elsewhere, Equation 4 gives only an approximate value of the screen curve but it has been shown that the approximation is in practice normally good.

8.3.2 Screen Design

Screening takes place through screen drums, perforated with holes or slits. The screen openings are all the time kept free from blockage by the passage of some type of clearing organ along the screen surface. The perforations consist of cylindrical holes, normally with a diameter of 1.2–2.5 mm, coned towards the accept side. They are placed in a certain *screen hole pattern,* and the *open screen area* amounts to 10–20 per cent. When the demand for a cleaning effect is very high, screen slits are used, with an opening of down to a few tenths of a mm. The open area and thereby the capacity is lower on a slitted screen drum than on a hole-perforated one.

It should be pointed out that not only particles larger than the screen openings are separated. Through orientation of long particles parallel with the screen surface, a considerable separation of long particles is obtained, even when their cross-section dimensions are considerably smaller than the screen openings. The flow through the screen drum can take place from outside inwards or the opposite, see *Figure 8.7.*

If the flow follows the previous case, one works against the centrifugal force, which is favourable for the separation of heavy particles. This gives an increased pressure drop, however. In flow inside - outwards, low-pressure drops are obtained, but there is a tendency to collect heavy contaminations (sand, stone etc) on the screen surface (on the inject side).

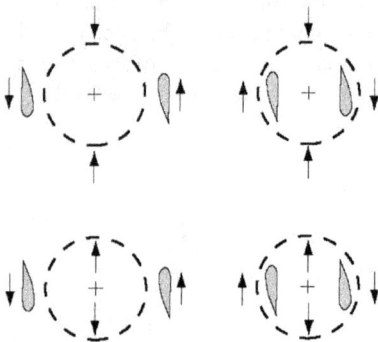

Figure 8.7. Screening and cleaning geometries. The arrows show the direction of movement of the inject and the cleaning organs.

The cleaning organs can be placed either on the inject or on the accept side. The cleaning organs are designed for sucking away (from the inject side) blocking material or removal by pressing (from the accept side). A suction effect is obtained in the form of an under/overpressure zone, generated by the cleaning organ, which moves at a high speed (at least 20 m/s) concentrically with the screen surface.

In some newly developed screens, the suction pulse is so extensive that water is sucked back through the cone-shaped screen holes (the fibres get stuck in the cones). This water then locally dilutes the inject, and then it is possible to screen with input inject concentrations of circa 3 per cent. Conventional screening takes place at a concentration of less than 1 per cent.

If the cleaning organs lie on the inject side, the generated pressure pulses will be dampened slightly by the screen drum. This procedure is therefore used in the input to headboxes, where the sensitivity to pressure variations is great.

8.3.3 Screen Coupling

Screening in one step does not normally give sufficient separation of contaminations, and also too high fibre losses with the reject. Several screens are therefore normally combined in the screen system. *Figure 8.8* gives examples of a screen system with double screening of the accept and cascade screening of the reject. Double screening means that the accept from a primary screen is screened in a secondary screen.

The reject is returned to a reject screen from which in turn the accept is returned to the primary screen. The fibre flow in the reject from the reject screen is considerably less than from the primary screen, at the same time as the accept from the secondary screen is considerably cleaner than that from the primary screen. In industrial screening systems, the reject is often post-treated in some way. For mechanical pulps, the reject is refined and the strongest pulp fraction is thus obtained.

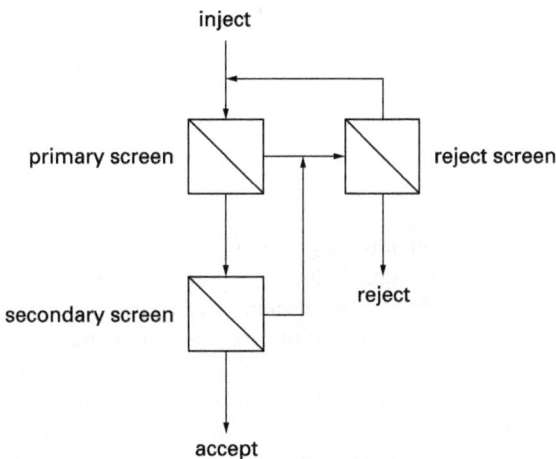

Figure 8.8. Screen system with double screening of accept and cascade screening of reject.

8.4 Centrifugal Separation

Besides screens, centrifugal separators are often included in the pulp preparation systems. These are cyclones in which a fibre suspension is exposed to high centrifugal forces, see *Figure 8.9*, and where the separation takes place according to particle density.

The inject is fed tangentially, and the accept is taken out in the central, upwards-directed pipe. The reject is flushed out at the bottom of the cone. To give a good cleaning effect, a vortex cleaner should have a small diameter, which means that a large number of vortex cleaners must be connected in parallel to manage normal pulp flows. The fibre concentration in a vortex cleaner should be below 1 per cent. At higher concentrations, fibre networks arise and hold back the contaminations to be separated. To avoid too great a thickening of the mix at the reject outlet and thereby plugging, the reject level must be high, at least 20 per cent.

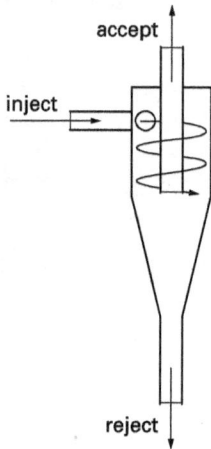

Figure 8.9. Basic principle of a centrifugal cleaner.

If vortex cleaning is necessary, special consideration must be given to the large reject quantities, which are necessary. To reduce the total reject to a suitably low level, the reject must therefore be treated in several vortex cleaning steps. With five vortex cleaning steps with 30 per cent reject in each case, the final fibre loss amounts to 0.24 per cent of the originally fibre input.

Traditionally, vortex cleaners have been used to separate e.g. shives and bark specks. Today, vortex cleaners are also used to separate fibre materials according to specific area, which e.g. for mechanical pulp means that the fibre material can be divided into fractions with different bonding properties. Another field of use is to separate light particles, e.g. plastic contaminations, in recycled paper preparation. In this "reversed vortex cleaning", the reject is taken out through the upwards-directed pipe.

8.5 Deaeration

Deaeration of the stock blend in the short circulation has the task of removing all air bubbles, even those that adhere to fibre surfaces. The stock blend is sprayed through a slice onto the roof of a deaeration tank under vacuum, as illustrated in *Figure 8.10*. Air bubbles are thus separated out and taken to the vacuum system. The deaerated stock blend then passes over the overflow under very constant flow conditions to the feed pump, which carries the stock blend to the headbox.

The presence of air bubbles leads to large problems particularly at high machine speeds, i.e. when the stock blend is added under high pressure in the headbox. This air then expands in the jet from the headbox so that the jet quality drops, and it subsequently disturbs the dewatering process in the wire section.

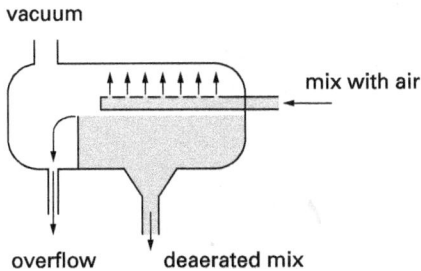

Figure 8.10. Deaeration tank.

8.6 Simplified Short Circulation

A proposal to improve the short circulation with regard to grammage stability compared with *Figure 8.1* is shown in *Figure 8.11*. The thick stock is here diluted with deaerated white water (the air content in the thick stock is normally very low), and there is no reject withdrawal of fibre material. The stock blend finally passes a junk trap, which is only intermittently emptied of impurities, and does not give a continuous reject flow.

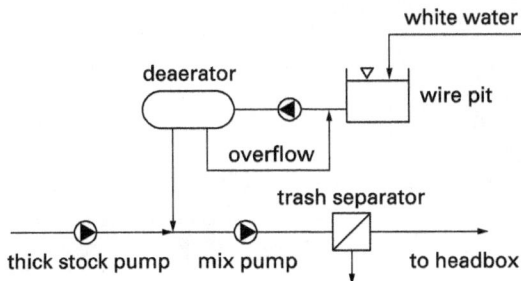

Figure 8.11. Proposal for a simplified short circulation.

The trash separator also has a dispersing effect on fibre flocs and stickies. The finished stock blend is fed to the headbox. A precondition for applying this alternative is that the thick stock is already screened, but this is today fully possible since there are several types of screen equipments, which work within a concentration range of 3–4 per cent. It is further assumed that vortex cleaning has already been carried out to the necessary extent, preferably in connection with the pulp preparation. A local dilution and thickening is then necessary. As mentioned earlier, vortex cleaning is not yet possible at concentrations above one per cent.

9 Polyelectrolyte Adsorption onto Cellulose Fibres

Lars Wågberg
Department of Fibre and Polymer Technology, KTH

9.1 Background

The field of polyelectrolyte adsorption onto cellulosic fibres has been studied in detail for around 40 years. Initially the studies focused on wet strength additives but later the focus has been changed towards retention and fixing agents in order to elucidate the working mechanism behind these additives. Today the knowledge about the equilibrium adsorption , in mg/g, of polyelectrolytes is rather good whereas the kinetics of adsorption is not as well investigated. Furthermore it has become obvious in many investigations during the last ten years that the conformation of the adsorbed polymers on the fibre surface is probably as important as the adsorbed amount for the action of these polyelectrolytes. Furthermore, fairly recent results indicate that the change of this conformation with time is also very important. These two latest issues, i.e. the conformation and its change with time are at the moment very poorly characterised.

With this as a background this chapter focuses upon the following questions

a) Why are the polyelectrolytes adsorbed onto cellulosic fibres? The easiest way to describe the adsorption is as an interaction between the charges on the fibres and the charges on the polyelectrolytes. However the entropy gain upon the release of counterions to the charges on the fibres and the polyelectrolytes gives a much larger contribution to the adsorption energy than the charge interaction. There are some investigations which clarify the influence of certain basic parameters on the fibres and the surfaces and these investigations will be reviewed. An attempt will also be made to create a link between published adsorption data and the Scheutjens-Fleer theory for polyelectrolyte adsorption. Another important question is how the polyelectrolytes are adsorbed on the fibres, i.e. are the segments only found in trains on the surface or will there be some loops and/or tails protruding into the solution? Very little is known about this but the topic will be discussed in some detail.

b) Where are the polyelectrolytes adsorbed- on the external surfaces of the fibres or within the fibre wall and how is this related to the molecular properties of the polyelectrolytes? There are a lot of definitions in the literature about fibre surfaces, external surfaces etc. but it is important to link the adsorption to molecular properties and the work in this area will be reviewed. The influence of fines will also be discussed.

c) What controls the kinetics of polyelectrolyte adsorption and are there models to describe this adsorption? It is still not entirely known how polymers that adsorb onto the internal surfaces of the fibres are then transported through the fibre wall. The work conducted in this area will be reviewed. It is also important to know the kinetics of the reconformation of the polymer on the fibre surface and this topic will also be discussed in detail. Finally the kinetics of desorption will be treated.

9.2 General

For most cationic polymeric additives used in the paper industry it is essential to have good adsorption of the polymer onto fibres, fines and fillers in order to have good efficiency of the polymer. The process of polyelectrolyte adsorption onto cellulose fibres has therefore attracted a considerable interest over the last 40–50 years and a number of reviews on the topic have already been published [1–4]. It is therefore not the purpose of the present chapter to give a complete review on all the literature available in the area but more to answer the following questions

- Why and how are the polyelectrolytes adsorbed on the fibres?
- Where are they adsorbed?
- How fast do they adsorb?

In order to create a link to the current knowledge of the theoretical understanding of polyelectrolyte adsorption the examples chosen from the literature will be linked to mainly the Scheutjens-Fleer theory for polyelectrolyte adsorption [5, 6]. It is furthermore the ambition of this chapter to highlight some future developments and research needs with this analysis.

9.3 Why and How are the Polyelectrolytes Adsorbed onto Cellulosic Fibres

In the theoretical treatment of the polyelectrolyte adsorption [5, 6] four major entities are used to discuss and explain polyelectrolyte adsorption and these are χ, χ_s, σ_0 and q_m. The first two correspond to the Flory-Huggins parameter of polymers in solution and the adsorption energy parameter respectively. The χ_s parameter is dependant on the difference between the adsorption energy of a polymer segment and a solvent molecule. Simply stated the adsorption is enhanced if the χ parameter has a value larger than 0.5 since the polymer then has a poor solubility in the solvent used .The adsorption is also enhanced the higher the value of the χ_s parameter.The σ_0 and the q_m correspond to the surface charge and the charge of a polymer segment respectively.

In [5] the theoretical predictions of the SF (Scheutjens-Fleer)-theory for polyelectrolyteadsorption were summarised with the following *Figure 9.1*.

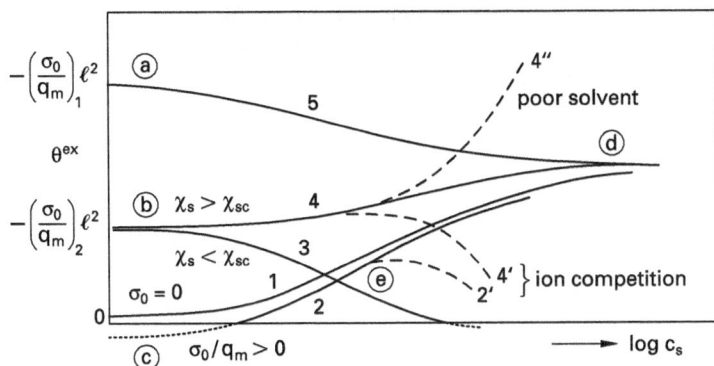

Figure 9.1. Schematic overview of polyelectrolyte adsorption. Different polyelectrolytes and different surfaces were used for the numerical simulations and the adsorption is shown as a function of salt concentration, c_s. From [5].

The curves in *Figure 9.1* that will be more thoroughly discussed are **3**, **4** (**4′** and **4″**) and **5**, which describe the adsorption of polyelectrolytes on oppositely charged surfaces. The ordinate in the figure represents the adsorbed amount of polymer on the surface. For pure charge compensation the sum of the charges on the surface and of the adsorbed polymer is zero, i.e. $q_m \cdot \theta_{ex} = \sigma_0 \cdot l^2$ (where $\sigma_0 \cdot l^2$ is the charge per site on the surface). This situation corresponds to $(\sigma_0/q_m)_2 \cdot l^2$ in the figure. For the situation where there is a non-electrostatic contribution to the adsorbed amount, θ_{ex} will both contain a fraction corresponding to charge compensation and a fraction corresponding to a positive χ_s value. This situation corresponds to $(\sigma_0/q_m)_1 \cdot l^2$ in *Figure 9.1*.

To repeat, the parameters of largest importance for the present discussion are :

- (σ_0/q_m) which is the ratio between the surface charge and the polymer segment charge
- χ_s which was defined above and it might be added that this is the non-electrostatic adsorption energy parameter
- χ_{sc} which is the critical value of the χ_s parameter above which adsorption will occur

Curve **3** in *Figure 9.1* corresponds to pure electrosorption, for polyelectrolytes with a high charge density. The only driving force for the adsorption is the charge interaction between the polymer segments and the charge on the surface. For this situation there is a charge balance between the charges on the surface and on the polymer and as the salt concentration is increased the interaction between the segments and the surface is decreased and the adsorption is hence decreased. For curve **4** there is also a non-electrostatic contribution to the adsorption of the polymer and as the electrolyte concentration is increased the repulsion between the polymer segments on the solid surface is decreased and the adsorption is increased.

Since many charged polymers loose some of their solubility, i.e. the efficient χ parameter is increased, as the repulsion between the segment decreases the adsorption will be even further enhanced. This means that the adsorption can increase drastically as indicated in the figure (**4″**). On the other hand if there is a specific interaction between the counterions and the surface, the

150

counterions will start to compete for the surface sites with the polymer segments, which will lead to a decrease in adsorption (curve **4'**).

In *Figure 9.1* there is also one curve (curve **5**), which corresponds to the situation with an initial, i.e. at low salt concentrations, contribution from both electrosorption and non-electrostatic interactions between the polymer segments and the surface. This situation is valid for polyelectrolytes with a low charge density and as the salt concentration is increased the adsorption approaches the adsorption for highly charged polyelectrolytes but with the same χ_s. A schematic representation of this situation is shown in *Figure 9.2* where cartoons of the adsorbed structure has been combined with *Figure 9.1* [5].

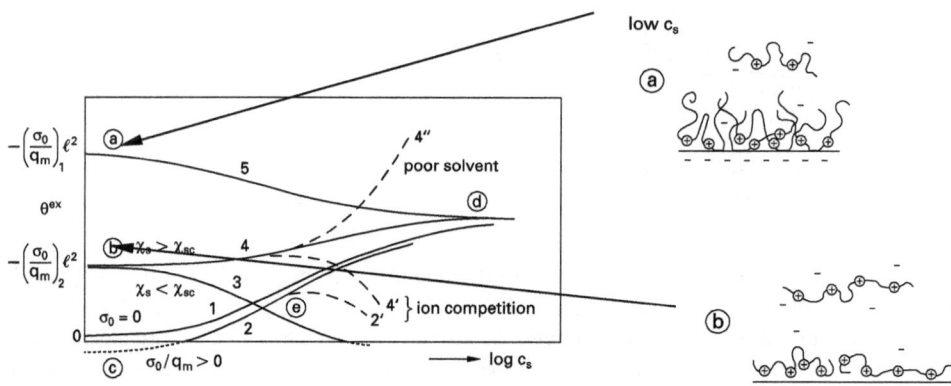

Figure 9.2. A schematic representation of the adsorbed polyelectrolyte structure at differentsalt concentrations and different polyelectrolyte surface interactions in Figure 9.1.

With this as a background it is now useful to try to review what has been done regarding polyelectyrolyte adsorption on cellulose fibres. It is naturally difficult to exactly fit the published data totally according to the categories mentioned above so the interpretations made in the following text have to be treated with caution. Van de Steeg [7, 8] has tried to create a link between the SF theory and the adsorption of cationic starch on micro-crystalline cellulose and for pure electrosorption the following results were achieved from calculations for adsorption of polyelectrolytes of low charge density to oppositely charged surfaces, *Figure 9.3a and Figure 9.3 b*.

There are two major features in *Figure 9.3 a*. First it can be noted that as the charge density, α, of the polymer increases there is a decrease in adsorption of the polyelectrolyte at low salt concentrations. Furthermore, as the salt concentration increases the polyelectrolytes with the lower charge densities will loose more of their adsorption compared with the polyelectrolytes of higher charge density. If the polyelectrolyte with a charge density of 0.015 is compared with the polylelctrolyte with a charge density of 0.2 it can be seen that at a salt concentration of around 0.03 M the adsorption of the polymer with the lower charge density is virtually zero whereas the adsorption of the polymer with a charge density of 0.2 is unchanged compared to deionised water. These data can also be plotted as a function of charge density, α, and then the results in *Figure 9.3b* are achieved. For the lower salt concentrations there is a characteristic maxima in adsorbed amount at increasing charge density that is a result of the assumption that there is no contribution to the interaction between the polyelectrolyte and the surface apart from pure electrostatic interaction.

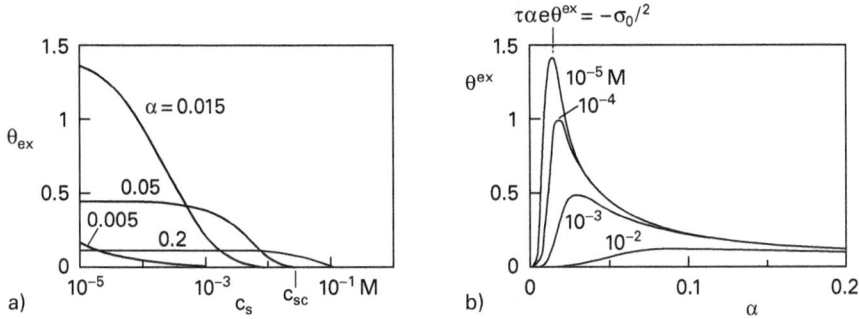

Figure 9.3. Pure electrosorption of polyelectrolytes with low charge density on oppositely charged surfaces. Figure a) shows the adsorbed amount as a function of salt concentration for polyelectrolytes of different charge densities(α) and figure b) shows the adsorption as a function of charge density (α) for different salt concentrations. The calculations were based on the following parameters in the SF theory [6]: Number of segments in the polymer = 100, lattice size (hexagonal lattice) = 0.6 nm, $\chi = 0.5$, $\chi_s = 0$, solution volume fraction of polyelectrolyte (φ_b) = 10^{-3}, surface charge (σ_0) = 10^{-3} mC/m^2) adapted from [8].

These kinds of results were partly presented in [7] but also found for the adsorption of cationic potato starch on peroxide bleached mechanical pulps [9] and for the adsorption of cationic polyacrylamides on bleached kraft pulps (50/50 mixture of hardwood and softwood)[10]. Examples of the results from [9] are shown in *Figure 9.4 a* and *Figure 9.4b* below

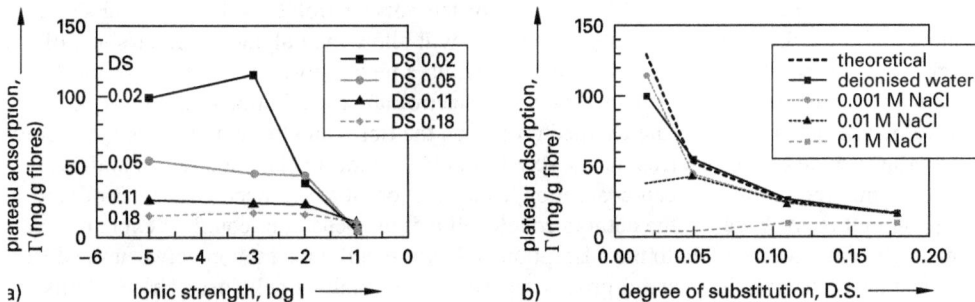

Figure 9.4. Plateau adsorption as a function of salt concentration for adsorption of cationic potato starch, of different charge densities, on bleached thermomechnical pulp and b. Plateau adsorption as a function of charge density (D.S.) of the starch for different salt concentrations. Adapted from [9].

These results suggest that there is no non-electrostatic contribution to the adsorption energy and that small ions can totally prevent the adsorption of the polymers onto the fibres. Similar results but with a different shape of the curve were also found in [11] (adsorption of cationic polyacrylamides on bleached softwood kraft pulp) and [12] (adsorption of cationic potato starch on bleached softwood kraft pulp). In these investigations it was found that the adsorption was initially increasing upon increased salt concentration and then again decreased when the salt concentration was increased above about 10^{-2} M NaCl. This is shown for adsorption of cationic polyacrylamide in *Figure 9.5* where the adsorption of cationic polyacrylamide on bleached softwood fibres is shown. In the figure the adsorption in % (polymer on fibre) is shown at a polymer addition of 1 % for different molar degrees of substitution of the polymer.

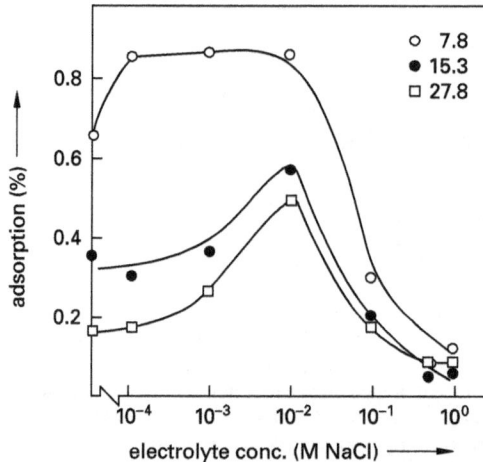

Figure 9.5. Adsorption of cationic polyacrylamides, with different degrees of substitution, on bleached softwood pulp as a function of salt, NaCl, concentration. The adsorption is shown in % of polymer on the fibres(i.e. 1 % = 10 mg/g). Polymer addition was 1 % (10 mg/g) and the pH during adsorption was 8.Adapted from [11].

The adsorption behaviour can most probably be ascribed to a pure electrosorption behaviour where the initial increase is linked to a coiling of the polyelectrolyte with increasing salt concentration. Since the fibres are highly porous this will allow the polymer to enter some of the larger pores in the fibre wall. This will in turn lead to a higher polyelectrolyte adsorption. The maxima could in turn also be explained by a non-electrostatic contribution to the adsorption and a specific interaction between the charged groups on the fibres and the sodium ions in the solution. However, since no such tendencies were found in [10] this explanation seems unlikely.

The same trend in the plateau-adsorption, as a function of salt concentration, was found in [13] and in this investigation the authors concluded that the behaviour could be explained by a non-electrostatic contribution to the adsorption and by a specific interaction between the different (counter) ions and the charged groups on the microcrystalline cellulose. These results are shown in *Figure 9.6* below and the experimental results were compared with theoretical predictions from the SF-theory [6]. The authors support their conclusion with the results from experiments with different monovalent counterions but it is still not clear how the porous nature of the microcrystalline cellulose substrate might have affected the results.

Hence it might be concluded that the available SF theory can be used to describe the adsorption cationic polymers to cellulosic fibres to a very large extent, despite the fact that some factors still need to be clarified.Long before theoretical modelling was a common tool for studying polyelectrolyte adsorption several investigators found a link between the charges on the polyelectrolyte and the charges on the adsorbed polyelectrolyte [14–18]. Trout [14] was apparently the first to introduce the term ion-exchange reaction where the counterions to the charged groups on the fibres were exchanged with the charged groups on the polyelectrolyte. An example of this is shown in *Figure 9.7* where the adsorption of polyethyleneimine onto different types of fibres is shown. Similar conclusions regarding the adsorption of wet strength resins onto fibres, i.e. a charge compensation between charges on the fibres and charges on the polyelectrolyte, were also found by Bates [15].

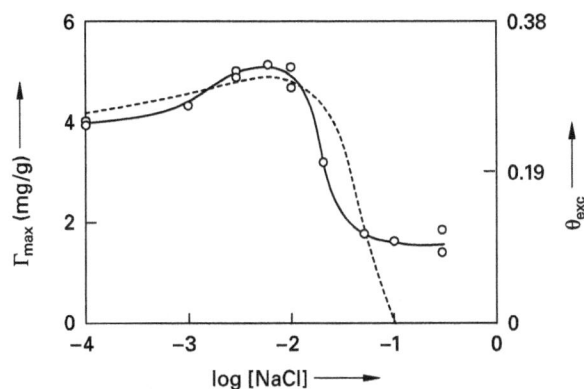

Figure 9.6. Adsorption of cationic amylopectin (D.S. = 0.035) on microcrystalline cellulose, and comparison between experimental results and theoretical calculations, as a function of salt concentration. Adapted from [13].

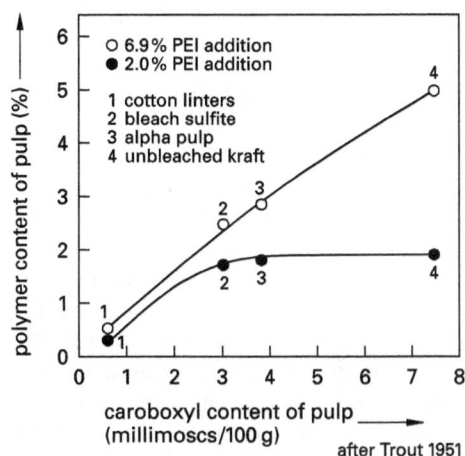

Figure 9.7. Adsorption of polyethyleneimine on different types of fibres as a function of the amount of charged groups on the fibres. The fibres were (according to the specification in [14]): cotton linters, bleached sulphite, alpha pulp and unbleached kraft. Adapted from [14].

It was not clear from these measurements whether there was a 1:1 combination of the charges on the polyelectrolyte with the charges on the fibres since only the total adsorption was measured. To resolve this question a different approach was used by Winter et.al [16] who measured the adsorption of a 3.6-ionene and poly DiMethyDiAllylAmmoniumChloride (DMDAAC) on both carboxymethylated rayon fibres and carboxymethylated bleached chemical softwood fibres. In addition to measuring of just the adsorption of polyelectrolyte the release of counterions was simultaneously estimated by measuring the conductivity in the solution. The results showed that there was a linear 1:1 relationship between the adsorption 3.6-ionene and ion-exchange capacity of both types of fibres, which is shown in *Figure 9.8.*

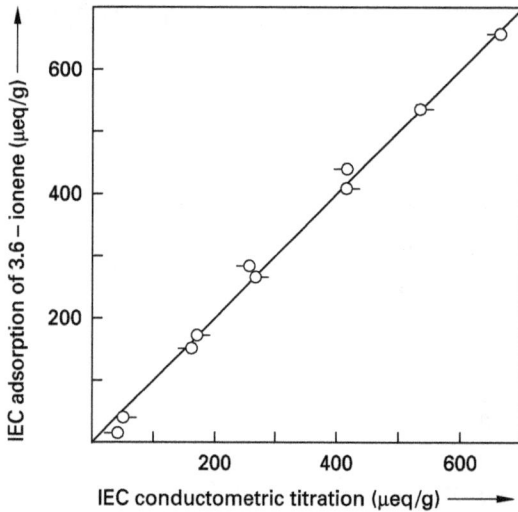

Figure 9.8. Comparison between ion-exchange capacities from adsorption isotherms of 3.6. ionene at pH = 8 in deionised water, and the ion-exchange capacity from conductometric titrations. The figure contains results from experiments with both rayon fibres and bleached softwood fibres Adapted from [16].

The results from the combination of both adsorption measurements and the measured release of counter ions showed that there indeed was a very close match between the charges on the fibre surface and the charges on the polyelectrolyte. The authors defined an adsorption stoichiometry as the relation between the released amount of counter ions and the adsorbed amount of polyelectrolyte charges and found a stoichiometry of 90 % regardless of surface charge of the fibres and the types of fibres used. This behaviour matches the behaviour of pure electrosorption but since the release of counterions could not be used to determine the adsorption stoichiometry at higher salt concentrations it was not possible to conclude if χ_s is zero. Wågberg et.al. [17] later refined the technique in [16] by specifically measuring the release of bromine ions, counter ions to the 3.6 ionene, and found that the adsorption stoichiometry decreased as the fibre surfaces were saturated with polyelectrolyte, as shown in *Figure 9.9.*This indicated that the polyelectrolyte was adsorbed with a larger fraction of the segments in loops and tails as the surfaces were completely saturated. This will be discussed in more detail later.

The results in [16, 17] also indicated that polyelectrolyte adsorption could be used for a fast determination of the charge of the fibres provided the pores of the fibres were large enough [18]. As shown in *Figure 9.9*, for the fibres with a D.S.(degree of substitution) of 0.064, there was no difference in adsorption results between the fibres and the microfibrils prepared from the same fibres. This showed that the polyelectrolyte could reach all the charges in the fibre wall for this specific type of fibre. By utilising polyelectrolytes with different molecular mass the adsorption on different structural levels in the fibres could be determined [19]. This will also be discussed under the headline "Where are the polyelectrolytes adsorbed" below.

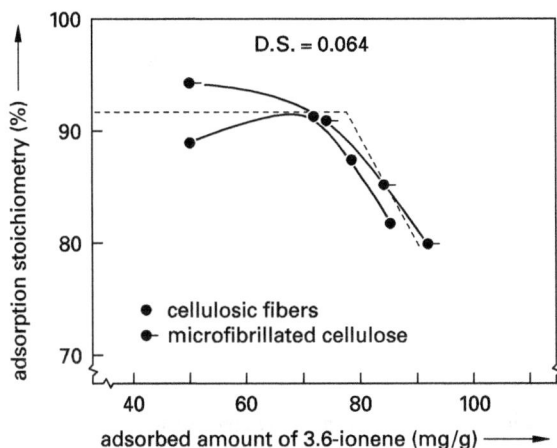

Figure 9.9. Adsorption stoichiometry, i.e. amount of released counter ions divided by the amount of adsorbed polyelectrolyte charges in %, as a function of polyelectrolyte adsorption. The charge of the fibres and fibrils was 381 µeq./g and the pH during the adsorption measurement was pH = 7 and only deionised water was used. Adapted from [17].

As discussed earlier the adsorption results in *Figure 9.9* indicated that the conformation of the adsorbed polyelectrolytes changed as the fibres start to become completely saturated with polymer. Naturally it is very hard to determine the exact conformation of the polyelectrolytes on the fibre surface and to the knowledge of the author there are no direct measurements of the conformation of cationic polyelectrolytes on cellulosic surfaces published. However, indirect measurements that are based on binding of anionic particles to cellulosic fibres that first have been pre-saturated to different degrees with cationic polyelectrolytes have been published [20]. These measurements are based on the quite simple assumption that polyelectrolytes protruding out from the fibre surface are more efficient to bind anionic particles than polyelectrolytes with a flat conformation on the fibre surface. An example of these measurements are shown in *Figure 9.10* where the binding of montmorillonite clay particles to bleached softwood fibres which have first been saturated with cationic polyacrylamide to different degrees. As can be seen in the figure the binding of the montmorillonite clay increases as the adsorption of C-PAM increases (note that the abscissa gives the added amount but this corresponds to the adsorbed amount in this special case since all added polyelectrolyte was adsorbed by the fibre [20]). It is interesting to note that despite a considerable increase in adsorption there is no more binding of the montmorillonite clay after a certain level of polyelectrolyte adsorption. This might also be due to a complete coverage of the fibre surface with montmorillonite clay.

The change in adsorption conformation and the ability of the adsorbed polyelectrolyte to adsorb montmorillonite is also schematically shown in *Figure 9.11*.

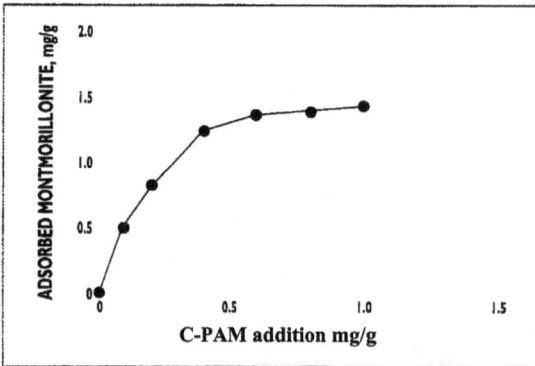

Figure 9.10. Adsorption of montmorillonite clay to cellulosic fibres which have been presaturated to different degrees with C-PAM. The preadsorption time was 3 minutes and the fibre concenbtration was 2 g/l during the preadsorption. The pH during the clay adsorption was pH = 8, the concentration of fibres was 20 g/l, addition of montmorillonite was 2 mg/g and the adsorption time was 5 minutes. Adapted from [20]

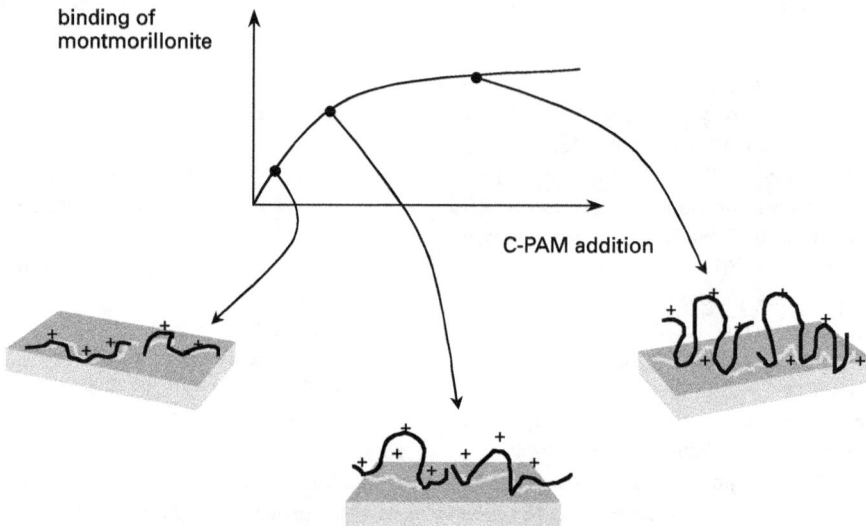

Figure 9.11. Schematic representation of how the adsorbed amount of C-PAM changes with addition of the polyelectrolyte and how this will change the binding ability of added montmorillonite.

Similar results are discussed in [5] regarding the increase of the hydrodynamic layer thickness of cationic polyacrylamide adsorbed onto silica surfaces upon increasing polymer adsorption. There is large similarity between the results shown in *Figure 9.10* and the results shown in [5]. In [5] it was suggested that this curve shape, i.e. a flattening off in thickness of the adsorbed layer at higher polyelectrolyte adsorption, was typical for diffuse chain conformation for pure electrosorption. In this mode the chains are captured in the electrical double layer and are hence not lying flat on the surface and the chain conformations are, according to [5], independent of the degree of ion exchange in the double layer as soon as 50 % of the small ions are replaced by polyelectrolyte chains. Since most polyelectrolytes in the paper industry are efficient through some kind of bridging between surfaces it must be important, for future research, to determine in detail the factors controlling the conformations of polyelectrolytes adsorbed to cellulose surfaces and to find ways of controlling this conformation. It will also be important to link simulations of how charged co-polymers are adsorbed to the different surfaces with similar measurements of model systems of polyelectrolytes and model cellulose surfaces.

Apart from studying the change of conformation of adsorbed polyelectrolytes with an increase of the adsorbed amount it is also important to study how the conformation is changed upon increasing salt concentration. Again, this has not, to the knowledge of the author, been directly examined experimentally for cellulose surfaces. However for model systems, i.e. adsorption of cationic polyacrylamides on silicon surfaces this has been examined by using ellipsometry [21]. As demonstrated in *Figure 9.12* there is a rather dramatic expansion of the adsorbed layer of C-PAM when changing the NaCl concentration from 10 mM to 100 mM and the time scale for the change is around 100 s. The concept of kinetics of adsorption and reconformation will though be treated separately under the headline "Kinetics of adsorption and reconformation" and this example is included here merely to demonstrate the change in conformation of an adsorbed polyelectrolyte when the salt concentration is changed.

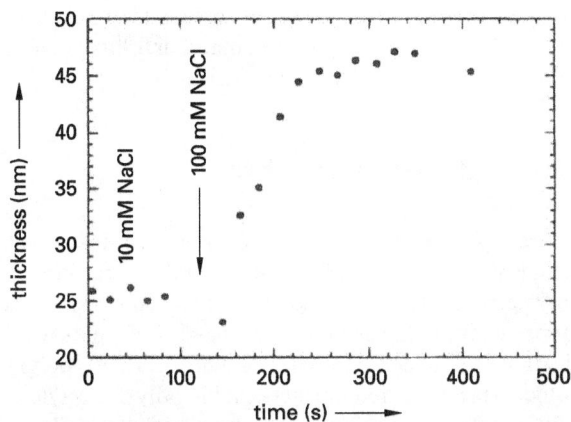

Figure 9.12. Change in polymer conformation upon change in salt concentration. The system consisted of cationic polyacrylamide adsorbed on silicon surfaces and 1 mM NaHCO$_3$ was used as a buffer throughout the experiments. Adapted from [21].

With an even higher salt concentration the polymer will have an even larger extension and eventually the polymer will be totally displaced by small ions as the salt concentration is increased as described by curve **3** in *Figure 9.1*. This behaviour has also been described theoretically by van de Steeg [13] and as shown in *Figure 9.13*, there is a dramatic change in conformation when the salt concentration is changed from 10 mM to 100 mM. In the figure the fraction of trains and loops decreases while the fraction of tails increases. Hence, by comparison of *Figure 9.12* and *Figure 9.13* it has once again been found that the SF-theory can be used to predict the adsorption of polyelectrolytes on cellulose and similar surfaces.

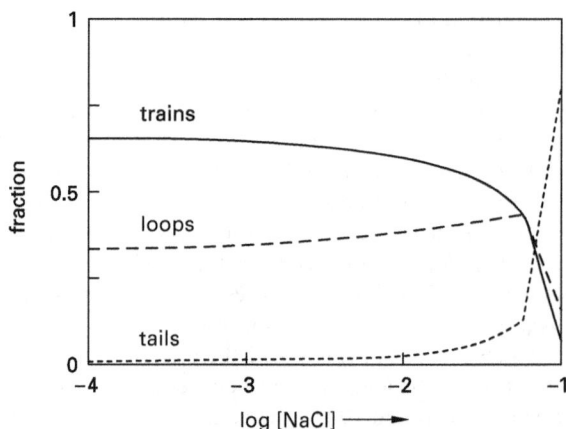

Figure 9.13. Structure of the adsorbed layer as a function of salt concentration. From model calculations with the SF-theory [6]. The amount of loops, tails and trains are given as fractions of the adsorbed amount. Chain length = 500, segment charge = 0.1, surface charge density −0.01 C/m², interaction parameter for the segments $\chi_{sP} = 0.6$ an interaction pareameter for the cations $\chi_{sC} = 4$. Adapted from [13]

It must be stressed once again, though, that there are no direct measurements of polyelectrolyte conformation on cellulose surfaces , to the knowledge of the author, and this is an area, which definitely has a large potential for future research and development.

9.4 Where are Polyelecrolytes Adsorbed on Fibres

Since cellulosic fibres are porous in nature it is obvious that the molecular mass of the polyelectrolyte will have a large impact on the adsorbed amount. It has also generally been found that when the molecular mass of the polymer decreases there is an increase in adsorption since more surfaces will be available for the polymer with the lower molecular mass. This holds true for polyetyleneimine [22–26], cationic dextran [27] cationic polyacrylamides [28, 29], polyDMDAAC [30] and possibly also many other types of cationic water-soluble polyelectrolytes. As an example of this behaviour the adsorption of cationic polyacrylamides on cellulosic fibres are shown in *Figure 9.14* [29].

All the different investigations mentioned above show the same result, i.e. that the polymers will reach different levels of the fibre wall, and different structural levels of the microcrystalline cellulose substrate, depending on their molecular mass or more correctly the size of the polyelectrolytes. Several investigators [17–19] have also claimed that the combination of this fact and the fact that highly charged polyelectrolytes are adsorbed through an ion-exchange mechanism makes it possible to determine the charge on different structural levels in the fibres.

In [17] the adsorption and the adsorption stoichiometry, as defined earlier, were determined on fibres and microfibrillated fibres in order to test how the "removal" of the porous nature of the fibre wall would affect the adsorption. It was shown in *Figure 9.9* that the adsorption and adsorption stoichiometry of polybrene was not at all affected by the porous structure of the fibre wall and that this polymer hence had access to all available surfaces of the fibres used in the ex-

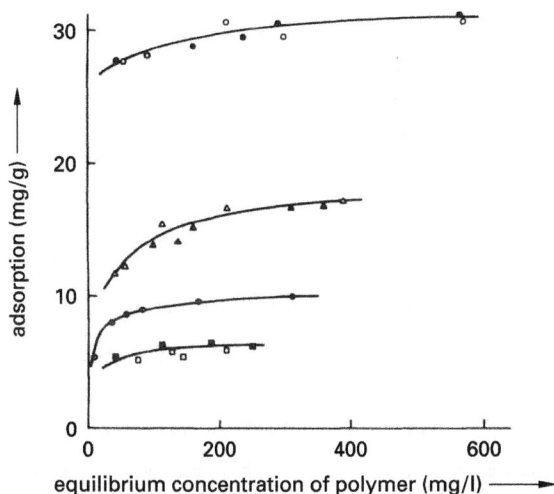

Figure 9.14. Adsorption of cationic polyacrylamide on cellulosic fibres for different molecular mass of the polyelectrolyte. The pH during the adsorption was pH = 4.5–5 , the fibre concentration was 4 g/l and the adsorption time was 15 000 minutes. $\bigcirc$ and $\bullet$ = 2.5 and 2.0 · 10^4, $\triangle$ $\blacktriangle$ = 5.2 and 5.3 · 10^5, $\blacksquare$ $\square$ = 9.0 and 10 · 10^6. The higher value of the molecular weight corresponds to fluorescently labelled C-PAM whereas the lower value corresponds to the non-labelled polymer. The half filled circle corresponds to polybrene adsorption. Adapted from [29].

periments. In [17] it was also concluded that this was true for the fibres that had been carboxymethylated to a degree of substitution of 0.064. There was naturally a large difference in adsorbed amount compared to the native fibres but the adsorption stoichiometry showed the same pattern as for the native fibres. This, in turn, showed that the polybrene molecule has a dimension that is close to the dimension of the pores of the fibre wall and when the fibre wall has a low degree of swelling the molecule will not be able to reach all the charges in the fibre wall. It would hence be of great importance to have an exact knowledge of the dimension of the polybrene molecule but unfortunaltely this is not available in the literature. In [18] the dimension, i.e. radius of gyration, was estimated to be somewhere between 26 Å and 225 Å which is in the same size range of pores that has been estimated from solute exclusion techniques [31]. However, due to the large uncertainty in the determination of the dimension of this molecule it is difficult to judge if the reported values agree.

Recently, the size-range of the pore size distribution of the fibre wall, as reported in [31] has been questioned in [32]. These authors compared the adsorption of polyethyleneimine (PEI) with different molecular mass on porous glass and non-porous glass and on cellulosic fibres. In order to suppress the electrostatic effect to as high degree as possible the adsorption experiments were performed at pH = 10. As is shown in *Figure 9.15* [32] there is a break in the adsorption (i.e. saturation adsorption in mg/g) at a polymer radius of around 13 nm. This means that the surfaces available for a polymer with a radius of 1 nm are also available for a polymer with a radius of gyration of 13 nm. In the figure is also included the specific adsorption for PEI on glass.

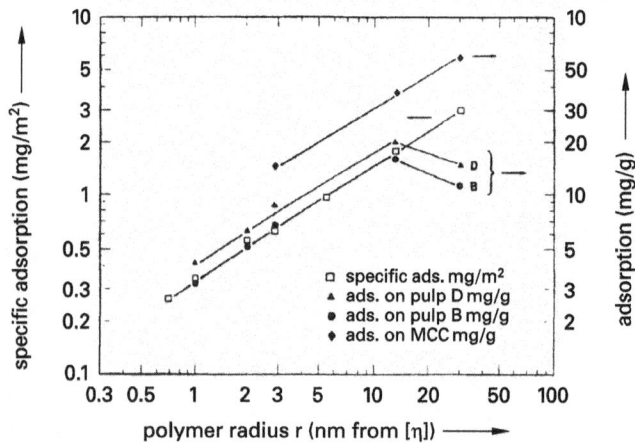

Figure 9.15. The saturation adsorption of PEI on bleached sulphite pulp (D), a bleached softwood kraft pulp(B) and microcrystalline pulp as a function of polymer size. The adsorption was conducted at pH = 10. The specific adsorption on non-porous glass in mg/m^2 (left scale) is included for comparison and the adsorption in mg/g for the different cellulose substrates are shown on the right scale. Adapted from [32].

Since the relationship between the pore size on porous glass and the radius of the polymer was 3–5, i.e. the pore has to be 3–5 times larger to allow the polymer to enter the pores, it was suggested that there was a rather narrow pore size distribution of the fibres around 40–65 nm which is in rather sharp contrast to the pore size range 0.25–5 nm as found in [31]. In [32] it was also argued that there are larger pores in the fibre wall but they are found in pockets that are only possible to reach via smaller pores. However, this statement needs further investigation. It can hence be concluded that there exists a relationship between the size of the polyelectrolyte molecules and the pore size of the fibre wall. This is also schematically shown in *Figure 9.16* where the measured dimension of a polyDiMethylDiAllylAmmonium Chloride with a molecular mass of 70.000 is drawn to scale to the dimension of the pores of the fibre wall as determined with deep etched freeze drying Scanning Electron Microscopy technique. For clarity the stretched dimension of the polyelecvtrolyte is also shown in order to show how the coiling of the molecule will allow for a penetration into some of the pores of the fibres wall.

It can hence be concluded that there is a need to further study this relationship using polyelectrolytes with a well-known size distribution and with a good characterization of how the dimension of the polyelectrolyte changes with salt concentration and the presence of a surface to which the polyelectrolytes can adsorb. Apart form being of purely academic interest these kind of investigations would give further information on the structure of the fibre wall and secondly it would give important information to paper chemists, among others, who would like to add polyelectrolytes to change the interior and the exterior of the fibre wall.

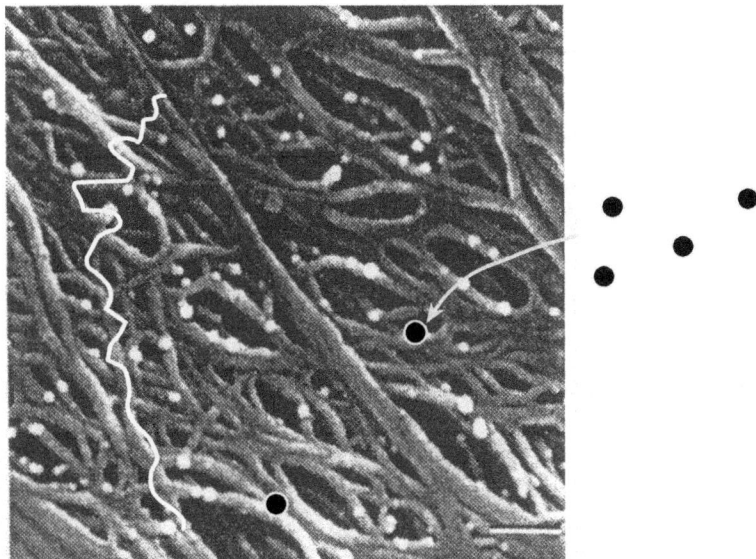

Figure 9.16. Schematic representation of how a poly-DiMethylDiAllylAmmonium Chloride with a molecular mass of 70.000,drawn to scale to the dimension of the pores of the fibre wall as determined with deep etched freeze drying Scanning Electron Microscopy technique, will have access to some parts of the fibre wall. For clarity the stretched dimension of the polyelectrolyte is also shown in order to show how the coiling of the molecule will allow for a penetration into some of the pores of the fibres wall. The bar in the figure corresponds to 100 nm.

Another factor, which is of great importance for the adsorption of polyelectrolytes to cellulosic fibres, is the presence of cellulosic fines from the fibres. Since these materials have large specific surface area, i.e. a large area available for high molecular mass polyelectrolytes, they will adsorb a high amount of added polyelectrolytes despite the fairly low amount of fines. Naturally this is valid only for polyelectrolytes with a molecular mass large enough to prevent the polymers for entering the fibre wall. The influence of fines has been discussed in a number of investigations [9, 33–35] and it was established that it is not only the amount of fines that is important but also the properties of the fines [35]. In *Figure 9.17a* and *Figure 9.17b* this effect is summarised [35]. In *Figure 9.17a* the plateau adsorption of cationic starch on different bleached softwood fibres is shown and in *Figure 9.17b* the adsorption of starch on the fines from the different fibres are shown. The adsorption on the long-fibre fraction from the different pulps is the same whereas there is a fairly large difference in adsorption on the total pulps.

Since the adsorption to the long-fibre fraction was the same the authors [35] concluded that the difference in adsorption could be ascribed to the fines fraction and as is shown in *Figure 9.17b* there was also a large difference between different fines. Microscopy investigations showed that there actually was a difference in the morphology of the fines that could explain the adsorption results. The fines from the beaten fibres had a fibrillar structure whereas the fines from the unbeaten fibres consisted to a large extent of non-fibrillar fibre fragments. The authors could not explain the levelling off at lower D.S. of the cationic starch in *Figure 17b* in [35] but with reference to the SF-theory [5, 6, 8] it may be suggested here that there is a maximum in the plateau adsorption as a function of the charge of the polyelectrolyte. The difference between the fines and the fibres might also be caused by a different χ_s parameter for these two materials and that the fibres have a higher χ_s value. This has to be taken as a suggestion but it is definitely worth further investigations. Again this shows the usefulness of comparing the experimental results with the existing theories.

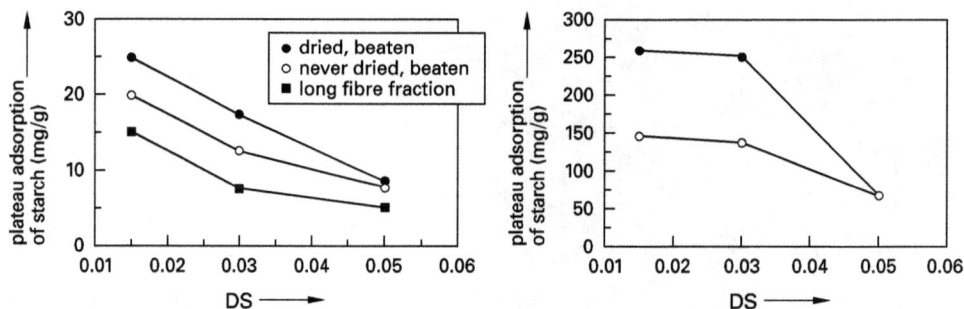

Figure 9.17. a) Plateau adsorption of cationic potato starch as a function of D.S. (degree of substitution) of the starch. A beaten, dried softwood kraft pulp($\bullet$) is compared with a never dried, beaten softwood kraft pulp ($\circ$) and the long fibre fraction of both pulps($\blacksquare$). pH during the adsorption was 7 and the adsorption time was 10 minutes. b) Plateau adsorption on the fines fraction from the pulps mentioned in Figure 9.14 a). The adsorbed amount was calculated from the adsorption data in Figure 9.14a and the amount of fines in the different pulps. Adapted from [35].

Concluding this section it is obvious that the porous structure of the fibres will influence the adsorption to a large extent and since both the dimensions of the polyelectrolytes and the dimensions of the pores in the fibres is not exactly known it will be hard to predict the adsorption of a polyelectrolyte on a certain type of fibre. Even if it might not be fully worth the effort to get an exact relationship for all types of polymers and fibres it is recommended to establish different adsorption regimes for certain fibre types and certain types of polyelectrolytes,. i.e. fixing agents, dry strength agents, wet strength agents and retention aids. It is also clear that the amount and types of fines are important parameters for the polyelectrolyte adsorption and from a practical point of view it is the opinion of the author that this area definitely needs more work. It is naturally a tough area to study since the preparation of the fines fraction will always take time-consuming sample preparation methodology development.

9.5 How Fast are the Polyelectrolytes Adsorbed to Cellulosic Fibres

In many industrial applications the effect of the added polyelectrolyte is expected only seconds after the polyelectrolyte has been added to the fibre/water mixture. This makes the adsorption kinetics very important and consequently this fact has given support to a fair amount of research on adsorption kinetics. As pointed out by for example van de Ven [36] and by Lindström [28] the following processes occur during the adsorption

- Transport of the polyelectrolytes from the solution to the fibre surface (the definition of the available fibre surface is dependant on the molecular mass of the polyelectrolyte)
- Attachment of the polyelectrolyte onto the fibre surface

- Reconformation of the polyelectrolyte on the fibre surface
- Detachment of the polyelectrolyte from the fibre surface

The general understanding, at present, is that as the polyelectrolytes collide with the fibres either by Brownian motion [36] or through turbulent transport [30, 37] and this process is equivalent to the first item mentioned above. Several attempts have also been made to model the adsorption kinetics of polyelectrolytes onto cellulosic fibres [22, 30, 36]. All these attempts use some kind of Langmuir model to describe the kinetics. Kindler et al. [22] introduced a stagnant layer around the fibres in order to use a mass-transfer model whereas van de Ven [36] and Wågberg [30] utilised a Brownian collision or a turbulent collision model where the fibres are viewed as rigid spheres, which collide with spherical polyelectrolyte molecules. In [36] a desorption process was also included in the proposed rate equation but in most modelling situations the time for desorption was set at infinity. In [30] only one fitting parameter was used to fit the data to the experimental results and that was the collision efficiency factor, α, and no desorption process was used. The equations can naturally be made more or less complicated and in [38] another factor was included and this was a factor to describe the migration of polyelectrolyte into the lumen of the fibres with the characteristic time τ_{lumen}.

All these models gave excellent fit to the experimental data but in [22] very thick stagnant layers were found (of the order of 4 cm) and the authors concluded that the stagnant layer approach was too simple and that the layer thickness, as defined in the equations, would also include diffusion and subsequent adsorption within the fibrous structure, entropy barriers, or solvent interactions". With the approach chosen in [22] it was not possible to separate these processes. Nevertheless the modelling showed that it was possible to get a good match with experimental results the only complication was that a more advanced model was needed to sort out the details. van de Ven [36,38] used a more elaborate model to get a more detailed description of the adsorption process. To give a few examples of the results the following characteristic timescales can be mentioned [36, 38]

- τ_{ads} is typical around 1 minute
- τ_{lumen} is typical of the order of 200 minutes

These values were obtained by fitting the adsorption model to the data shown in *Figure 9.18* below. The full lines in the figure correspond to the fit of the model and the symbols correspond to actually measured values. As can be seen the fit is very good and the model can hence be used to predict the adsorption of polyethyleneimine onto cellulosic fibres under certain conditions. Naturally the calculated timescales are dependant on the experimental conditions but the times mentioned above are representative of typical conditions occurring in practice.

In the turbulent model [30] it was found that there was a very good agreement between the experimental data for the adsorption of the low molecular mass polyelectrolyte and the simulated values assuming a collision efficiency factor of 0.25, i.e. every fourth collision led to an adsorption. For the high molecular mass polyelectrolyte there was not at all as good agreement and it was obvious that a more elaborate approach was needed but on the other hand this kind of model also needs further input parameters which in turn makes it necessary to find timescales for polyelectrolyte reconformation on the surfaces and polymer diffusion onto fibre surfaces. These data are not available today and further work should be definitely devoted to determine these.

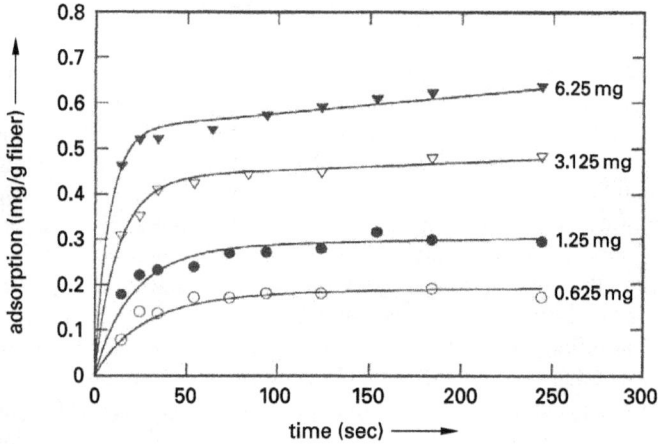

Figure 9.18. Adsorption of polyethyleneimine as a function of time for different polymer additions. Fibre concentration was 2 g/l, stirring speed 80 rev./min, KCl concentration was $2.5 \cdot 10^{-4}$ and the pH was 6. The desorption time was set to infinity. Adapted from [38].

Another interesting feature was found in [30] and that was that the adsorption of the low molecular polyDMDAAC was not at all affected by the presence of a preasdorbed saturated layer of high molecular mass polyDMDAAC on the surface of the fibres. This is shown in *Figure 9.19* below.

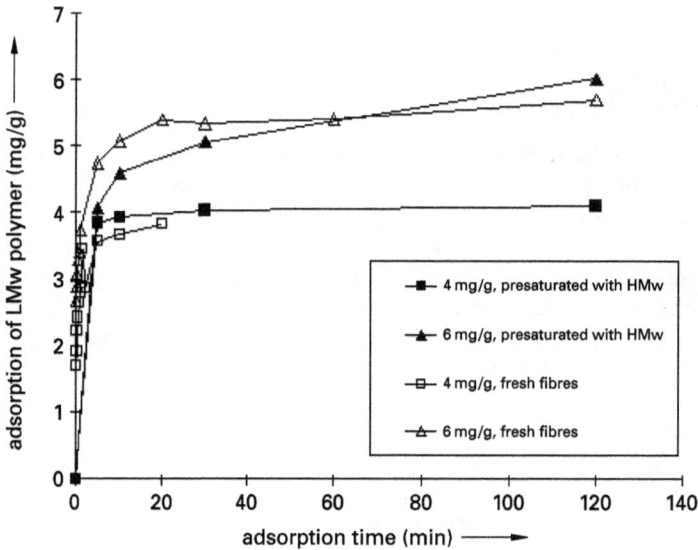

Figure 9.19. Kintetics of adsorption of LM_w polymer on fresh fibres and fibres precovered with HM_w polymer.. The fibre concentration was 5 g/l and the pH = 8 in all measurements. Adapted from [30].

This indicates that the polyelectrolytes diffuse into the fibre structure before they are adsorbed onto the cellulose surface and that they do not first adsorb on the external surface and then migrate into the fibre wall by a reptation process. Needless to say this is an area for future research and again new techniques, with high enough resolution are needed to study these processes.

The results presented in [37] clearly indicated that the polyelectrolytes adsorb in a state very similar to the one they have in solution and with time they reconform to a more flat conformation according to the theories for polyelectrolyte adsorption. Since almost all types of actions of the polyelectrolytes depend on some kind of bridging between surfaces it is essential to have some information about the kinetics of this reconformation. As was mentioned earlier in this paper the literature in this area is a very limited and there are only some indirect measurements available on this topic [39, 40]. In [39] the reconformation time was estimated from the release of counterions from the polyelectrolyte and in [40] it was estimated from binding of anionic polyacrylamide at different times after adsorption of a high molecular mass cationic polyacrylamide to the fibres. An example from [40] is shown in *Figure 9.20*. The authors in [40] concluded that as the adsorption of anionic polyacrylamide levels off the cationic polymer would have reached its equilibrium conformation. The time taken to reach this levelling off was taken as a measure of the polymer reconformation time.

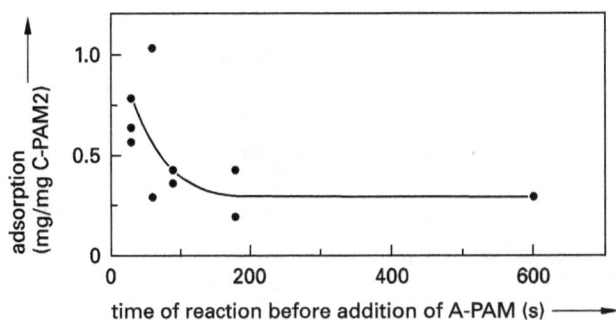

Figure 9.20. Adsorption of anionic polyacrylamide (A-PAM) onto a fines-free, bleached softwood kraft pulp to which a high molecular mass cationic polyacrylamide (C-PAM) first had been pre-adsorbed. A-PAM was added at different times after addition of the C-PAM. pH during the measurements was pH = 7.5. Adopted from [40].

It can be seen in the figure that the reconformation times are rather short and after 100 s little is changing. A simplified cartoon of this process is also shown in *Figure 9.21*.

In *Table 9.1* a short summary of the data in [39] and [40] are shown and the overall conclusion is that the polymer reconforms very fast to a flat conformation for the polyelectrolytes used in these investigations. The D.S.-values given in the table corresponds to the D.S. of the C-PAM and that the 3.6.-ionene is fully charged in the entire pH-range.

Interpretation

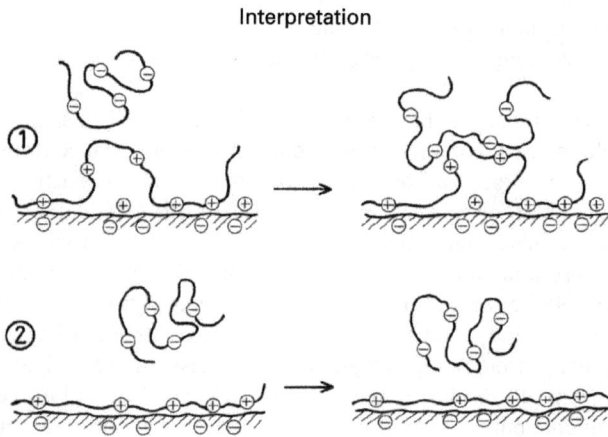

Figure 9.21. Simplified interpretation of the reconformation indirectly detected in Figure 9.20. Adopted from [40].

Table 9.1. Summary of the indirect estimation of reconformation times of cationic polyelectrolytes adsorbed onto cellulosic fibres. The D.S.-values correspond to an average number of charged groups on the polyelectrolytes given as mole fractions of the total number of groups in the polymer backbone.

System investigated	Fibres	D.S.-value (Fraction by mole)	Time to Equilibrium (s)	Reference
A-PAM to pre-adsorbed C-PAM	Bleached softwood fibres	A-PAM = 0.16 C-PAM = 0.3	100–200	[40]
C-PAM	Carboxymethylated bleached softwood D.S. = 0.065	0.3	60–180	[39]
3.6. -ionene	- " -	0.18	≈ 60	[39]

It should be mentioned that the reconformation times are rather short compared to reconformation times achieved for adsorption of polyvinylpyridine on anionic latex particles [41]. These authors found that two different types of reconformation processes occurred. When the polymer was added to the latex at a very low feeding rate there was one type of relaxation on the surface and this relaxation process was of the order of 2000 minutes. When the polymer was added at a higher feed rate there was first an oversaturation of the surface by the polyelectrolyte and when the polymers started to reconform there was also a desorption process taking place. The time for the reconformation in this case was of the order of 10 minutes. It is today not clear how the reconformation of polyelectrolytes on fibre surfaces should be categorised and how the relatively short reconformation times on the fibre surfaces should be explained. It might be suggested that there is a different type of reconformation on a macroscopically rough surface than on a flat surface. This in turn might be very important and the reconformation process and the factors controlling this should definitely be studied further. With instrumentation such as ellipsometry and with the new developments of smooth cellulose surfaces [42,43] this will be a very interesting future research area.

The last process mentioned above, i.e. the exchange of adsorbed polyelectrolytes has been an area for large controversy over the years for a number of reasons. First of all the general concept of the polymer adsorption as an equilibrium process has been very important for the development of different theoretical models [5] based on thermodynamic relations. This discussion was spurred by the fact that it is very hard to find any desorbed polymer when diluting a system with adsorbed polymers with pure solvent. This was thoroughly discussed in [5] and on a segment basis the adsorption can definitely be regarded as an equilibrium process but it will naturally be difficult for a large macromolecule to desorb due to its large number of attachment points. For polyelectrolytes with opposite sign to the surface, the desorption will be even more difficult and careful investigations of the desorption process have been conducted by Tanaka et.al [3, 44, 45] by using an elegant technique with fluorescent labelling. By first treating the fibres with cationic polyacrylamides containing a fluorescent label, washing and then treating the fibres with a non-labelled polyacrylamide it was possible to detect the desorption simply by measuring the fluorescence of the water phase surrounding the fibres. The non-labelled polymer was denoted as a displacer since it was used to test if the pre-adsorbed polymer could be displaced or not.

It was found in these investigations that;

1. No desorption took place without displacing polymer present
2. The desorption was very slow and virtually non-existent for high molecular mass polyelectrolytes and increased with decreasing molecular mass and for a C-PAM with a molecular mass of $2.9 \cdot 10^4$ (1.43 meq./g) a final desorption of 50 % was finished within 2 days. For a C-PAM with a molecular mass of $5.2 \cdot 10^5$ (1.38 meq./g) about 25 % could be desorbed and this process was finished within 12 days.
3. With a decreasing charge density the exchange was faster. By using a C-PAM with a molecular mass of $5.2 \cdot 10^5$ and a charge density of 0.66 meq./g the fraction of desorbed C-PAM could be increased to around 50 % and this process was finished within 1 day.

An example of these results is shown in *Figure 9.22* and is representative of the results discussed under 2) above.

The reason why only fractions of the polyelectrolyte can be desorbed despite a ten-fold excess of displacing polyelectrolyte is not exactly known but probably this can be explained by either heterogeneity in the adsorbed conformation of the preadsorbed C-PAM or a very slow migration of the polyelectrolyte into the outer layers of the fibres. Probably both these suggestions are valid but the fact that the amount of C-PAM that could be desorbed was very dependent on the time between adsorption and desorption suggested that the penetration into the external part of the fibre wall was indeed important [44]. It must be stressed though that this penetration was found for very long times and hence this finding should not be mixed up with the findings mentioned in conjunction with *Figure 9.19*.

Apart from these simple model systems there are some interesting aspects on polyelectrolyte adsorption kinetics that need some further comments. In practice it is very common to add different types of polyelectrolytes very close to each other in the papermaking system. Since most of the polyelectrolytes, as mentioned previously, have their effect while being adsorbed to the fibres or fines surfaces it is important to know which of the polyelectrolytes in a mixture that will adsorb to the fibres first and how this will affect the effect of the polyelectrolyte. In [20] it was for example found that an addition of a highly charged cationic polyelectrolyte could sig-

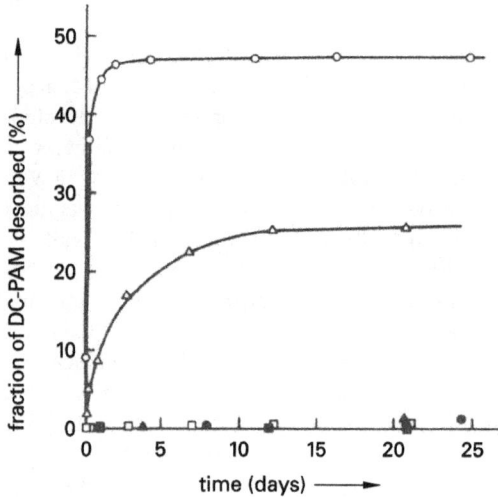

Figure 9.22. The fraction of desorbed fluorescently labelled C-PAM as a function of displacement time. Filled symbols correspond to the situation with no displacer present. The molecular mass of the polyelectrolytes was $O = 2.9 \cdot 10^4, \triangle = 5.2 \cdot 10^5$ and $\square = 9.0 \cdot 10^6$. Adsorption time was 10 days and the pH was 4.5–5 during the experiments. Adapted from (45)

nificantly enhance the flocculating ability of micro-particle based retention agents. Theoretically it might be postulated that small molecules should diffuse faster to the fibre surface, provided that Brownian diffusion across some kind of stagnant liquid layer on the surfaces of the fibres is a rate-limiting step. However since turbulent collisions might be the dominating collision mechanism [30, 37] it is not so clear how the molecular mass might affect the adsorption process since the cellulosic fibre will dominate this process due to their size. In [46] this was investigated for C-PAM with different molecular weight by utilising the fluorescent labelling technique as mentioned earlier. The results showed that in a 1:1 mixture of high and low molecular weights C-PAM very close to equal amounts were adsorbed to the cellulosic fibres. This indicates that Brownian diffusion is not the mechanism responsible for the adsorption of polyelectrolytes on cellulosic fibres under realistic conditions.

Finally it should be briefly discussed how the presence of different components in a furnish might affect the equilibrium adsorption and the adsorption kinetics. In [28] it was detected that the fine material from a bleached sulphite pulp showed a transient surface potential, from microelectrophoresis measurements, in the presence of fibres [2]. The authors concluded that this was due to a more rapid adsorption of the C-PAM to the fines material and at low polyelectrolyte additions some polyelectrolyte could be transferred to the long fibre fraction. This conclusion was based on the fact that in the absence of long fibres this transfer could not be detected [2]. The reverse process, a transfer of C-PAM from fibres to polystyrene latex was detected in [47] and it was found that the amount of segments from the polymer protruding into solution was important for the transfer reaction. It was detected that the transfer was easiest for the medium molecular mass polymer and most difficult for the low molecular mass polymer leaving the high molecular mass polymer in between. A larger number of anchoring points for this polymer can explain the fact that the high molecular mass polyelectrolyte transferred with a lower rate than the medium molecular mass polyelectrolyte. The reason to the low transfer rate of the low molecular mass polyelectrolyte can be explained by the fact that it is rapidly adsorbed into the interior of the fibre wall and is hence not as easily transferred due to this. The importance of the extension of the polymer segments into solution could also bet detected as a dependence of delay time between adsorption of C-PAM and addition of polystyrene latex. When the delay time was increased there was a lower amount of C-PAM transferred to the latex.

9.6 Recommendations for Future Work

It has been indicated in the text where more research is needed, according to the author, but in summary the following research areas can be mentioned:

- Tailormaking of polyelectrolytes to give them certain χ_s and χ parameters so that they can tolerate systems with high salt concentrations and still give high adsorption to fibres and fines. By using the information in *Figure 9.1* it should be possible to produce suitable polyelectrolytes.Furthermore, by using non-charged adsorbing species it should also be possible to create polyelectrolytes which could tolerate high amounts of dissoloved and colloidal substances.
- Prepare block-copolymers, which have one type of segments with a high affinity for the fibre surface and one type of segments that have a high affinity for other type of materials. These latter materials can be microparticles, fillers, dissolved and/or colloidal materials. Generally the block-copolymers are very interesting for several applications but today their cost might be a bit too high. This might however change very fast and therefore they are of high interest for research purposes.
- There is a need to find a better correlation between polyelectrolyte size and the pore size of the fibre wall and to determine the rate constants for penetration of polyelectrolytes into the fibre wall. This is for example of importance when trying to modify both the fibre surface and the interior of the fibre wall.
- Determination of the conformation of polyelectrolytes on fibre surfaces. Clarification of the importance
- Surface structure of the fibres
- Surface charge of the fibres
- Surface structure of the fibres
- Polyelectrolyte structure
- Polyelectrolytre charge
- Salt concentration and type of salt
- Time after polyelectrolyte addition
- Studies of some of these items are underway and here the different types of smooth cellulosic surfaces that have been developed during the last years will be of large strategic research importance.
- Creation of a better link between the more recent development of polyelectrolyte adsorption theories and the adsorption of different types of polyelectrolytes onto cellulosic fibres is essential. In this way more efficient experiments will be conducted and furthermore a theo-

retical model for adsorption of polyelectrolytes onto fibres can be developed including both statics and dynamics. The dynamic aspect is very important and has often been neglected.

9.7 References

[1]   Wågberg, L., and Ödberg, L.: Polymer adsorption on cellulosic fibres. *Nordic Pulp Paper Res. J.* 4,2(1989)135.

[2]   Lindström, T.: Some fundamental chemical aspects on paper forming. In: C.F. Baker and V.W. Punton (Eds.) *Fundamentals of Papermaking.* Cambridge Mechanical Eng. Publ. Ltd., London, 1989, p. 309.

[3]   Ödberg, L., Swerin A., and Tanaka, H.: Kinetic apects of the adsorption of polymers on cellulosic fibres. *Nordic Pulp Paper Res. J.* 8,1(1993)6.

[4]   Swerin, A., and Ödberg, L.: Some aspects of retention aids. In: C.F. Baker (Ed) *The fundamentals of papermaking materials.* Pira International, Leatherhead, UK, 1997, p. 265.

[5]   Fleer, G.J., Cohen-Stuart, M.A., Scheutjens, J.M.H.M., Cosgrove, T., and Vincent, B. (Eds.): *Polymers at Interfaces.* Chapman&Hall, London, 1993, pp. 343-375.

[6]   Böhmer, M.R., Evers, O.A., and Scheutjens, J.M.H.M.: Weak polyelectrolytes between two surfaces: Adsorption and stabilization. *Macromolecules* 23,8(1990)2288.

[7]   van de Steeg, H.G.M.: *Cationic starches on cellulose surfaces (PhD Thesis).* University of Wageningen 1992.

[8]   van de Steeg, H.G.M., Cohen Stuart, M.A., de Kaizer, A., and Bijsterbosch, B.: Polyelectrolyte adsorption – A subtle balance of forces. *Langmuir* 8(1992)2538.

[9]   Wågberg, L., and Kolar, K.: Adsorption of cationic starch on fibres from mechanical pulps. *Ber. Bunsenges. Phys. Chem.* 100,6(1996)984.

[10] Pelton, R.H.: Electrolyte effects in the adsorption and desorption of a cationic polyacrylamide on cellulosic fibres. *J. Colloid Interface Sci.* 111,2(1986)475.

[11] Lindström, T., and Wågberg, L.: Effects of pH and electrolyte concentration on the adsorption of cationic polyacrylamides on cellulose. *Tappi J.* 66,6(1983)83.

[12] Hedborg, F.and Lindström, T.: Adsorption of cationic potato starch onto bleached softwood cellulosic fibres. *Nordic Pulp Paper Res. J.* 8,2(1993)258.

[13] van de Steeg, H.G.M., de Kaizer, A., Cohen Stuart, M.A., and Bijsterbosch, B.H.: Adsorption of cationic amylopectin on microcrystalline cellulose. *Colloids Surfaces A: Physicochemical Eng. Aspects* 70(1993)77.

[14] Trout, P.E.: The mechanism of the improvement of wet strength of paper by polyethyleneimine. *Tappi* 34,12(1951)539.

[15] Bates, N.A.: Polyamide-Epichlorihydrin wet-strength resin. I. Retention by pulp. *Tappi* 52,6(1969).

[16] Winter, L., Wågberg, Ödberg, L., and Lindström, T.: Polyelectrolytes adsorbed on the surface of cellulosic materials. *J Colloid Interf. Sci.* 111,2(1986)537.

[17] Wågberg, L., Winter, L., Ödberg, L., and Lindström, T.: On the charge stoichiometry upon adsorption of a cationic polyelectrolyte on cellulosic materials. *Colloids Surfaces* 27(1987)163.

[18] Wågberg, L., Winter, L., and Lindström, T.: Determination of ion-exchange capacity of carboxymethylated cellulose fibres using colloid and conductometric titrations. In: V.Punton (Ed.) *Papermaking Raw Materials*. Mech. Eng. Publ. Ltd, London, 1985, p. 917.

[19] Wågberg, L., Ödberg, L., and Glad-Nordmark, G.: Charge determination of porous substrates by polyelectrolyte adsorption. Part 1. Carboxymethylated, bleached cellulosic fibres. *Nordic Pulp Paper Res. J.* 4,2(1989)71.

[20] Wågberg, L., Björklund, M., Åsell, I., and Swerin, A.: On the mechanism of flocculation by microparticle retention aid systems. *Tappi J.* 79,6(1996)157.

[21] Ödberg, L., Sandberg, S., Welin-Klintström , S., and Arwin, H.: Thickness of adsorbed layers of high molecular weigth polyelectrolytes studied by ellipsometry. *Langmuir* 11,7(1995)2621.

[22] Kindler, W.A., and Swanson, J.W.: Adsorption kinetics in the polyethyleneimine-cellulose fiber system. *J. Polym. Sci.* A-2,9(1971)853.

[23] Horn, D., and Melzer, J.: Electrostatic and steric effects of cationic polymers adsorbed on cellulose fibres. In: V. Punton (Ed.) *Fibre-Water interactions in papermaking, Vol.1*. Oxford University Press, Oxford, 1977, p 135.

[24] Alince, B.: Polyethyleneimine adsorption on cellulose. *Cell. Chem. Techn.* 8(1974)573.

[25] Alince, B.: The role of porosity in polyethyleneimine adsorption onto cellulosic fibres. *J. Appl. Polym. Sci.* 39(1990)355.

[26] Alince, B., Vanerek, A., and van de Ven, T.G.M.: Effects of surface topography, pH and salt on the adsorption of polydisperse polyetyleneimine onto pulp fibres. *Ber.Bunsenges.Phys. Chem.* 100,6(1996)954.

[27] Heinegård, C., Lindström, T., and Söremark, C.: Reversal of charge and flocculation of microcrystalline cellulose with cationic dextranes. In: *Colloid and Interface Sci. Vol. IV*. Academic Press, New York, London, San Fransisco, 1976, p. 139.

[28] Lindström, T., and Söremark, K.: Adsorption of cationic polyacrylamides on cellulose. *J. Colloid Interface Sci.* 55(1976)305.

[29] Tanaka, H., Ödberg, L., Wågberg, L.and Lindström, T.: Adsorption of cationic polyacrylamides onto monodisperse polystyrene latices and cellulose fibres: Effect of molecular weight and charge density of cationic polyacrylamides. *J. Colloid Interface Sci.* 134,1(1990)219.

[30] Wågberg, L., and Hägglund, R.: Kinetics Of Polyelectrolyte Adsorption On Cellulosic Fibres. *Langmuir* 17,4(2001)1096-1103.

[31] Stone, J.E., and Scallan, A.M.: A structural model for the cell wall of water-swollen wood pulp fibres based o their accessibility to macromolecules. *Cell. Chem. Techn.* 2(1968)343.

[32] Alince, B., and van de Ven, T.G.M.: Porosity of swollen pulp fibres evaluated by polymer adsorption. In: C.F. Baker (Ed) *The fundamentals of papermaking materials*. Pira International, Leatherhead, UK, 1997, p. 771.

[33] Marton, J., and Marton, T.: Wet end starch: adsorption of starch on cellulosic fibres. *Tappi* 59,12(1976)121.

[34] Marton, J.: The role of surface chemistry in fines-cationic starch interactions. *Tappi* 63, 4(1980)87.

[35] Wågberg, L., and Björklund, M.: Adsorption of cationic potato starch on cellulosic fibres. *Nordic Pulp Paper Res. J.* 8,4(1993)399.

172

[36] van de Ven, T.G.M.: Kinetic aspects of polymer and polyelectrolyte adsorption on surfaces. *Adv. Colloid Interface Sci.* 48(1994)121.

[37] Falk, M., Ödberg, L., Wågberg, L., and Risinger, G.: Adsorption kinetics for cationic polyelectrolytes onto pulp fibres in turbulent flow. *Colloids Surfaces* 40(1989)115.

[38] Petlicki, J., and van de Ven, T.G.M.: Adsorption of polyethyleneimine onto cellulosic fibres. *Colloids Surfaces A: Physicochem. Eng. Aspects* 83(1994)9.

[39] Wågberg, L., Ödberg, L., Lindström, T., and Aksberg, R.: Kinetics of adsorption and ion-exchange reactions during adsorption of cationic polyelectrolytes onto cellulosic fibres. *J. Colloid Interface Sci.* 123,1(1988)287.

[40] Aksberg, R., and Ödberg, L.: Sequential adsorption of cationic and anionic polyelectrolytes on bleached cellulosic fibres. *Nordic Pulp Paper Res. J.* 5(1990)168.

[41] Pfefferkorn, E., and Elaissari, A.: Adsorption-desorption processes in charged polymer/colloid systems; Structural relaxation of adsorbed macromolecules. *J. Colloid Interface Sci.* 138(1990)187.

[42] Buchholz, V., Wegner, G., Stemme, S., and Ödberg, L.: Regeneration, Derivatization and Utilization of Cellulose in Ultrathin Films. *Adv. Mater.* 8,5(1996)399.

[43] Gunnars, S., Wågberg, L, and Cohen Stuart, M.A.: Preparation and characterization of thin, smooth cellulose surfaces by spincoating. *Manuscript in preparation* (2000).

[44] Tanaka, H., and Ödberg, L.: Exchange of cationic polymers adsorbed on cellulose fibres and on polystyrene latex. In: C.F. Baker and V.W. Punton (Eds.) *Fundamentals of Papermaking Cambridge.* Mechanical Eng. Publ. Ltd., London, 1989, p. 453.

[45] Tanaka, H., Ödberg, L, Wågberg, L., and Lindström, T.: Exchange of cationic polyacrylamides adsorbed on monodisperse polystyrene latex and cellulose fibres: Effect of molecular weight. *J. Colloid Interface Sci.* 134,1(1990)229.

[46] Tanaka, H., Swerin, A., Ödberg, A., and Park, S.B.: Competitive adsorption of cationic polyacrylamides with different molecular weights onto polystyrene latex, cellulose beads and cellulose fibres. *J. Pulp Paper Sci.* 23,8(1997)J359.

[47] Tanaka, H., and Ödberg, L.: Transfer of cationic polymers from cellulose fibres to polystyrene latex. *J. Colloid Interface Sci.* 149,1(1992)40.

10 Web Forming

Bo Norman
Department of Fibre and Polymer Technology, KTH

10.1 Introduction

The complete forming process really includes everything from the input of the thick stock into the short circulation to the delivery of a wet paper web from the wire section. Although this includes also the short circulation, in which the mix is prepared, this part is treated in Chapter 8. A headbox feeds the mix to the wire section in the form of a thin jet across the full width of the wire. The sheet build-up takes place through dewatering of the mix in the wire section, which may be of single or twin wire design. The final fibre distribution in the paper web is determined during the forming, and is characterised by formation and fibre orientation parameters. The fibre distribution at different levels in the z-direction of a paper web is of increasing importance.

10.2 Paper Formation and Fibre Orientation Parameters

The formation of a paper sheet is the local grammage distribution determined by the material distribution in the plane of the sheet in the floc size range, ie up to 30–50 mm. It has a great influence on many sheet properties and it is therefore desirable to be able to quantify this distribution. In the field of paper process technology, care should be taken not to confuse the words "forming", which refers to the process, and "formation" which refers to the structure. To characterize the formation, a signal is required which represents the local grammage. An approximate measurement can sometimes be made with light absorption. Beta-radiation absorption gives a correct result, however, and a laboratory method has been developed using beta-radiography to measure formation.

Formation is quantified in terms of

- The formation number F,
- The small scale and large scale formation numbers F_1 and F_2 respectively or, most completely, by
- The formation spectrum, covering a range from ca 0.1 mm to 30–50 mm.

Fibre orientation is characterised by

- Fibre orientation anisotropy; machine direction/cross direction (MD/CD)
- Fibre orientation misalignment angle α (relative to MD).

10.2.1 Light and Beta-Radiation Absorption

Traditionally, paper is studied against a background light and the variations in light transmission are judged in order to assess the formation. This simplified method can, however, give misleading results especially in two cases:

- when the paper consists of components with different optical properties
- when the paper is calendered.

Light transmission through a paper can be described by the Kubelka-Munk equations. These show that it is the product of the light absorption coefficient k and the grammage w and the product of the light scattering coefficient s and the grammage w which determine the degree of light transmission. For a filler-containing paper where both k and s differ between fibres and filler, it is impossible to determine whether e.g. a local reduction in light transmission depends on an increase in the fibre amount or in a slightly smaller increase in the local filler amount. During calendering, the thickest regions (high local grammage) will be subjected to an extra high pressure and the light scattering coefficient may decrease locally. This will increase the light transmission without influencing the local grammage. All formation measurement with the help of light transmission should therefore take place before calendering.

Beta-radiation absorption is normally used on-line to determine the grammage of the paper at the dry end of the machine. The use of beta-radiation is especially suitable for grammage determination on paper since beta-radiation has a practically constant absorption coefficient for all relevant chemical elements and also a negligible scattering. Unfortunately, it is not possible to measure formation on-line in the same way as grammage since the very small measurement area gives inadequate amounts of transmitted radiation.

Laboratory measurement of formation is carried out at e.g. STFI with the help of beta-radiography. A beta-radiogram is first produced by recording on an X-ray film the radiation transmitted through the sample from a radioactive (C-14) source with an extension of 100×1500 mm. A calibration scale with known grammages is placed along one edge of the sample. After about 30 minutes' exposure, the film is developed. This radiogram is then analysed with a desktop scanner, with a resolution of 300 dpi (dots per inch), which corresponds to circa 0.08 mm. The calibration scale with known grammages makes it possible automatically to translate the blackening of the film into local grammage data. *Figure 10.1* shows variations in the local grammage of a sample of newsprint, measured with beta-radiation absorption. The measurement was made with a circular measurement area with a diameter of 0.1 mm.

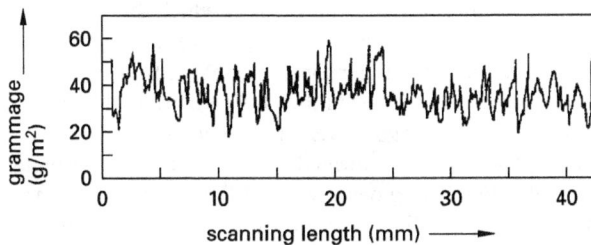

Figure 10.1. Local grammage along a sample of newsprint, with 0.1 mm measuring area.

10.2.2 Formation Number and Formation Spectrum

The grammage variations can be characterised by the formation number F, which is defined as the coefficient of variation of the local grammage, i.e. the standard deviation $\sigma(w)$ divided by the average grammage w_m:

$$F = \sigma(w) / w_m \tag{10.1}$$

The geometrical resolution of the grammage measurement is of decisive importance for the size of the formation number. The smaller the measurement area, the more small-scale variations can be detected and the greater is the formation number. The formation number is grammage-dependent and is statistically inversely proportional to the square root of the mean grammage.

A more complete characterization of the formation is given by an effect spectrum for the local grammage normalized with respect to the mean grammage. The most common type of effect spectrum is the frequency spectrum which can be obtained directly by frequency analysis of a measurement signal. In order to transfer to a variable which is independent of the scanning speed and to characterize the floc occurrence in a paper more directly, the frequency f is transformed to the corresponding wavelength l where u is the speed of scanning:

$$l = u / j \qquad\qquad (10.2)$$

The wavelength spectrum thus obtained describes the occurrence of flocs of different sizes. The wavelength spectrum of the grammage is called the formation spectrum. The area beneath an effect spectrum is equal to the variance of the studied dimension, and the area beneath a formation spectrum is thus equal to the square of the formation number F^2. The complete information in the form of a formation spectrum can be simplified with the small-scale value F_1 and the large-scale value F_2. In the STFI method, a small-scale range of 0.3–3 mm and a large-scale range of 3–30 mm are used. The calculation of these two values involves integration of the wavelength spectrum over the corresponding wavelength intervals.

Because of the logarithmic wavelength scale, spectra presented as variance per unit wavelength have the disadvantage that it is difficult directly to perceive the importance of large flocs. To give a more correct picture of the contribution to the real area beneath the spectrum, in spite of the logarithmic scale, a slightly modified form is now used at STFI, in which the variance is presented within wavelength intervals proportional to the wavelength. This means in practice that the spectral density in a wavelength spectrum on a logarithmic wavelength axis is multiplied by the wavelength (and a constant which is dependent on the chosen band width) so that the level at small wavelengths decreases while that at large wavelengths increases.

Figure 10.2 shows radiograms of two paper samples, formed at different mix concentrations.The image to the right shows clearly the dominating occurrence of large flocs in the paper with the higher forming concentration. The corresponding formation spectra are shown in *Figure 10.3*. These formation spectra also clearly demonstrate the domination of large flocs at the higher forming concentration.

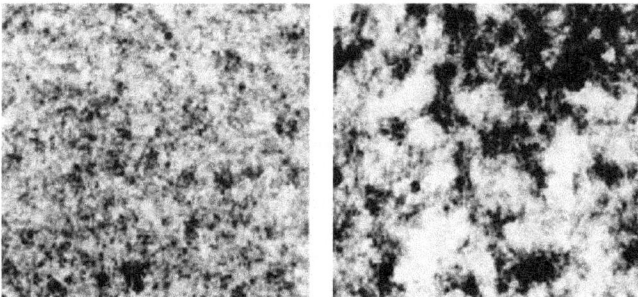

Figure 10.2. Beta radiograms for the two kraft paper samples. Softwood kraft pulp, 60 g/m², blade forming at different consistencies (FEX trial). In the radiogram, high grey level corresponds to low grammage. Left: Forming concentration 4 g/l Right: Forming concentration 10 g/l.

10.2.3 Fibre Orientation Anisotropy and Misalignment

The number of fibres is measured in all directions, and the result is presented in a polar diagram, see *Figure 10.4*. Often an ellipse is fitted to the results, to describe the fibre orientation distribution.

Two important parameters are evaluation from the shape and slope of this ellipse:

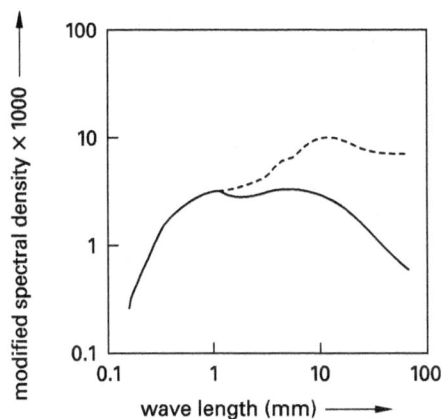

Figure 10.3. Modified formation spectra for the two samples in figure 37.2 Unbroken line: Concentration 4 g/l, F = 13.3, F_1 = 9.3, F_2 = 9.5 Broken line: Concentration 10 g/l, F19.1, F_1 = 9.8, F_2 = 16.4.

- Fibre orientation *anisotropy ratio* MD/CD
- Fibre orientation *misalignment angle* α.

The MD/CD ratio is often approximated by the ratio of the long axis (a) and short axis (b) of the ellipse.

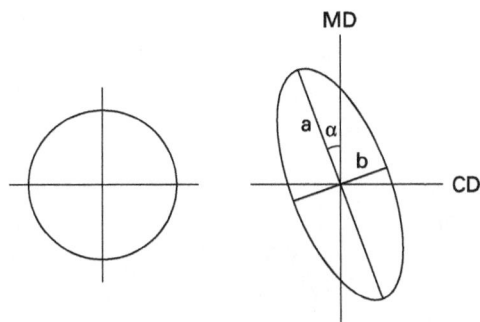

Figure 10.4. Fibre orientation distribution, polar diagram Left: Isotropic distribution. Right: Anisotropy ratio a/b and misalignment angle α.

10.3 The Headbox

The main function of the headbox is to distribute the mix evenly across the wire section. This means, for example, that the flow from a pipe with a diameter of 800 mm shall be transformed into a 10 mm high and 10 000 mm wide jet, with absolutely the same flow rate and flow direction at any point, as indicated in *Figure 10.5*. This transformation takes place mainly in three steps:

- The transverse distributor makes a first distribution of the mix across the machine
- One or two internal pipe sections improve the CD profile
- The headbox slice forms the final jet.

Between the transverse distributor and the slice there is an intermediate chamber.

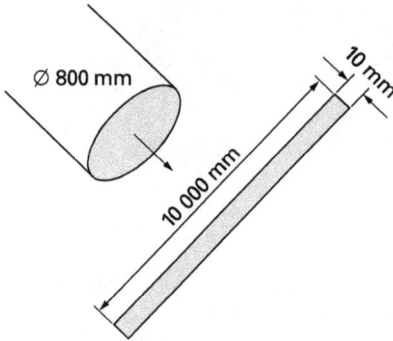

Figure 10.5. Feed pipe for mix flow and cross-section of jet from headbox (not to scale).

10.3.1 Cross Direction (CD) Distribution

A modern transverse distributor consists of a channel, which runs across the whole headbox, and from which discharge takes place successively through a hole in the wall, see *Figure 10.6,* and a pipe package, which leads to the headbox slice (or to an intermediate chamber).

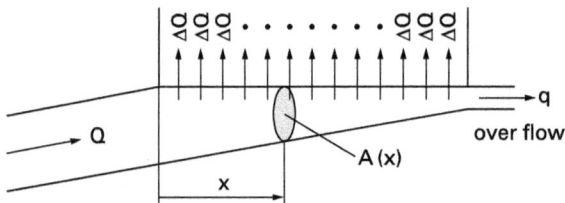

Figure 10.6. Transverse distributor with main flow Q, discharges ΔQ and overflow q.

If a perfect headbox jet is to be delivered to the wire section, the mix flow ΔQ, the same through each pipe, must be the same. This means that the static pressure along the transverse distributor channel must be kept constant. The friction pressure drop along the channel must therefore be compensated for by a corresponding pressure rise. This can be achieved by transforming a part of the velocity energy in the flow along the channel to static pressure by successively reducing the flow rate. The channel then functions in principle as a diffusor, even though the cross-sectional area $A(x)$ decreases in practice along the channel in contrast to a conventional diffusor. This is possible since the volume flow along the channel decreases continuously because of the discharges ΔQ.

$$u(x) = \frac{Q - n(x)\Delta Q}{A(x)} \tag{10.3}$$

Equation (10.3) describes the flow rate $u(x)$ along the transverse distributor, where $n(x)$ is the number of part flows ΔQ discharged before position x. The cross-sectional area $A(x)$ along the channel changes so that for a given dimensioning of the input flow Q and the overflow q, a constant static pressure is obtained along the whole transverse distributor. It will not however be possible to maintain a constant pressure along the whole channel under other flow conditions. If there is a pressure difference between the inlet and outlet, the overflow q can be adjusted so that pressure agreement is attained. Some pressure deviations along the channel can nevertheless still remain. The distance between the discharge holes should be so large that bridging of fibres across two adjacent holes is avoided. Such "fibre piling" would lead to the build-up of disturbing fibre flocs.

The discharge holes lead to the headbox slice and the intermediate chamber via a pipe package. The higher the pressure drop across this pipe package, the smaller are the differences between the individual discharge flows ΔQ caused by variations in static pressure along the transverse distributor channel. To achieve these high pressure drops, the flow rate must be high which means that the flow area in the pipe package must be relatively small. This means that the open area at the inlet of the pipe package, i.e. the ratio of the cross-sectional area of the flow to the total cross-sectional area, is small.

To improve the stability of the flow at the transfer from pipe package to subsequent slice or intermediate chamber, the pipes are expanded with a gradual or sudden increase in area, so that the flow rate is successively reduced. The output from the headbox slice can then take place with a large open area, see also Section 10.3.3.

10.3.2 Intermediate Chamber

If there is an intermediate chamber, its function is to equalize the speed profile of the flow from the pipe package of the transverse distributor so that the headbox slice is fed with an even better distributed flow. Depending on the construction of the intermediate chamber, a headbox is called either an air-cushion box or a hydraulic box. The air-cushion box contains an air volume above the mix in the intermediate chamber whereas the hydraulic box is completely filled with liquid. Some hydraulic headboxes lack an intermediate chamber, and the pipe package from the transverse distributor then feeds the headbox slice directly.

The **air-cushion headboxes** are a further development of the original, completely open, headboxes where gravity was the only driving force for the outflow through the slice. At high machine speeds, however, too high a mix height is required in an open headbox to achieve the required jet speed. A jet speed of 250 m/min requires, for example, a mix height of 0.9 m. At high speeds, it was therefore better to close the headbox and place a pressurized air-cushion over the mix in the intermediate chamber to create the conditions for a higher outflow speed. The function of the compressible air cushion was also to have a damping effect on pressure fluctuations in the mix flow to the headbox.

A traditional way of equalizing speed profiles which is applied in e.g. wind tunnels is to let the flow pass a number of nets which cause pressure drops and thus generate lateral flow from areas with high flow rates. Unfortunately, nets cannot be applied in a headbox since fibres would very rapidly clog them. Instead, perforated distributor rolls having circular holes with an open area of at least 50 per cent are instead placed in the intermediate chamber. Their function is to generate a pressure drop and thereby even out the speed profile across the headbox. They are kept in slow rotation to reduce the tendency for fibre flocs to form at the edges of the holes. Earlier perforated roll boxes could be equipped with up to five perforated rolls, but a modern headbox is normally equipped with two, see *Figure 10.7*.

The mix height in the intermediate chamber is normally 500–1000 mm, and the mix is fed directly from this chamber to the headbox slice. This means that the contraction in the headbox slice, i.e. the ratio of the input to the output cross-sectional area of the slice, is large. Air-cushion boxes are now used mainly at low machine speeds and in the manufacture of different types of kraft paper, which require large slices.

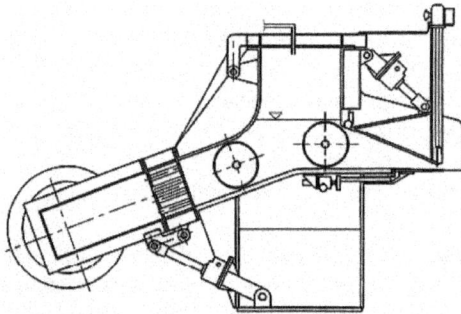

Figure 10.7. Air cusion headbox with two perforated rolls.

Hydraulic headboxes were constructed especially for twin-wire forming. One of the main requirements was small dimensions to facilitate a short free jet from the headbox between the wires. Hydraulic boxes lack the conventional air cushion and are available either with or without an intermediate chamber. If pressure pulsations occur in the mix flow, it may be necessary to install a separate damper before a hydraulic box, which in turn contains an air cushion. In a hydraulic box with an intermediate chamber, the perforated rolls in the air-cushion box are replaced with a fixed pipe package between the intermediate chamber and the headbox slice to equalize the profile. This is possible since the turbulence level is higher, so that floc formation on the upstream side of the pipe package is avoided. The total mix height at the entrance to the outlet slice is normally less than in the air cushion boxes and, compared with these, a hydraulic box therefore often has a smaller slice contraction.

10.3.3 Headbox Slice

In the outlet slice of the headbox, the mix is accelerated and the degree of acceleration is equal to the slice contraction. The greater the acceleration, the smaller are the relative speed variations in the jet. Turbulence must also be included in the harmful speed variations, except for the most fine-scaled turbulence, which is favourable for floc decomposition. The turbulence energy, which occurs at the beginning of the slice because of the mixing process between the single jets, will also remain in the jet, which leaves the headbox, since no essential turbulence decay takes place. The higher this turbulence energy is, the more rapidly is the jet broken in the free flow process from headbox to wire section. To generate a jet with the smallest possible speed variations, feeding with a large open area and a high slice contraction is therefore favourable. The lower mix heights in hydraulic headboxes than in air-cushion headboxes often results in a smaller degree of large-scale turbulence in the jet, however, and this in turn leads to less large-scale grammage variations in the finished paper.

Manufacturers of headboxes often set an empirically determined upper limit for the input speed of the mix when dimensioning the headbox slice. It should be pointed out, however, that the speed in itself is not harmful but rather the speed-related turbulence energy in the flow, which also depends on the particular feed geometry.

The size of the speed defects fed to the outlet slice is determined to a high degree by the size of **the open area of the feed**. A low open area (thinly distributed jets) gives large local speed variations, whereas a 100 per cent open area could yield a feed with small local speed variations. In air-cushion boxes, the open area of the feed is determined by the dimensions of the last perforated roll and, as mentioned earlier, the open area of the perforated rolls is normally about 50 per cent. In the first hydraulic boxes, the outlet slice was fed by pipe packages with a fairly low open area, which gave relatively high turbulence levels. To counteract the generation of too large-scale speed fluctuations, Beloit divides the channel into several, thin channels with the help of thin, flexible middle walls, see *Figure 10.8*, a technique which was introduced at the end of the 1960's.

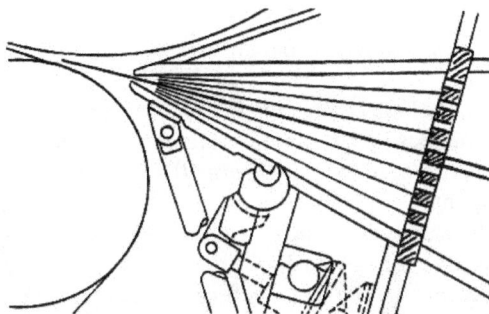

Figure 10.8. Headbox slice with thin, flexible middle walls. For two-layer forming, see Section 10.3.5 (Beloit).

Voith developed another solution during the 1970s, which modified the outlet section of the pipe package, which fed the slice. At the inlet side, the pipe package consisted of widely spaced cylindrical pipes (low open area). Towards the outlet of the pipes, the shape was modified so that they were expanded to a hexagonal shape. These hexagons could then be packed close to each other, so that a very high open area was attained.

A similar principle but with a rectangular discharge shape is now also used by Valmet. Along the walls of the pipes, a border layer of water develops, and the vertical walls can generate streaks in the paper if several are placed above each other. Valmet has therefore chosen to use a certain displacement between the different rows, although this in turn causes problems at the edges of the headbox.

Escher Wyss produced an alternative pipe package in the 1970's in so-called step-diffuser headboxes, see *Figure 10.9.*

velocity profiles

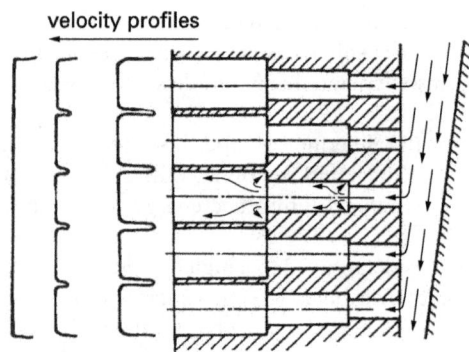

Figure 10.9. The pipe package of the transverse distributor has been equipped with pipes with stepwise increasing flow area, with circular cross-sections except in the discharge section where they are square shaped (Escher Wyss).

An area increase takes place through sudden expansions of the circular pipe section. Only the final stage has a square section in order to permit close packing. The sudden increase in area means that the border layers are broken up, which reduces the problems of grammage streaks when several units are stacked vertically.

During the 1990's, Beloit has moved on to a combination of step diffusers and a final gentle transition from a circular to a rectangular section.

Slice contraction. It has been stated earlier that a high contraction in the discharge region of the headbox is a good way of reducing speed variations in the output jet. *Figure 10.10* shows results from studies of a channel in Beloit's Converflo headbox (cf. *Figure 10.8*), which was fed with relatively widely spaced pipes. It is evident in the figure that the smallest discharge height, i.e. the largest slice contraction, gives the most even speed profile of the jet.

As mentioned previously, air-cushion boxes have a large stock height at the entrance to the slice. This gives a high contraction, and also permits the use of large slices with acceptable jet quality. On e.g. sack paper machines, slices of up to 80 m can be desirable. It has not been possible to use such large slices on hydraulic boxes, which in the first place have been developed for printing paper forming in twin-wire machines. The slice opening then seldom exceeds 20 mm. Recently, hydraulic boxes have been developed also for e.g. sack paper, in which the number of pipe layers in the pipe package which feeds the slice has been increased to obtain an acceptable jet quality.

High contraction in the slice of the headbox causes a considerable orienting of the fibres in the machine direction. It can therefore be impossible to make sheets with a low anisotropy ratio using such a headbox. The fibre orientation in the finished sheet is, however, often determined primarily by the dewatering ratio in the wire section.

The accelerating flow in the headbox slice is good for the state of flocculation of the fibres. The flocs are stretched in the flow direction and they can be broken down into smaller units, especially in mechanical pulps. The influence of slice contraction on the anisotropy and formation is demonstrated in Section 10.5.3 "Twin-wire forming".

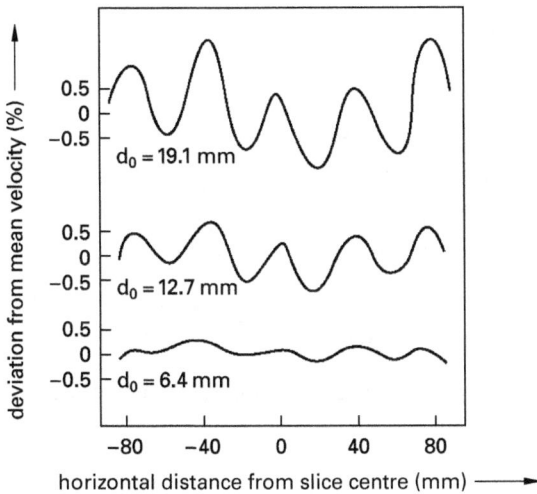

Figure 10.10. Velocity CD–profiles at different slice openings d_0 (Beloit).

10.3.4 Mix Jet

As mentioned previously, a good mix jet requires that the speed and flow direction shall be the same across the whole width of the machine and that the jet has a constant thickness. Even if the thickness of the jet is constant at the exit from the headbox, it will be changed during the journey from headbox to wire, because of crooked flow and inherent turbulence. The lower the degree of turbulence, the longer can the free jet exist without harmful deformations.

Jet contraction and angle. *Figure 10.11* shows the discharge from a headbox and the free mix jet. The upper wall often ends with a "lip slice". This has two functions, on the one hand to facilitate a local change in the slice height (see later in this Section), and on the other hand to reduce, through local acceleration, the influence of the border layer formed along the upper wall. Because of the jet contraction, the jet thickness h at "vena contracta" is always less than the geometrical slice opening h_0 between the upper and lower lips. The ratio is called the contraction coefficient μ of the jet; Equation (10.4).

$$\mu = \frac{h}{h_0} \tag{10.4}$$

The contraction coefficient μ and jet angle α depend on the angle of inclination of the front wall, the relative downwards protrusion Y/h_0 and the relative protrusion of the lower lip x/h_0. The slice h_0 is adjusted by a vertical displacement of the upper lip. The angle α and thereby the point of impact in the wire section is normally changed by changing the protrusion x of the lower lip.

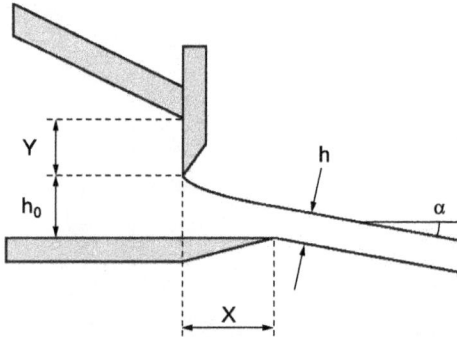

Figure 10.11. Outflow from a headbox with sloping upper wall and a vertical lip ruler.

For a headbox, the **jet speed** at vena contracta can in the general case be calculated with the aid of the energy equation (10.5):

$$p + \frac{\rho}{2} u^2 + \rho g H_l + U = \text{constant} \tag{10.5}$$

Here, p is pressure, ρ is density, u is flow speed, H_l is height level and U internal energy

The friction pressure drop from the pressure measurement position, i.e. the positive sign downwards in the box in *Figure 10.11*, to the discharge is Δp_f. It is transformed e.g. through viscous dissipation into heat and thereby increases the internal energy U. The pressure drop Δp_f and the overpressure p_l in the box can be expressed as "pressure heights H" by inserting Equations (10.6) and (10.7).

$$\Delta p_f = \rho g \Delta H_f \tag{10.6}$$

$$p_1 = \rho g H_l \tag{10.7}$$

The mass equation for the volume flow Q further gives Equation (10.8).

$$Q = h_1 u_1 = \mu h_0 u \tag{10.8}$$

If equations (10.5)–(10.8) are combined, Equation (10.9) is obtained.

$$u = \sqrt{2 g H_1 \frac{1 - \dfrac{\Delta H_1}{H_1} - \dfrac{\Delta H_f}{H_1}}{1 - \left[\dfrac{\mu h_0}{h_1}\right]^2}} \tag{10.9}$$

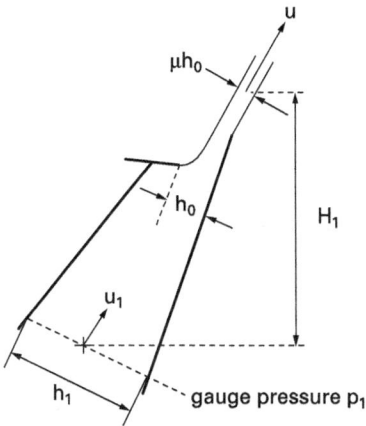

Figure 10.12. Outflow conditions from a hydraulic headbox.

This equation is thus valid for a hydraulic box (*Figure 10.12*), in which the friction pressure drop in the box, the dynamic pressure at the pressure measurement points and the height differences cannot be neglected. In air-cushion boxes, the mix speed inside the box is often so low that both the friction pressure drop and the dynamic pressure can be neglected. An air-cushion box also normally works horizontally. Equation (10.9) then reduces to the well-known Equation (10) for height-generated outflow.

$$u = \sqrt{2gH_1} \tag{10.10}$$

H_1 is here the sum of the mix height in the middle chamber and the pressure height in the air cushion. At very low jet speeds, it may be necessary to use an under pressure in the air cushion.

Control of grammage profile. In the final stage before the outflow from the headbox, the transverse distribution of the mix can be influenced by the slice profile, through a local deformation of the lip slice (cf. *Figure 10.11*). A local reduction in the slice opening reduces the mix outflow in this position, but at the same time certain lateral speed components are generated. These sideways components cause quality problems in the finished paper since the fibre direction distribution within the sheet is no longer symmetrical around the machine direction.

The size of the lateral flow components depends to a high degree on the design of the discharge slice of the headbox. With the very rapid mix acceleration obtained in the earlier air-cushion boxes with a vertical front wall, the lateral flow was relatively small, but the more pointed slices on modern headboxes gives larger lateral flow components. Local adjustments of the slice may be necessary because of an uneven feed from the transverse distributor, because of transverse effects in the wire section or drying section or because a total optimization of product properties often requires a grammage profile with certain deviations from constant grammage.

In those cases where the feed from the transverse distributor is uneven, the lateral flow effects will be much smaller than if the headbox is equipped with an intermediate chamber. As mentioned earlier, this makes it possible to achieve a certain degree of equalization through lateral flow already in the intermediate chamber, and consequently there will be fewer tendencies to a crooked flow in the output jet.

The position of the lip ruler is controlled by "lip screws", which are placed across the whole width of the headbox at intervals of 100–150 mm. On modern machines, the lip ruler is equipped with position indicators so that the slice profile can easily be read. Together with remote-controlled lip screws, this also makes automatic control of the lip profile possible.

A new way of controlling the grammage profile is through profile dilution, where white water is added locally in different transverse positions to compensate for streaks with locally high grammage. According to one method, the dilution water is added in the pipes from the transverse distributor channel (Voith), see *Figure 10.13*, while in another (Metso) the addition takes place it a step in a step diffuser pipe package.

The important advantage of the dilution method compared with lip deformation is that local CD-flow flow in the headbox jet is avoided and it is also possible to influence comparatively thin grammage streaks. The first two new Swedish paper machines equipped with profile dilution were started up during 1996, and today dilution CD-profile control is standard on all new headboxes.

Figure 10.13. Design for addition of dilution fluid to pipes feeding headbox (Voith).

10.3.5 Multi-Layer Headboxes

For the manufacture of soft crêpe paper, two-layer headboxes are used at several mills in USA. From a single headbox, a layered jet is then fed out which contains two different types of stock. This provides better possibilities of optimizing both the crêping process and the softness of the final product. Liner is another product where two-layer manufacture is applied industrially, especially in new machines for test liner.

In the manufacture of three-layered printing paper, there is a great potential for both raw material saving and quality improvement. It is possible to use different raw materials in the surface layer and middle layer, it is possible to fractionate a given stock and to place the different components in a suitable way in the thickness direction, and it is possible to place different types and different amounts of filler on the surfaces and in the middle of the paper. Several machine manufacturers are carrying out development work on multi-layer headboxes.

The main principle of a multi-layer headbox is a separate transverse distributor for each layer and that the mix flows are then kept separate in the intermediate chamber and in the outlet slice with the help of separation vanes. They can be thin and flexible, as in the headbox shown in *Figure 10.11*. They can also be made stiffer through reinforcement with e.g. kevlar fibre, which has been found to be favourable from a stability viewpoint.

The technical design of the slice outlet and the end shape and length of the separation vanes are very critical. Much development work still remains here. A new idea, which is under development on the FEX machine at STFI, is to reduce the mixing between the individual layers by adding white water at the end of the separation vanes. A suitable flow velocity of the added water can counteract the instabilities generated by the boundary layers on the separation vanes. With the help of the new technology for profile dilution, it is possible to control the grammage profile of the individual layers in multi-layer forming.

10.3.6 High-Concentration Headboxes

Conventional forming takes place at consistencies of less than one per cent. Forming at high consistencies means smaller volume flows and thus smaller pipe diameters, pump sizes etc. High-concentration forming thus has a potential for both reduced investment costs and lower energy consumption.

A high-concentration headbox has been developed at STFI, which works at a concentration of about 3 per cent. The mix is fluidized by being pumped past bends, which cause a pressure drop and thus generate turbulence. The turbulence then decays in narrow channels with a smallest height of circa 2.5 mm, see *Figure 10.14*. The size of the fibre flocs, which are re-formed as the turbulence decays, is thus limited geometrically by the available space.

The forming process at high concentration can be likened to an extrusion procedure. The stock jet which leaving the headbox consists of a connected fibre network, so that the final sheet structure has already to some extent been formed in the headbox. The process therefore differs considerably from a conventional forming process where the sheet is built up in the wire section. The high-concentration formed sheet has a high z-strength and also as higher sheet strength than conventional sheets. The principle is therefore used industrially for forming middle layers in carton board.

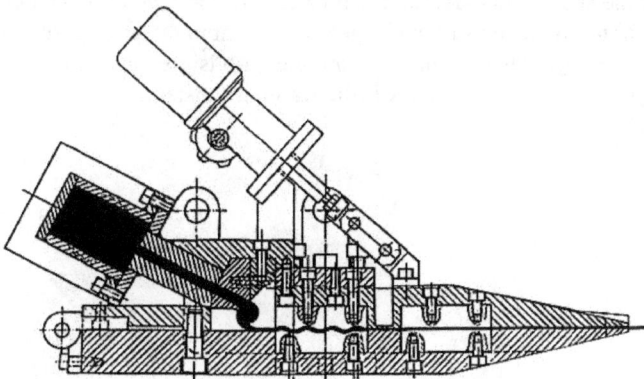

Figure 10.14. Headbox for high-concentration forming (STFI).

10.4 The Dewatering Process

The water is drained from the mix in the wire section and a wet sheet is formed. The dewatering process can thus in principle be of two different kinds, filtration or thickening, see *Figure 10.15.*

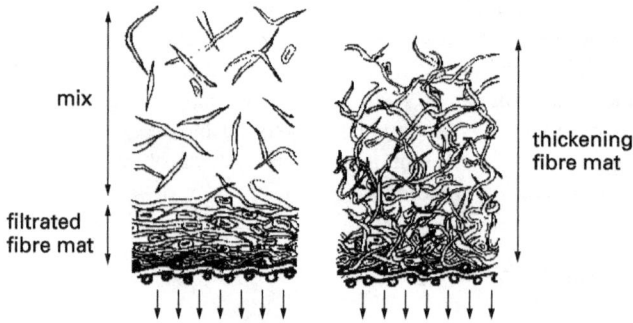

Figure 10.15. Dewatering through filtration (left) or thickening (to right), (Parker).

The filtration case is valid in conventional sheet forming, i.e. the fibres are placed successively on the wire in a wet sheet as the suspending water is removed. The strongly layered sheet structure characteristic of conventional paper is then obtained. Above the wet sheet there is a mix with the same concentration as in the jet from the headbox. At the dry line (sometimes called the wet line, depending on the context), dewatering according to the filtration principle is completed and the remainder of the dewatering takes place as a thickening process. The wet sheet is thus progressively compressed as its dry content is increased.

In high-concentration forming, the mix has so high a concentration (ca 3 per cent) that a connected fibre network with considerable network strength is formed already in the headbox (Section 10.3). The dewatering in this case is purely a thickening process and the resulting sheet structure is considerably more "three-dimensional" than that of the more layered "two-dimensional" sheets.

The dewatering resistance in filtration dewatering depends on the properties of the fibres (dimensions, elastic properties), the structure of the formed wet web, the penetration of the sheet into the wire, the compression of the sheet by the dewatering pressure etc. Only empirical methods have therefore been used so far to predict dewatering capacity. A commonly used version is the Koszeny-Karman equation where dQ/dt is flow rate per unit area, Δp is pressure drop, C is concentration, S is surface area, μ is fluid viscosity and K is the Karman constant.

Porosity-concentration:

$$\frac{dQ}{dt} = \frac{1}{K} \frac{(1-C)^3}{S^2 C^2} \frac{1}{\mu} \Delta p \qquad (10.11)$$

$$t = \frac{G}{C} w^\alpha (\Delta p)^n \qquad (10.12)$$

Here, t is dewatering time, C is mix concentration, w is deposited grammage and Δp is dewatering pressure. G, a and n are empirical constants.

It is important that the fibre distribution is as uniform as possible so that the best formation is attained. Traditionally, there are three main ways of improving the fibre distribution, viz. dewatering, directed shearing and turbulence, see *Figure 10.16.*

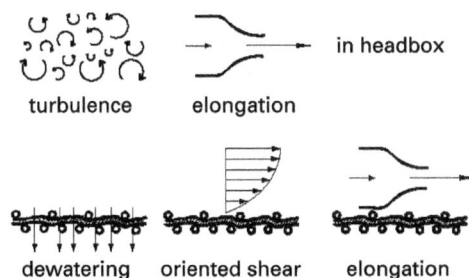

Figure 10.16. Principles for improving fibre distribution: Turbulence, elongation, dewatering and oriented shear.

Dewatering in itself has a self-healing effect on the fibre distribution. This can be explained by the fact that a local low fibre amount leads to a local low dewatering resistance, which in turn increases the local dewatering and thus tends to increase the amount of deposited fibres, thus having an equalizing effect on the fibre distribution. A sheet with a random fibre distribution would therefore show a poorer formation than a real sheet formed at a low concentration, e.g. a laboratory sheet, especially regarding small-scale unevenness (Norman, doctoral thesis 1974).

At more normal forming consistencies, mechanically connected fibre flocs are also formed, with or without network strength. The weakest of these can be broken up during the dewatering process through directed shearing. The directed shearing can also contribute to removing fibres so that they can be deposited in areas with local low dewatering resistance. Directed shearing also influences the fibre orientation. The orientation of the fibres increases since the fibres, which are deposited on the surface of the wet sheet, are "combed out" in the shearing direction. In this way, the fibre orientation can be increased in the machine direction by giving the mix a speed in the machine direction, which deviates from the speed of the wire. The jet-to-wire discharge ratio is the ratio of the speed of the mix to the speed of the wire.

By generating turbulence in the mix, it is possible to break fibre flocs in the dewatering process itself and thereby to improve the formation of the paper. On a fourdrinier wire, the activity level, i.e. the degree of disturbance of the free liquid surface, usually quantifies the degree of turbulence in the mix. In dewatering on a fourdrinier wire, the possibility of increasing the turbulence is limited since too high a degree of activity leads to harmful disturbances of the paper formation. In twin-wire dewatering on the other hand, higher turbulence levels can be used since there is no free liquid surface.

It has recently been shown that fibre flocs can be broken down through a tearing-apart effect with the help of an extensional flow. Extensional flow is developed already by the acceleration in the headbox, and a considerable improvement in formation can be obtained using a high slice contraction, (Nordström, doctoral thesis 1995, see also Section 10.4.3). Extensional flow can also be achieved through pressure pulses in twin-wire dewatering.

10.4.1 Forming Wires (Fabrics)

The wire through which the mix is dewatered consists of a fabric spliced into an endless band. The fabric is characterized by the threads (material, dimensions) and the fabric pattern. The traditional bronze wire was single-layered with one layer of lengthwise threads and one layer of transverse threads, and with a 2-thread symmetry where the fabric pattern is continuously repeated after two threads. The fabric pattern was thus the most simple imaginable; each thread in the length direction and in the cross-direction ran alternatingly above and below the crossing threads.

During the 1970's, the plastic wire replaced the bronze wire since the increasing machine speeds gave the bronze wire too short a lifespan, due to the poor wear resistance of the bronze threads. The lifespan was increased radically by a transition to wear-resistant polymer threads. The 2-thread plastic wires had such a high stretchability that they could not be tightened sufficiently with the existing wire-stretching equipment. This depended on the low bending stiffness of the polymer threads in combination with the high average angle formed by the threads with the plane of the wire in this fabric pattern. A transfer then took place to multi-phase fabric patterns, in which the average angle of the threads to the plane of the wire is smaller and the tensile stiffness is increased. A 5-thread pattern, for example, means that a thread runs over four crossing threads and below a fifth.

A wire should be constructed so that the lengthwise threads are protected from wear since the force, which drags the wire through the wire section, loads these. The wear should therefore in the first place be absorbed by the cross-threads, which should lie on the wear side to the greatest possible extent. The paper side of a wire should give the best possible support in the sheet forming, so that the smallest possible wire marking is generated in the paper and so that the wire openings are not blocked by fibres and the dewatering reduced. Especially for the latter reason, it is an advantage if the cross-threads lie on the paper side to the greatest possible extent. Fibres which lie predominately in the length direction can then more easily bridge the wire openings without penetrating down and hindering the dewatering.

For a single-layered wire, the geometry of the paper side is by definition opposite to that of the wear side. It is therefore impossible to construct a single-layered wire so that both the paper side and the wear side have mainly transversely running threads in the surface. To facilitate an optimum construction of both the paper and wear side of the wire, the double-layered wire was first introduced (Nordiskafilt, 1970s) and subsequently the multi-layered wire, see *Figure 10.17*.

With a multi-layered construction, it is in principle possible to start from a coarse lower wire, which is dimensioned for maximum resistance to wear, and a fine upper wire, which is optimized for sheet support in the dewatering. These wires are then woven together into one unit. Development regarding wires is still in progress and includes thread materials, thread dimensions and thread treatment, and fabric patterns.

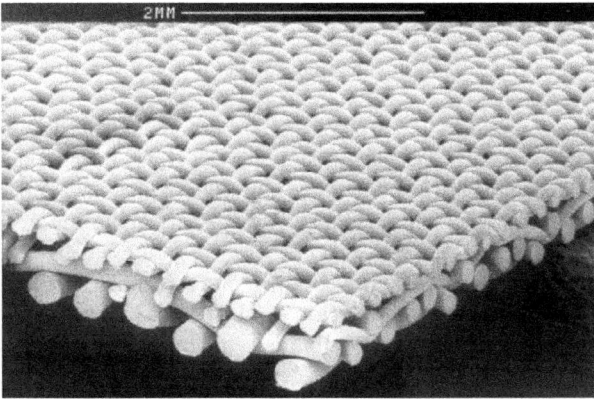

Figure 10.17. Multi-layered wire (Albany International, Halmstad).

10.4.2 Fourdrinier Forming

With the development of the continuous paper machine in the beginning of the 19th century, the fourdrinier wire section was introduced. This was formed by a number of rolls, which supported the wire within its horizontal dewatering section. To reduce the friction against the wire, the rolls could be rotated. The dewatering took place completely with the help of gravity, but with the development towards an increased production capacity, a demand for special dewatering elements gradually arose. To improve the formation of the manufactured paper, the wire was shaken horizontally sideways, generating a mixture of "directed shearing" and "turbulence" in the mix on the wire (*Figure 10.18*).

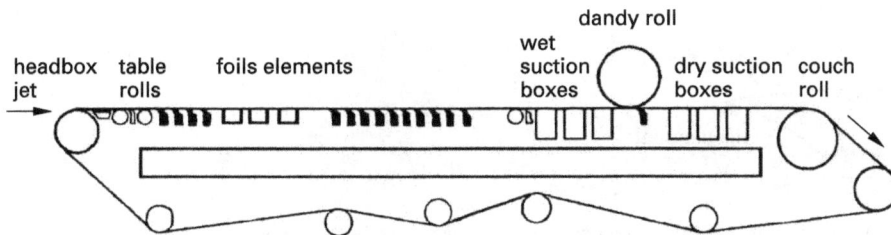

Figure 10.18. Fourdrinier machine.

As the machine speed was gradually increased, a speed range was entered where the rotating support rolls, the register rolls, developed a certain dewatering ability. This arose due to a local under pressure between wire and roll with increasing separation distance in the exit nip. The maximum attainable under pressure corresponds to the dynamic pressure $\rho \bar{v}^2/2$ of the mix on the wire. The under pressure at the register roll deflects the wire slightly downwards before it returns to its original level at the next register roll. This vertical shaking increases the degree of activity on the wire and has a favourable effect on the formation of the paper being made. At machine speeds of above about 500 m/min, however, the shaking becomes so strong that the degree of activity on the wire becomes too high. At high machine speeds, the register roll can therefore no longer be used as a dewatering element.

Before a register roll, the wire normally holds a water layer on the bottom side and this is pressed up through the wire before the above-mentioned suction pulse generates dewatering. The washing effect, which is thus, introduced results in a low fines (and filler) content in the paper on the wire side, i.e. on the side of the paper, which has lain against the wire in the dewatering.

During the 1960's, **the foil** was introduced as a dewatering element on fourdrinier wire machines. This consists of a strip fitted below the wire, the upper side of which forms a grazing angle with the wire which generates an under pressure. The amplitude of the suction pulse is determined by the foil angle, which is normally about one degree. As in register roll dewatering, the under pressure means that the wire is deflected vertically which means that turbulence is generated on the wire. A suitable foil angle is chosen to attain the desired degree of activity on the wire and this angle is smaller the higher the machine speed. A number of foils are often fitted onto a vacuum box, which means that the dewatering capacity is increased since dewatering can also take place between the strips. *Figure 10.19* shows the difference in under pressure pulse between a register roll and a foil.

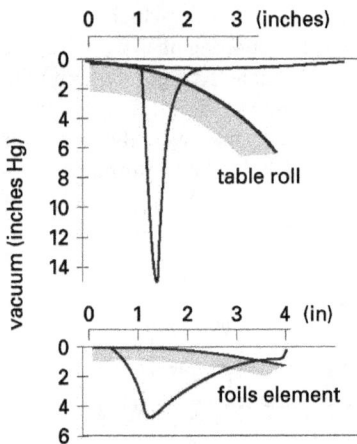

Figure 10.19. Comparison between the suction pulses of a table roll (above) and of foils element (below) at 2000 ft/min (Wrist).

A **step foil** consists of a flat, short supporting surface and a lower surface, which is parallel with this, between which there is a gap of a few millimetres. A step foil unit consists of a number of step foils fitted onto a vacuum box. To change the degree of activity on a wire without having to change to a foil list with a different angle, a step foil unit can be used and the vacuum level on this can be changed. The higher the vacuum, the greater is the downward suction and thereby the greater the degree of activity on the wire.

With the **Iso-Flo foil** arrangement from Johnson Wire, a number of horizontal supporting strips are fitted on a vacuum box. Every alternate strip is slightly lowered so that already at a moderate vacuum, the wire is sucked down towards the lower supporting strips and thereby receives a vertical pulsating movement. The vacuum level and thereby the dewatering capacity can here then be increased without influencing the vertical agitation.

The **fibre direction anisotropy** in the sheet, which leaves the wire section, depends partly on the fibre direction distribution in the mix jet from the headbox and partly on the change in the fibre direction distribution during the dewatering process. The fibre orientation in the length direction cannot be reduced during the dewatering process; directed shearing can only increase it. This is produced as a result of a speed difference between mix and wire, i.e. as a result of a jet-to-wire discharge ratio, which is not 1. The greater the speed difference, the greater is the directional effect (*Figure 10.20*). Only relatively small speed differences can be used in practice, however, since large differences result in disturbances in the mix and thereby a poorer formation. In addition to the effects of fibre orientation, the anisotropy of the finished paper is also affected to a great extent by the web tension conditions in the drying section.

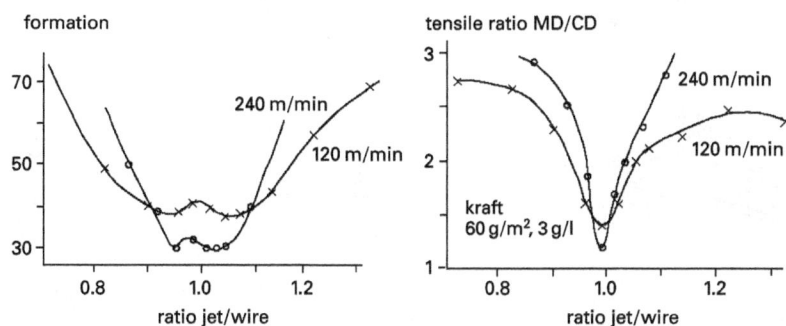

Figure 10.20. Influence of jet -to-wire speed difference on formation and anisotropy ratio (Svensson and Österberg).

10.5 Twin-Wire Forming

The greater the machine speed, the longer is the wire section required to dewater a given product. At machine speeds above about 1000 m/min, the problems increase with an unstable liquid surface on a fourdrinier wire section, e.g. depending on a considerable friction between air and mix. This means that there are practical difficulties in applying fourdrinier wire forming at high machine speeds. Attention was given to these problems already during the 1950's and experiments then began in which the mix was ejected between two wires, so-called twin-wire forming. This new principle had the two advantages of increasing the dewatering capacity by allowing dewatering through two wires and of avoiding the free surface between mix and air in the dewatering process. During the 1970's, twin-wire forming became the dominating principle for newly installed printing paper machines. It was soon found that two-sided dewatering gave a more symmetrical paper than could be attained on the fourdrinier wire machines. This also led to a modernization of many printing paper machines, which were rebuilt for twin-wire forming. A forming unit in which the dewatering starts directly in a twin-wire nip is usually called a gap-former.

In a hybrid-former, on the other hand, the initial dewatering takes place on a fourdrinier wire section and the rest of the dewatering (from an average concentration on the wire of about 2 per cent) according to the twin-wire principle. Most hybrid machines were originally fourdrinier wire machines complemented with an upper wire for dewatering upwards at the end of the wire section. Thus, at a relatively low cost, the following advantages have been attained with twin-wire dewatering:

- high dewatering capacity
- good formation
- symmetrical sheet
- low linting (fibre release in printing)

It should be pointed out that twin-wire dewatering does not double, but rather quadruples the dewatering capacity compared with one-sided equipment, due to the fact that dewatering takes place through two wires at the same time as the dewatering resistance is halved since half the grammage is formed on each wire, see *Figure 10.21*.

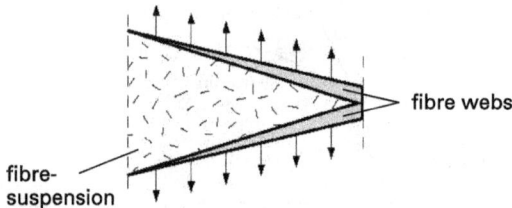

Figure 10.21. Two-sided dewatering.

10.5.1 Roll Forming

In the simplest form of twin-wire forming, the two wires are stretched around the periphery of a rotating roll. If a tensile stress T (force per unit width) is applied in the outer wire, the dewatering pressure p generated in the mix between the wires gives Equation (10.13) where R is the radius of curvature of the outer wire.

$$p = T / R \tag{10.13}$$

To a good approximation, the radius of curvature of the wire becomes equal to the radius of the roll. The dewatering pressure then becomes constant during the dewatering process, except in the first stage during which the pressure is built up. To facilitate two-sided dewatering, the surface of the roll is open so that it can receive the water, which passes through the inner wire, see *Figure 10.22*.

The water, which is pressed into the roll, is expelled after the internal wire with the sheet has left the roll. It is thereafter led to the wire pit. The rest of the water passes through the external wire and is also led to the wire pit. The surface of the roll usually contains several suction zones so that the pressure below the wire can be controlled. This is necessary if symmetrical dewatering is to be obtained. It is in fact necessary to compensate for the centrifugal pressure p_c, which influences the mix between the wires according to Equation (10.14) where ρ is the density of the mix, h is the distance between the wires, v is the mix velocity and R is the radius of curvature of the roll.

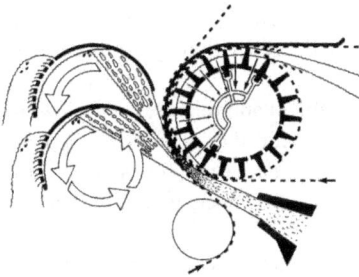

Figure 10.22. Twin-wire dewatering according to the roll forming principle.

$$p_c = \rho h v^2 / R \qquad (10.14)$$

Because of the influence of the centrifugal force, the pressure drop across the internal wire decreases by p_c. To compensate for this and to maintain the same pressure drop across the external and the internal wires, a corresponding under pressure is created with the aid of the suction zones in the roll.

Fibre orientation anisotropy. When the mix jet from the headbox enters between the wires, it is met by a dewatering pressure of $p = T/R$, which means that the jet is decelerated. If the energy equation (Bernoulli's Equation) is applied to the mix from the free jet and to the mix between the wires, we obtain Equation {10.15} where u is the jet velocity, y is the mix velocity after the deceleration and T/R is the dewatering pressure.

$$\rho \frac{u^2}{2} + 0 = \rho \frac{y^2}{2} + \frac{T}{R} \qquad (10.15)$$

If it is required that the sheet shall have the lowest possible fibre orientation, the mix shall, through the braking at the inlet to the forming zone, assume the same speed as the wires. The combing-out effect of directed shearing is then minimized during the rest of the dewatering process. By inserting $y = v$ in Equation (10.15), where v is the wire speed and then solving the discharge ratio $(u/v)_0$ for minimum anisotropy, we obtain Equation (10.16).

$$\left(\frac{u}{v} \right)_0 = \sqrt{1 + \frac{2}{\rho v^2} \frac{T}{R}} \qquad (10.16)$$

This means that if a paper with the least possible fibre orientation is desired on a twin-wire machine, a jet-to-wire discharge ratio greater than 1 must always be used. This differs in principle from the fourdrinier wire case, where the least fibre orientation is attained at a jet to wire velocity ratio equal to 1.

Influence of headbox. As described above, the pressure is constant during most of the dewatering process in roll forming. This means that it is possible to introduce directed shearing or turbulence during the forming and thus improve the formation. The only variable by which the formation can be influenced is the jet speed or in reality the jet-to-wire speed difference. At a speed difference of zero, the state in the jet from the headbox will mainly be frozen, but a slight speed difference will increase the fibre orientation and improve the formation by directed shearing.

Roll forming can be characterized as an easily run (few variables) and reliable process which gives a slightly flocculated paper. The formation is not acceptable for high-class printing paper, and pure roll forming is not therefore used today on new printing paper machines. The undisturbed dewatering process gives a relatively high retention level, however, which reduces the need for a retention agent.

Figure 10.23 shows how the fibre orientation of the paper measured as the tensile stiffness anisotropy depends on the speed difference between the mix and the wires. The traditional result is shown by the square symbols in the figure, which represent a headbox with a low slice contraction. The anisotropy in the jet itself can be estimated to be about 1.6 and increases rapidly with the increasing speed difference between mix and wires. The filled circles represent a headbox with high slice contraction and a jet anisotropy of about 2.4. Also in this case, the anisotropy of the paper increases with increasing speed difference. The unfilled circles represent a headbox with a high headbox contraction where the orientation conditions remain completely unchanged. Already in the jet, the anisotropy is above 4 and only a small decrease can be achieved by altering the speed difference between mix and wires. The conclusion is that, at a sufficiently high contraction in the headbox slice, the jet develops a high degree of fibre orientation, which cannot then be further influenced in the dewatering process (Nordström, doctoral thesis 1995).

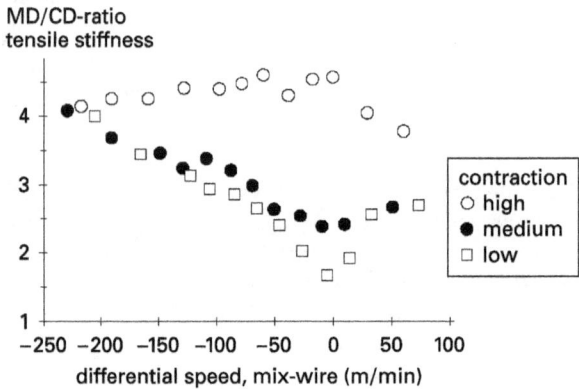

Figure 10.23. The influence on anisotropy of the speed difference between a decelerated mix jet and wires for different contraction in the headbox slice.

Figure 10.24 shows the corresponding influence on the small-scale formation of the paper, i.e. within a wavelength range of 0.3–3 mm. The square symbols represent the traditional case here also, and the formation is improved when the speed difference between mix and wires increases from zero. This formation improvement has earlier been described as being the result of a directed shearing. A further increase in the speed difference leads to a poorer formation. The figure also shows that at a high headbox contraction (unfilled circles), the process is changed. The best formation is instead obtained at a speed difference of zero. This must be due to an improved state of flocculation in the headbox jet at the high contraction. The formation is then worsened by the application of directed shearing through a speed difference.

small scale formation F(0.3–3) (%)

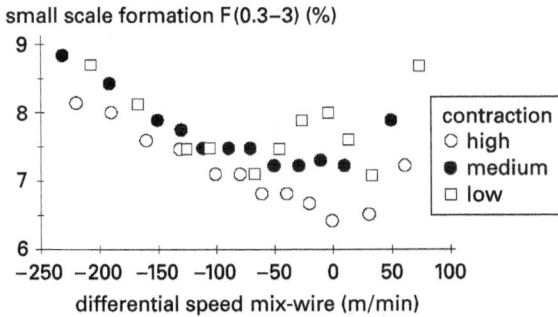

Figure 10.24. The influence on small-scale formation of the speed difference between a braked mix jet and wires for different contractions in the headbox slice. Roll forming at 700 m/min, TMP pulp (FEX-test).

10.5.2 Blade Forming

In twin-wire forming, as opposed to pure roll forming, it is possible to generate a pulsating dewatering pressure. This can be produced by deflection of the wires across a deflector blade, see *Figure 10.25*. The figure demonstrates how a force F on the outer wire is needed, to balance the vertical downwards components of the pulling forces T (left and right) on the tensioned outer wire.

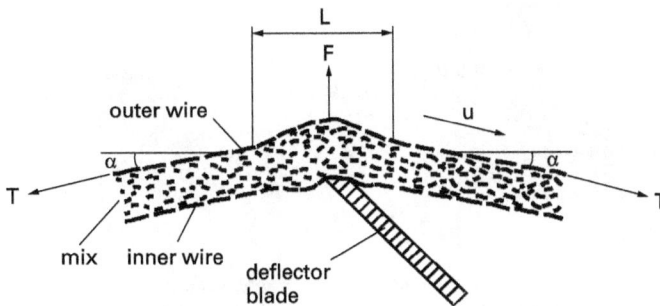

Figure 10.25. Dewatering force F during wire deflection across a deflector blade (Norman 1978).

The following vertical force balance is then valid for the outer wire where F (kN/m) is the force on the outer wire, T (kN/m) is wire tension and 2α is the total angle of outer wire deflection (Equation 10.17).

$$F = 2T \sin \alpha \qquad (10.17)$$

When the two wires pass the blade without deflection, no dewatering pressure is generated. If the deflector blade were pushed upwards against the inner wire, it would deflect. The space between the wires would then decrease. The mix would then lift the outer wire, to try to restore the original space between the wires. In doing so, it would bend the outer wire, and below the now curved outer wire, a pressure zone would then be generated according to the basic Equation {10.13} $p = T/R$. This pressure zone would transfer the force F to the deflector blade and generates dewatering through the wires.

The distribution of the force F in the form of a pressure pulse with extension L in the mix between the wires depends on several factors, e.g.:

• the shape of the deflector
• the wire speed
• the mix thickness between the wires

A major affect of a pressure pulse on fibre flocs in the mix between the wires is acceleration in the downstream end of the pulse. This will stretch the flocs and eventually break them apart. *Figure 10.26* shows an example of pressure pulse shape. The pressure has been measured through pressure tappings drilled in a V-shaped blade with a local angle change of 2.3° at the middle of the blade.

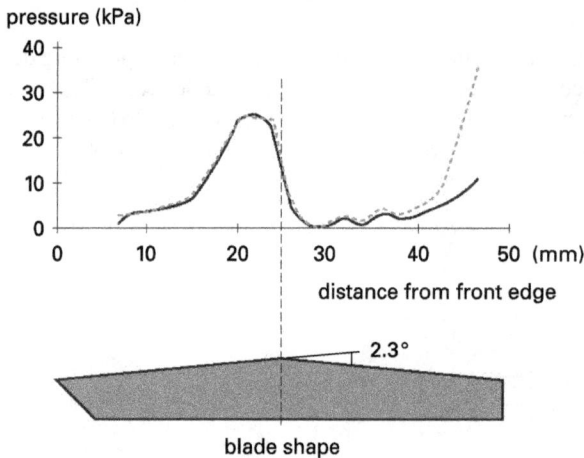

Figure 10.26. Pressure distribution along a blade with an angular change at the centre. Broken curve: Higher deflection angle over downstream edge. Machine speed 1200 m/min, wire tension 6 kN/m, wire distance circa 2 mm (FEX trial, 1995).

The figure shows that the pressure pulse at blade centre is developed almost completely before the angular change. This means that if a flat blade is used in a blade former, with outer wire deflection at the blade front edge, the pressure pulse will lie almost completely before the blade. This explains why approximately the same dewatering can be obtained through both wires in blade forming.

If all deflector blades are placed on the same side of the wires, the pressure event is practically unchanged if the distance between the wires is changed, because the outer wire is freely mobile in the thickness direction. This means that the dewatering pressure, like in a roll former, is self-adjusting at changes in e.g. slice opening or mix dewatering properties. If a deflector is also added on the opposite sides of the wires, however, the wire deflection angles over the blades will become strongly dependent on the mix thickness between the wires, see *Figure 10.27*.

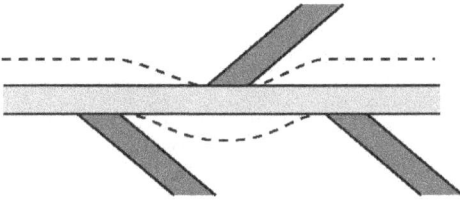

Figure 10.27. Wire shape with deflector blades on both sides. Unbroken lines: Original wire shape Broken lines: Doubled distance between the wires.

Originally, the wires could pass along straight paths between the three blades, whereas a doubling of the wire distance means a sick-sack wire run. To maintain a given pressure with a change in the thickness of the headbox jet or of the dewatering properties of the mix, the positions of the deflectors on one side must therefore be adjusted if these are placed on both sides of the wires. This was not, however, possible with the earlier designs of blade former units.

Forming and retention. The pressure pulses generated by the deflector blades cause a directed shearing and elongational flow in the mix between the wires. This contributes to an improved formation in the paper compared with the constant dewatering pressure in roll forming. The disturbed flow conditions during the dewatering in a blade former leads, however, to greater material losses through the wires, ie to a lower retention level.

10.5.3 Roll Blade Formers

For printing paper machines during the late 1980s, dewatering with combinations of constant and pulsating dewatering pressure was applied. The first design according to this principle was the Bel Baie III, see *Figure 10.28*. The advantages of this arrangement are that the initial dewatering takes place under pulsating conditions, which gives good formation, while the later dewatering takes place under constant pressure, which is favourable for the retention. Valmet in their Speedformer HS used the reversed combination, i.e. roll followed by blades.

In connection with board manufacture, which takes place at relatively low speeds, it has been difficult to avoid flow instabilities in a twin-wire nip. A development was introduced by Dörries (now part of Voith) in the form of an upwards-dewatering wire, placed on top of a fourdrinier wire. Blades located below the wire could be pressed upwards with individually adjustable loads, see *Figure 10.29*.

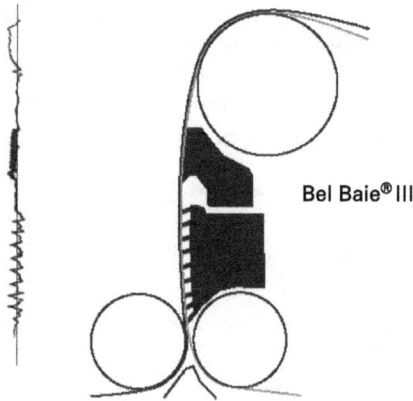

Figure 10.28. Bel Baie III, twin-wire forming with blades and roll. The pressure event during dewatering is shown to the left (Beloit).

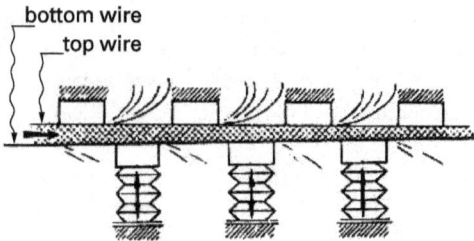

Figure 10.29. Duoformer D, twin-wire top forming with unit with loadable blades (Dörries patent, 1986).

This was the first twin-wire arrangement where the distribution of the dewatering pressure along the forming zone could be varied during operation. However, it was not realised at the time, that the pressure event was in reality pulsating, with the wires following a sick-sack shape (see *Figure 10.27*).

In 1991, on the FEX machine at STFI, initial roll forming was combined with blade forming according to the Dörries principle, with loadable blades on one side, see *Figure 10.30*. The headbox and the outer wire lead-in roll are movable around the centre of the forming roll and the roll cover angle can thus be adjusted during operation. After the forming roll there is an arrangement of deflectors. To the right, a number of fixed blades are fitted along a vertical line. To the left are movable blades, equipped with position indicators, and with individually adjustable load forces.

It is important that the movable blades are on the same side as the movable, outer wire on the roll. This will generate self-adjusting pressure pulses, even when the thickness of the mix layer, between the wires leaving the roll, is changed This design arrangement, with a symmetrical forming process at all stages, makes it possible to select an suitable amount of the dewatering at constant pressure on the roll, and thus leave an optimum amount of water for the blade section, permitting formation improvement with the adjustable pressure pulses, while maintaining a high retention level. The basic principles of the STFI-former is since the middle 1990s also applied by the two major machine manufacturers Voith and Valmet (later Metso). The design of the Voith CFD is shown in *Figure 10.31*.

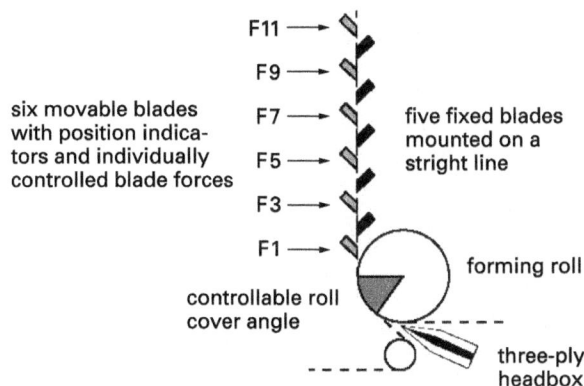

Figure 10.30. The STFI-former: Initial dewatering with constant pressure, followed by adjustable pressure pulses (FEX 1991).

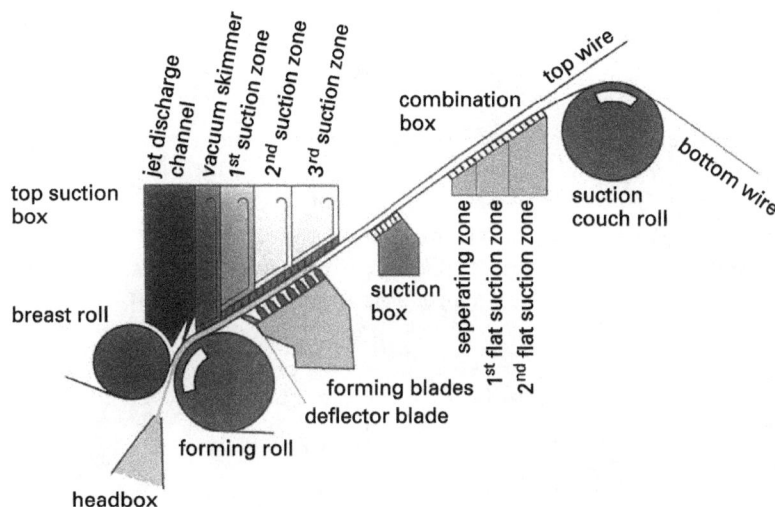

Figure 10.31. Voith CFD former with initial roll forming, followed by blade forming with adjustable pressure pulses.

In the early 2000s, the design was slightly changed when the Douformer TQv was introduced. The blade section is then vertical, just like in the STFI-Former. It should be pointed out that the adjustable blades in the Voith designs always have been mounted against the inner wire of the preceding roll.

Valmet developed the SPEED-FORMER MB according the principles of the STFI-Former, but with the loadable blades against the roll inner wire. It was around year 2000 modified to the Valmet (later Metso) OptiFormer, now with the loadable blades on the same side as in the STFI-Former. The basic design is shown in *Figure 10.32*.

Figure 10.32. Metso OptiFormer with initial roll forming, followed by blade forming with adjustable pressure pulses.

10.6 Literature

Norman, B. (1989) *Overview of the physics of forming. 9th Fundamental Research Symposium, Vol 3*. Oxford, pp. 73–149.

Norman, B., and Söderberg, D. (2001) *Overview of forming literature 1990-2000. 12th Fundamental Research Symposium, Part 1*. Cambridge, pp. 431–558.

Paulapuro, H. (2000) *Web forming. Papermaking Science and Technology, Book 8: Papermaking Part 1, Chapter 6*. Finnish Paper Engineer's Association, pp. 191–250.

11 Grammage Variability

Bo Norman
Department of Fibre and Polymer Technology, KTH

11.1 Introduction

The demand that local variations in different paper properties be kept small is increasing steadily, as a result of a need both to increase the efficiency of the production process and to improve the convertibility, e.g. in more and more rapid printing presses and packaging machines. This makes increased demands both on the design of the manufacturing process and on the measurement of important paper properties on-line in order to give greater control possibilities.

Measurement equipment for on-line measurement of grammage and moisture content was introduced in the 1960s and, to begin with, it was then possible to measure and control the variations in time, i.e. the longitudinal variations. When traversing meters were installed, it also became possible to record the cross-machine variations. Besides the longitudinal and cross-machine variations, the random, uncontrollable variations, usually called residual variations, can be evaluated with variance analysis. A further way of characterizing the variations is with the help of a frequency spectrum.

Today, there is also measurement equipment for web thickness, which gives greater possibilities of controlling the roll quality, and for ash content, optical properties, including colour and "optical formation". Equipment for the measurement of different mechanical properties is also being developed. Progress is also being made towards the measurement of the contents of different components in white water, as a basis for retention control.

11.2 Characterization of Variations

The characterization of variations is here exemplified with grammage variations, but the method can of course be applied to the variations in an arbitrary property.

11.2.1 Distribution Curve

The easiest way of describing grammage variations is to plot the number of single samples within different grammage intervals in a distribution curve of the type shown in *Figure 11.1*. From this distribution curve, the average grammage wm can be calculated, and the width of the distribution is a measure of the spread of the grammage. A quality specification for a given product can include the average grammage and how large a proportion of the product may be lower than the grammage wmin.

number of samples

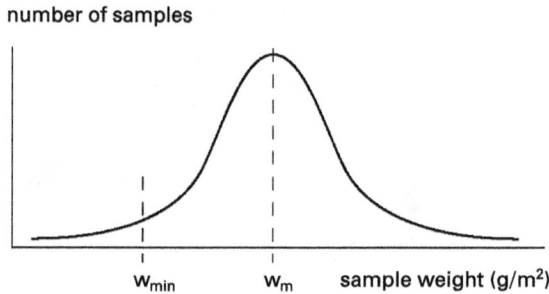

Figure 11.1. Distribution curve for local grammage.

Variance, standard deviation and variation coefficient. The grammage scatter is normally described with the help of the variance $\sigma^2(w)$. This is defined as the mean value of the square of the local deviation from the average grammage wm:

$$\sigma^2(w) = \frac{1}{n}\sum_{i=1}^{n}\left(w_i - w_m\right)^2 = \left(w_i^2\right)_m - \left(w_m\right)^2 \tag{11.1}$$

The validity of the equation can be shown by development of the middle line. The square root of the variance, $\sigma(w)$, is called the standard deviation and it is often expressed in relation to the average grammage, and is then called the coefficient of variation $V(w)$:

$$V(w) = \frac{\sigma(w)}{w_m} \tag{11.2}$$

(Cf. the definition of formation number, Equation (10.1))

The size of the measurement surface. The grammage variations which can be measured on a given paper web depend strongly on the size of the measurement area. The larger the measurement area, the smaller are the variations recorded. Consider a surface with an area A m^2, of a paper with an average of N fibres per m^2. The average number of fibres within the surface is then AN. If we assume that the number of fibres within the area A is Poisson-distributed, then the variance in the number of fibres is equal to the average number of fibres. The coefficient of variation of the grammage $V(w)$ according to the Poisson distribution is then given by Equation (11.3).

$$V(w) = \frac{\sqrt{mAN}}{mAN} = k\frac{1}{\sqrt{A}} \tag{11.3}$$

(m = the weight per fibre; k = constant)

This equation shows that the recorded variations decrease in proportion to the inverse of the square root of the area of the measurement region. For randomly spread fibres, the number of fibres within a given measurement area is Poisson-distributed. This distribution is only approximately valid for paper in the measurement areas concerned but the argument demonstrates that there is a relationship between the coefficient of variation of the grammage and the size of the measurement area.

11.2.2 Variance Analysis

For a variable composed of a number of independent components, the variance of the variable is equal to the sum of the variances of the components. For an analysis of the process reasons for the total variance in grammage, a division into variance components is productive. *Figure 11.2* shows two of the components. These are the grammage variations in the machine direction (MD) and in the cross-machine direction (CD).

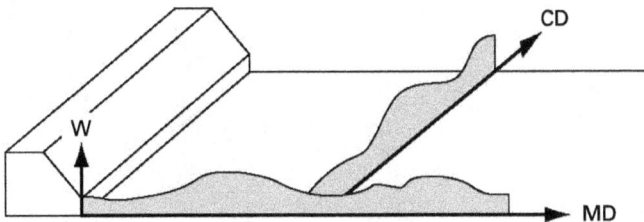

Figure 11.2. The length profile (MD) and cross-machine profile (CD) of the grammage.

The whole paper web is divided into a number of equally large sample areas and the grammage is determined for each sample. From these values, the total variance σ^2_{Tot} and longitudinal variance σ^2_{MD} for the average longitudinal profile of the grammage and the cross-machine variance σ^2_{CD} for the average cross-machine profile of the grammage are calculated. The following definition is then valid for the residual variance $\sigma^2_{Residual}$:

$$\sigma^2_{Residual} = \sigma^2_{Tot} - \sigma^2_{MD} - \sigma^2_{CD} \tag{11.4}$$

The residual variance is a measure of the random variations in grammage, which cannot be avoided. This is thus the most serious kind of grammage variation.

11.2.3 Traversing Measurement

To evaluate the properties of the paper across the whole web width, traversing measurement equipment is required. The measurement area then moves at an angle α to the direction of movement of the web, see *Figure 11.3*

For a newsprint paper machine, the following data can apply:

- Machine rate 1200 m/min = 20 m/s
- Machine width 9 m
- Traversing time 10 s
- Resultant angle $\alpha = \arctan 9/(10 \cdot 20) = 2.6°$

Since the angle α is small, the scanning line becomes almost parallel to the longitudinal direction of the paper web, so that only a very small part of the paper is in fact analysed. Shorter traversing times, a small measurement area and a short time between the measurements are therefore striven for.

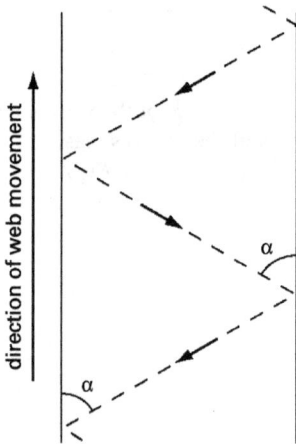

Figure 11.3. The movement of the measurement surface in traversing measurement.

It should be pointed out that the evaluation of average grammage profiles in the machine and cross-machine directions is not simple but requires fairly extensive computer treatment. As an example, *Figure 11.4* shows how an uneven grammage profile influences the recorded "length variations". With modern computer technology, it is now possible to evaluate all four variance components *Tot, MD, CD,* and *Residual* from on-line measurements with traversing measurement heads.

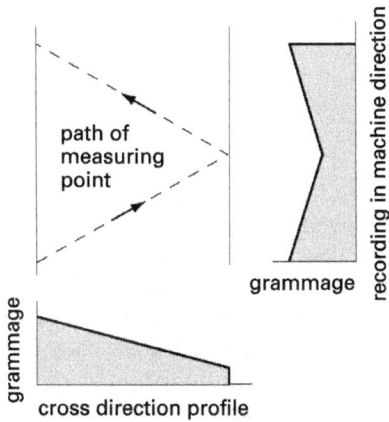

Figure 11.4. Left: The path of the measurement sensor and the cross direction profile. Right: Recorded variation in the machine direction.

11.2.4 Spectral Analysis

Another way of describing the variance is by an effect spectrum. One such spectrum is the frequency spectrum, which describes the contribution to the variance in different frequency ranges. In a frequency spectrum, periodic variations at a certain frequency appear as a "spike" in the spectrum, and it is possible to quantify the contribution to the variance of this periodic variation. With knowledge of the frequency of the disturbance, it is often possible to trace the source of the disturbance. A frequency spectrum can be transformed into a wavelength spectrum, from which the geometric extension of the disturbances can be read.

11.3 Grammage Variations

As mentioned initially, the grammage was the first web property to be measured on-line. The absorption of β-radiation from different isotopes (Promethium-147, Krypton-85 and others) was used as measurement principle. β-radiation is especially useful since its absorption properties are only mass dependent and independent of the composition of the absorbed pulp.

The radiation source is placed on one side of the web and the detector on the other. It is essential for the accuracy of the measurement that the distance between these units and their relative positions are kept constant during the traversing across the machine. This requires that the equipment be fitted in very stable frames, which have e.g. temperature control through water flow. At the end positions, the measurement heads can be driven out to the side of the web, which means that it is in principle possible to calibrate the meter, with no sample present between each traverse. Compensation for several error sources is necessary. As an example, it can be mentioned that in the measurement of the grammage of soft crêpe paper, the mass of the atmospheric moisture in the measurement gap can be of the same order of magnitude as the fibre mass. To separate the fibre mass from the total grammage, a measurement of the moisture content is also required, see further Section 11.4.

11.3.1 Length Variations

The length variations in grammage are in the first place due to concentration variations in the incoming pulp. To reduce these variations, an equalization chest is used, as was indicated already in the chapter "flow balances". The equalization takes place by mixing in the chest. The magnitude of the damping depends on the frequency, or on its inverse value the period, of the concentration oscillation, see *Figure 11.5*. Rapid variations with a short period are damped effectively, whereas variations which are slower than the average residence time in the chest V/Q (V = chest volume, Q = volumetric flow through chest) remain largely unaffected.

Division into several chests obtains an efficient equalization. The last equalization chest is the machine chest, which is dimensioned with consideration to the transport time from the tub to the measurement position for grammage at the dry end of the machine. Since this time amounts to a few minutes, the chest is dimensioned so that the emptying time is slightly longer. Periodic pressure variations generate grammage variations in the machine direction, especially when hydraulic headboxes are used. Pressure pulses can be generated by e.g. filters and pumps.

outgoing concentration variations

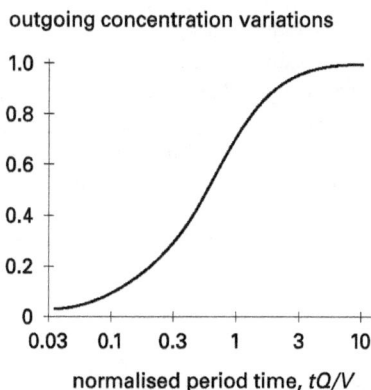

Figure 11.5. Damping in a chest of sinusoidal concentration variations with different periods, normalized with respect to the filling time of the chest V/Q.

Different grammage disturbances and their disturbance sources are listed in *Table 11.1*.

Table 11.1. Grammage variations in the machine direction within different frequency intervals and wavelength regions. Valid for a machine speed of 600 m/min.

Frequency 1/s	Wavelength m	Random Variations	Periodic Variations	Equalizing Measures
0.001	10,000	Stock Variations		Control of Thick Stock Dosage
0.01	1,000			
0.1	100			
1	10	Instability in Head-box	Pressure Pulsa-tions	Equalization in Machine Chest
10	1			
100	0,1			
1,000	0.01	Fibre Flocculation	Wire Mark	
10,000	0.001			

11.3.2 Cross-machine variations

The cross-machine profile of the grammage can be controlled with the slice profile. A local increase in the lip opening delivers a locally extra thick mix jet in this position but during the dewatering process, a considerable lateral redistribution then takes place. The grammage thus increases over a fairly wide area, the thicker the mix jet is the wider the area. A local increase in outflow takes place at the expense of the outflow on both sides, and this in turn leads to local grammage reductions around a central grammage increase.

The total profile response accompanying a change in the lip profile is thus complex. Computerized systems are therefore required to calculate the optimum change in the lip profile to attain a desired change in the grammage profile. It should be pointed out that the desired profile, the target profile, is not always constant across the whole wire width. To attain the desired dry content profile, a higher grammage may e.g. be required at the edges. It should also be pointed out that any local change in the lip opening leads to certain lateral flows, which influence the fibre direction distribution. Areas are created with a skew fibre direction distribution, which is a matter of increasing concern in the manufacture of e.g. copy paper and liner.

To avoid a skew fibre direction distribution, a method has now been introduced to control the concentration profile, see Chapter 10. In general, it is easier to adjust the grammage profile on a twin-wire machine than on a Fourdrinier machine, since considerably less lateral flow occurs. On twin-wire machines, on the other hand, it is more important that the mechanical properties of the wires are constant across the whole machine width since it is the wire tension, which generates the dewatering pressure.

11.3.3 Residual Variations

The residual variance is calculated, as has earlier been described using Equation 4. Since the residual variations are random and not controllable, only rebuilding of the high-concentration pulp system, short circulation, headbox or wire section can reduce them. With regard to the headbox, the stability is greater in a modern hydraulic box than in a conventional air-cushion headbox. This is because the volumes of the flowing mix are smaller, so that instabilities with a large spatial extension can be avoided. Normally, the residual variations increase with increasing machine speed. This is especially true in dewatering on Fourdrinier wire sections, where the mix on the open wire is exposed to flow disturbances in the lateral and longitudinal directions. In pure twin-wire dewatering, the mix is held between two wires and this means that the residual variations are about half the size of those in Fourdrinier wire dewatering.

11.4 Moisture Content Variations

The moisture content is often measured with radiation absorption, the transmission at two wavelengths within the infrared area being compared. The measurement equipment is placed on the same measurement head as the grammage meter. This means that the moisture content meter performs the same traversing movements as the grammage meter and that the evaluation therefore follows the same pattern.

11.4.1 Longitudinal Variations

Longitudinal variations in moisture content are controlled with the steam pressure in the drying cylinders. It should be pointed out that it is necessary to know the change in both fibre mass and moisture content to be able to control the steam pressure at a reasonable level. For example, a temporary increase in the fibre mass gives an increase in the moisture content but it is in this case the fibre mass, which must be altered, and not the drying capacity. It is not unusual for the drying capacity to be the production-limiting factor on a paper machine. Speed optimization can then be applied, which means that the steam pressure in the cylinders is kept at maximum all the time, and the dry content is instead controlled by changing the machine speed. Reducing the machine speed and thereby also the required drying capacity can compensate for too high a moisture content.

11.4.2 Cross-Machine Variations

Variations in the moisture content profile need not depend only on an uneven drying energy distribution. They can also be due to an uneven solids content profile after the wire section or an uneven pressing efficiency across the machine. However, it is very unusual to measure the dry content profile already after the press section, so it is difficult in practice to determine where in the process moisture streaks arise.

A common way of achieving a more even moisture profile is to add water in the streaks where the moisture content is low. It may seem uneconomic from an energy viewpoint to first over-dry the web and then add water, but it has been shown to be a very efficient control method. A large number of water nozzles are fitted across the machine, normally at the beginning of the dryer section. Each nozzle is equipped with a valve and these are controlled by a computerized system.

An alternative way of controlling the moisture profile is with the help of infrared radiators. At the end of the dryer, an infrared dryer is placed consisting of a large number of radiation elements in the cross direction. The power to the individual elements is controlled on the basis of the desired change in moisture profile.

11.5 Thickness Profiles

The roll quality is decisive for the runnability in a coming conversion stage, such as a printing press, sack machine etc. The thickness of the web is decisive for the roll quality and its profile must therefore be controlled. Varying the surface pressure across the machine calender does this. The local diameter of the smoothing rolls is dependent on the temperature. A locally too thick web requires a higher local surface pressure, and the local glazing temperature is therefore increased, with a greater diameter and reduced output web thickness as a consequence. The temperature is adjusted with the help of nozzles blowing hot or cold air. There are also systems where the temperature profile is controlled with induction spools.

11.6 Retention Control

To be able to maintain a constant composition of the web, with regard to e.g. filler content, the content of the corresponding component in the white water system is measured. Centrally placed measurement equipment is connected via test pipelines with different positions in the system, such as headbox and wire pit. Measurements are then made successively on samples from the different measurement positions, with a time interval of a few minutes. On the basis of these measurements, the retention can be calculated and the retention agent dosage can be controlled so that constant retention is maintained. The runnability is further improved if the high-concentration pulp is characterized with regard to the surface properties before the retention agent addition. This makes it possible to compensate for changes in the incoming fibre material before they have had time to influence the retention of a given component. The surface properties can be characterized by e.g. colloid titration and the retention agent requirement can be predicted.

12 Wet Pressing

Bo Norman
Department of Fibre and Polymer Technology, KTH

12.1 Introduction

Before reaching the dryer section, the paper web is pressed against one or between two felts in 2–4 press nips in which the dry solids content increases from ca. 20 per cent after the wire section to ca. 40–50 per cent. High dry solids content after the press section is desirable since it is less energy consuming to press out water than to vaporize it. The paper web is pressed either between a smooth surface and a single press felt: one-sided or single-felt pressing, or between two press felts: two-sided or double-felt pressing. The flow is thus mainly transverse, i.e. the water flows in the z-direction of the sheet, into the blanket, and then out of the felt through its reverse side. In the pressing of a moving web, there is also a certain superposed longitudinal lengthwise flow in the felt and sheet.

Increasing demands are being made on the pressing operation with regard to the properties of the final product. Surface defects, which worsen the printability of the paper, must not be introduced. Surface density is important for printing. For products where the bending stiffness is an important property, the desire for a high dry content level after the press section must be balanced against the need for high bending stiffness in the final product. The water squeezing can be altered by changing in the properties of the press equipment (pressure pulse, felts) and by changes in the paper web (the water content of the fibres, temperature, sheet structure).

This chapter presents the theory of pressing, on one hand plane pressing of sheets, and on the other hand the pressing of a moving paper web in a press nip. Finally, the design of some press sections is briefly described.

12.2 Plane Pressing

Both theoretically and experimentally, it is much easier to study plane pressing of a sheet between two parallel surfaces than the pressing of a moving web in a press nip. Most basic studies of the pressing process have therefore considered the plane pressing case.

If a given press force is applied, the force in the sheet will be balanced by two components:

- The mechanical elastic force in the fibre network
- The liquid pressure due to friction forces from the flowing liquid

The properties of the sheet during the pressing process can be described in a very simplified way by a so-called Kelvin model, *Figure 12.1*. The spring represents the network forces and the damper represents the hydraulic forces. According to the simplified model, the sum of the network force and the hydraulic forces is balanced by the applied press force. If the greater part of the press force is taken up by the fibre network, the process is said to be compression-limited. To increase the squeezing-out of water in a compression-limited process, the press force must be increased so that the "spring" can be further compressed. When the hydraulic forces dominate over the network force, the process is said to be flow-limited. In such a process, the squeezing of water increases if either the press force or the press time is increased.

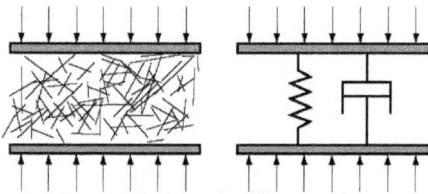

Figure 12.1. Sheet under compression between two plates, (left) and Kelvin model with spring and dashpot (right).

Water leaves the sheet from cavities between the fibres. The size of the cavities is characterized by the permeability of the sheet. Some water also comes from within the fibres, depending on the degree of swelling (Water Retention Value = WRV). The permeability of the fibres is considerably lower than that of the sheet.

12.2.1 Compression-Limited Press Nips

Water in the cavities between the fibres is squeezed out if the hydraulic pressure inside the sheet is higher than at its surface. When the network force of the compressed fibre network is so large that it resists the applied press force without further compression, no further water is squeezed out of the cavities. If the dry content after the pressing process depends only on the applied surface pressure, the process is said to be compression-limited, as mentioned previously. Assume that the pressure is increased, so that further compression of the sheet takes place and so that a corresponding volume of water is pressed out.

$$\Delta Q = A\Delta l \tag{12.1}$$

ΔQ is water volume squeezed out of the paper
A is the area of the sheet
Δl is the thickness decrease of the sheet

The change in the sheet thickness is determined by the modulus of elasticity E in the z-direction according to Hooke's law, which inserted in Equation 1 gives:

$$\frac{\Delta Q}{A} = \frac{l\Delta}{E} \tag{12.2}$$

l is the sheet thickness
Δp is the applied pressure increase

To calculate the effect of a pressure increase, it is thus necessary to know the elasticity modulus E. The higher the value of the elasticity modulus, the more difficult is the sheet to compress and the more compression-limited is the pressing process.

If a sheet is pressed for a long time between smooth but at the same time permeable surfaces, it is possible to attain high dry contents with moderate surface pressures. *Figure 12.2* shows the results from plane pressing tests in which the dry content has been calculated from the thickness of the sheet. This is a pure case of compression-limited pressing, since the press time was so long that all flow had ceased.

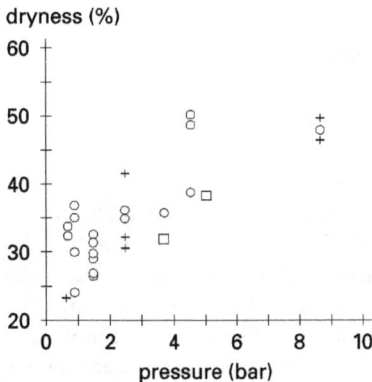

Figure 12.2. Dry content in static plane pressing of different pulps. (Carlsson and Lindström, STFI).

In the figure, it can be seen that already at a surface pressure of 5 bar (0.5 MPa), a dry solids content of 50 per cent can be reached. In the press nip of a paper machine, the maximum surface pressure amounts to circa 5 MPa. A surface pressure one magnitude higher, i.e. circa 50 MPa, gives dry contents of 75–85 per cent.

12.2.2 Flow-Limited Press Nips

In flow-limited press nips, the friction resistance in the outflow is decisive for the amount of water squeezed out. *Figure 12.3* shows the principle for the determination of the **permeability** of a sheet as a function of its dry content. The sheet is compressed between two permeable plates to the desired thickness and thus to the desired dry content, by a press force *F*. A hydraulic pressure difference *p* between the two sides of the sheet then generates a flow through the sheet, which is measured.

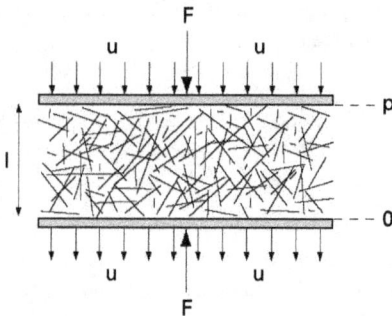

Figure 12.3. Permeability measurement on a compressed sheet.

The flow is normally laminar and can be described by the following Equation:

$$u = \frac{K\,p}{l\,\eta} \tag{12.3}$$

u is the average rate of flow across the whole surface of the sheet
K is the permeability of the sheet
p is the static pressure at the upper side of the sheet
l is the thickness of the sheet, and
η is the viscosity of the water.

The permeability *K* has the dimension m2 and is a measure of the effective, available area for water flow. There are two practical aspects on the evaluation of *K* according to *Figure 12.3*.

The surface structure of the permeable plates must be smooth enough to give an even pressure distribution on the sheet and at the same time the surfaces must be permeable. However, too small holes in the surface will make it easily sealed by the sheet fibres, which can cause a considerable sealing affect. In extreme cases, a low permeability of the interfaces between the permeable plates and the web surfaces may dominate the total permeability.

The flow across the sheet will generate fibre friction forces in the flow direction, and thus compressive forces on the sheet. This will in turn cause a compression force, increasing from top to bottom, and a corresponding gradient in sheet density. The evaluated permeability then represents a mean value within the actual sheet density range.

During the time Δt, a volume ΔQ of water passes through the sheet of area *A*, according to Equation 4.

$$\frac{\Delta Q}{A} = u\Delta t = \frac{K}{l\eta}p\Delta t \qquad (12.4)$$

The water flow is thus proportional to the product of pressure and time, which is called the *pressure impulse, I*. In the more general case with a varying pressure during the press time *T*, the following definition applies:

$$I = \int_0^T p(t)\,dt \qquad (12.5)$$

The press impulse is thus equal to the area below the pressure curve in a pressure-time diagram, as in e.g. *Figure 12.10*. In a flow-limited pressing process, the permeability *K* of the sheet is not constant but varies in time and in the *z*-direction of the sheet. A one-sided pressing process of this type is shown in *Figure 12.4*.

It is characteristic that the fibre network is compressed most on the side where the water leaves the sheet. Since the liquid pressure decreases to almost zero in the interface between the sheet and the felt, i.e. where the water leaves the sheet, the dominating part of the press force will compress the network at this level. The result of this phenomenon is that, during pressing in a single-felt press nip, paper develops a higher density and thus also higher surface strength on the side turned towards the felt. The effect becomes more marked the higher the pressure and the shorter the press time that is used. he opposite side will instead be smoother, du to contact with the even press surface.

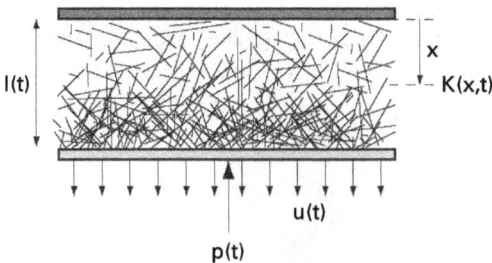

Figure 12.4. One-sided, flow-limited pressing process.

A **two-sided pressing** process consists of two one-sided processes, with half the grammage in each case, and with the impermeable surfaces turned towards each other at the dewatering centre, as shown in *Figure 12.5*. This means that the dewatering capacity is four times that in one-sided pressing, due to half the flow rate and twice the permeability in each half of the sheet. In a double-felted press nip, the two sides of the sheet become more densified than the centre. This is favourable both from a printability viewpoint and from a bending stiffness viewpoint.

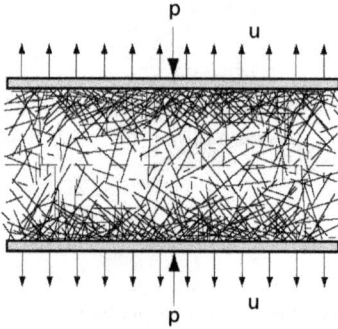

Figure 12.5. A two-sided, flow-limited pressing process.

Fibre-swelling. In the description above, no consideration has been given to the fact that water is also pressed out from within the fibres. For this to take place, the individual fibres must be compressed so that the liquid pressure inside them increases. The pores through which water can leave the fibres are very small, which means that fibres with a high *WRV* give a strongly flow-limited pressing process. The highly swollen fibres are also more plastic and therefore form a sheet structure with a low permeability.

In an extreme case, highly swollen fibres can be imagined as elastic, water-filled hoses. If such a hose is squeezed only locally, the water can in the first place flow along the hose. Efficient water removal then requires a very even pressure distribution, see further Section 12.3.

In summary, the degree of swelling of the fibres has a decisive importance for the dry content attained in a pressing process. Data are shown in *Figure 12.6* as an example, where it should be pointed out that they refer to laboratory roll pressing. WRV is also changed by the addition of polymers.

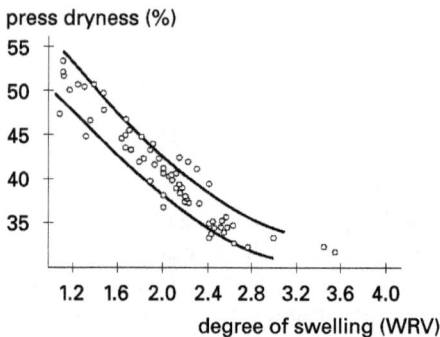

Figure 12.6. Press dry content as a function of *WRV*. Roll pressing. *pmax* = 3.45 MPa (Busker, Beloit).

The influence of pH. The swellability of a pulp depends e.g. on the content of dissociable groups (carboxyl-groups, sulphonic acid groups) and the ionic state in which these groups exist. At a high pH, the groups are dissociated and the repulsion between the charged groups leads to a swelling of the fibre. If a salt is added, the swelling decreases, since the charged groups are screened. If the charged groups are converted to their proton state through treatment at low pH, there are no longer any charged groups in the structure, and this means that the swelling of the fibre decreases. The degree of swelling change also depends upon the stiffness of the fibre wall.

The viscosity of water. An increase in the temperature of the sheet during the pressing process means that the temperature of the water is increased and that its viscosity is therefore reduced, see *Figure 12.7*. According to Equation 3, a reduction in the viscosity lowers the friction resistance in the outflow of the water, so that the dry content after pressing increases (unless the process is purely compression-limited).

viscosity ($10^3 \cdot \eta$ kg/ms)

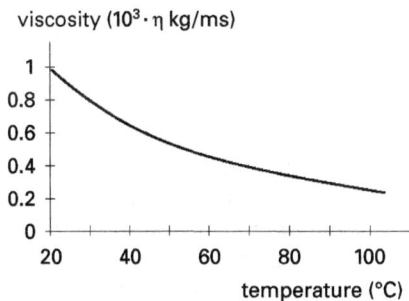

Figure 12.7. The viscosity of water as a function of temperature.

From experience, a temperature increase of 20°C usually gives a dry content increase after the press of about one per cent. If the web temperature can be increased to almost 100°C, pulps with a high lignin content, e.g. mechanical pulps, will also soften. The stiffness of the fibres thus decreases, and a dry content increase becomes possible as a result of the lower elasticity modulus of the fibre network (see Equation 2).

12.3 Press Felts

The task of the press felt in a press nip is to receive the water which has been pressed out of the paper web, to hold it when the web and felt separate after the press nip, and to carry it to the white water system. The felt thus helps the water in the press nip to flow out from the sheet in the z-direction.

Until the 1960's, felts were woven from wool and were "felted" to a more even structure through mechanical treatment in hot water. This felting process was based on the fact that the surface of the wool fibre is covered by small hooks, which can interact and thereby lock the fibres in a felted structure.

During the 1960's, wool was replaced as raw material in press felts by polymer fibres, in the first place to give the felts a longer lifespan. Because of their smooth surfaces, polymer fibres cannot be felted but they are instead needled onto a base fabric. Simplified, the base fabric can be described as a coarse wire in which the threads are usually of a monofilament type (homogeneous threads) or in certain cases multi filaments (spun threads). A preformed layer of batt fibres is placed on the base fabric, and barbed needles anchor these with each other and with the basic fabric by being thrust perpendicular to the felt a large number of times. *Figure 12.8* shows a cross-section of a felt with a multi-layer base fabric of monofilaments and needled batt fibres. Note that some fibres run vertically down into the felt, a result of the needling process.

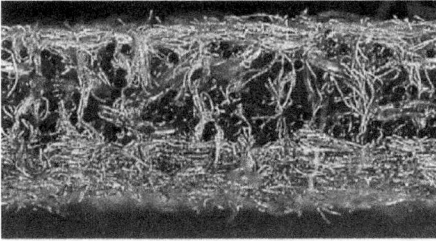

Figure 12.8. Needled press felt, 5 mm thick, with multi-layered base fabric (Albany Int, Halmstad).

The base fabric has two tasks, to provide a mechanically connecting frame for the felt and to create an incompressible volume, which can hold water inside the press nip. The base fabric today often consists of a double-layered fabric at the bottom and a finer single-layered fabric on top. To further reduce the unevenness of the base fabric, there is a development in which parallel thread layers are used in the lengthwise and crosswise directions, which are then fixed by the needling of batt fibres.

To bridge the unevenness of the basic fabric, coarse batt fibres are used lowermost and the surface is covered with fine batt fibres to give a smoother surface. The requirements regarding mechanical properties of the batt fibres are high, since it is their durability, which decides the lifespan of the press felt. The lifespan of the felt is determined by how long the structure can resist permanent compression. If the structure does not spring back sufficiently between two press pulses, it is no longer possible to keep it free from clogging. The permeability then decreases and thus also the potential for the felt to take up water from the web in the nip.

A common felt lifespan could be 2-3 months. *Figure 12.9* shows a press felt under normal pressure in contact with a newsprint sheet. It is clear that the surface of the felt is relatively uneven and that the pressure is therefore only applied to sparsely distributed areas. A consequence of this is probably that the water is only moved along the sheet to a certain extent. At the same

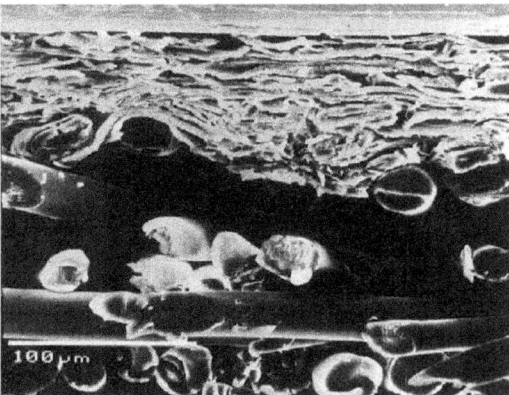

Figure 12.9. Press roll (top), newsprint sheet (middle) and press felt (bottom), (Beloit).

time, the felt can leave considerable amounts of water on the paper web at the time of separation; see further in Section "12.4.3 Rewetting".

Since both uneven pressure distribution and rewetting are surface effects, they increase in importance with decreasing grammage and for double felted in comparison to single felted pressing.

12.4 Press Nip

In Section 12.2, Plane pressing, the properties of a paper sheet were described in a one-dimensional pressing process in which all variations in the plane of the sheet were neglected. A paper web, which moves through a press nip together with one or two press, felts experiences strong pressure gradients in the machine direction and variations in the plane of the web because of the local unevenness of the felt.

The dry content of a press felt is as a rule so low that it becomes saturated with water in the press nip. To be able to remove water from the paper web, it is then necessary that water can be transferred from the felt to the press surface behind it. This is achieved in press roll surfaces by equipping them with either thin, parallel grooves (ventanip rolls) or small shallow indentations (blind-drilled rolls). Washing water is added to the press felts between the press nips, often by high-pressure spraying against the paper side. Some of this water is sucked up by suction boxes ("Uhle boxes"), but normally the dry content of the felt is still determined by the degree of compression in the press nips

12.4.1 Roll Pressing

In roll pressing, the sheet is exposed to a pressure pulse according to *Figure 12.10* where, as has been pointed out earlier, the area below the pressure curve is equal to the press impulse. The total press force on a press roll per metre roll width is called the line load and is usually expressed in kN/m. The maximum line load on a conventional roll press is 200–300 kN/m.

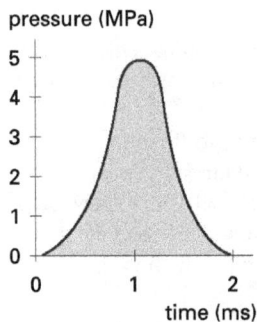

Figure 12.10. Example of a roll press pulse in the pressing of newsprint.

A simple relationship between press impulse I, the line load L of the press nip and the machine speed u can be derived as follows:

$$I = pt = \frac{F}{bl}t = \frac{F}{b}\frac{t}{l} = \frac{L}{u} \qquad (12.6)$$

I is press impulse (kPas), p (average) pressure (kPa), t press time (s), F press force (kN), b press nip width in CD (m), l press nip length in MD (m), L nip line load (kN/m) and u machine speed (m/s).

If the press pulse is to be maintained with increasing machine speed, the line load must thus be increased to the same extent.

In 1968, a model was developed to describe the process in roll pressing, still useful in its fundamental features, see *Figure 12.11*. The authors divide the pressing process into four main zones:

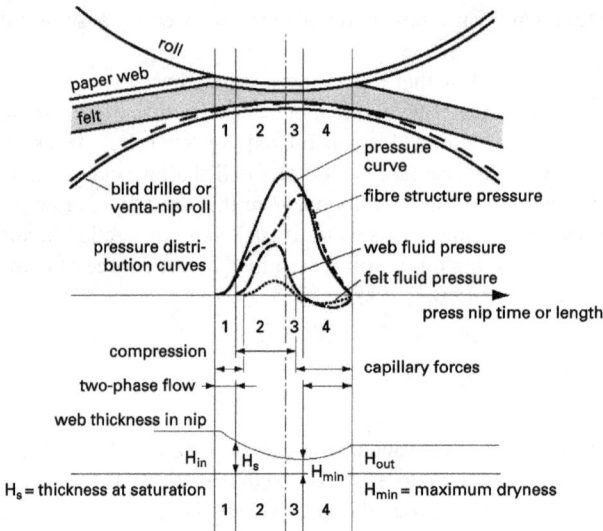

Figure 12.11. Roll pressing model (Wahlström, Larsson and Nilsson, KMW).

Zone 1: The whole applied pressure is taken up by the fibre network. Both the sheet and the felt contain air which is pressed out in this zone. The upper zone limit is the position where the sheet has become water-filled.

Zone 2: The hydraulic pressure is built up in the sheet and water begins to flow. In this zone, the felt also becomes saturated and its hydraulic pressure begins to build up. It never attains the pressure level in the sheet, however, since the roll surface behind the felt will let in surplus water into holes or grooves. A low felt pressure is required for water to flow out of the sheet. The zone ends at the position where the applied pressure reaches its maximum, which is assumed to be in the middle of the nip.

Zone 3: The water here continues to flow out of the sheet. The hydraulic pressure in the sheet decreases and, at the upper limit of the zone, it again reaches a value of zero. Note that the pressure on the fibre network continues to rise and reaches its maximum at the zone limit, not in the middle of the nip.

Zone 4: Negative values are here obtained for the hydraulic pressure, which suggests that water could be absorbed into the sheet again, so-called rewetting (see further Section 12.4.3). The zone is completed when the total pressure again reaches a value of zero.

To this original description of the four zones, a few complementary viewpoints can today be added:

The model does not take into consideration gradients in the z-direction of the sheet. It is further not certain that all air leaves the sheet in zone 1. Part of the air can be enclosed in the fibre network when the structure begins to compress. The volume of the air will of course decrease strongly when the pressure rises. No consideration has been taken to the flow in the plane of the sheet. The strong rewetting indicated in zone 4, where the thickness of the sheet increases from H_{min} to H_{out}, is now considered to be unrealistic.

12.4.2 Extended Press Nips

An increase in the press impulse leads to a higher pressing efficiency. In a conventional roll press, however, the maximum felt pressure allowed limits the applicable press force. If this is exceeded, the lifespan of the felt decreases radically. Further, the pressure increase along the web at press nip entrance can become so strong that the web is crushed, which means that it is locally pulled apart by flow in the plane of the sheet. A soft press nip gives a low maximum pressure. This can be achieved by double felting or by soft covers on the press rolls. Another but more expensive method is to use press rolls with extra large diameters.

To achieve any considerable increase in the press impulse, an extended press nip is required. This is based in principle on a non-exploited East-German patent, which treats a shoe-press, see *Figure 12.12*.

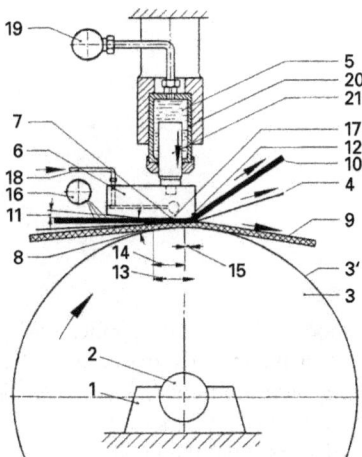

Figure 12.12. The principle for a shoe-press. (Jahn and Kretzschmar, Patent 1970).

Independently of the German invention, an extended press nip was developed by Beloit in the mid-1970s, see *Figure 12.13*. The lower press roll has here been replaced with a stationary, concave press-shoe. A belt runs over the shoe and between these, an oil film lubricates. Above the belt, press felts and paper web then run in a conventional way.

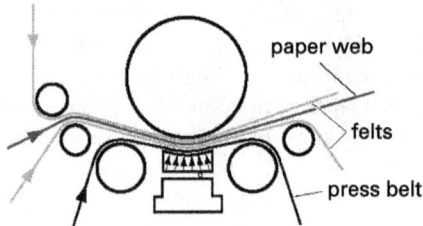

Figure 12.13. Extended press nip with oil-lubricated press-shoe. (Beloit).

The length of the press-shoe can be chosen so that the press nip becomes considerably longer (normal length circa 250 mm) than in roll pressing. In the extended press nips, which are used industrially today, line loads of more than 1000 kN/m can be used. The pressure distribution along the press nip can be influenced by the position of the support point of the press-shoe. If this is placed centrally, a symmetrical press curve is obtained. If instead the support is placed closer to the rear edge, a gradually increasing pressure is obtained which is favourable for the pressing-out of water. Pressure curves for a roll press and a shoe-press are compared in *Figure 12.14*.

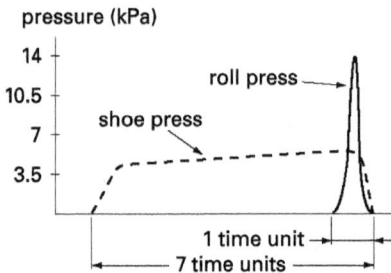

Figure 12.14. Pressure curves for roll press and shoe-press (Beloit).

Several machine manufacturers have now developed shoe-presses. Voith applies hydralic pressure to apply the press force. The pressure distribution is then controlled by the shoe shape. In Voith's version, the belt forms the periphery of a cylinder with normal press roll size. This means a compact structure with simplified control of oil mist leakage. Valmet-KMW's version is similar to that of Voith, but the press-shoe has been modified so that the press forces in the beginning, in the middle and at the end of the shoe can be controlled individually, on-line.

12.4.3 Rewetting

As "proof" that a considerable rewetting takes place in zone 4 (see *Figure 12.11*), experiments are usually referred to where dry solids contents of 65–70 per cent in the middle nip have been calculated, at the same time as the outgoing dry solids content of the sheet has been measured to 40 per cent. The dry content in the middle nip has then been calculated from measurements of the distance between the press rolls in the presence and absence of the paper web. There are of course two considerable sources of error, which make it impossible to calculate the dry content of the paper web from this type of thickness measurement.

•   Firstly, the water in the paper web in the middle of the press nip has a higher speed than the fibre network. This means that there is actually more water present than corresponds to the free space in the sheet. The excess speed of the water in a press felt has been studied experimentally and the water in the paper web is exposed to similar conditions in a press nip.
•   Secondly, the minimum thickness of the web is evaluated and not the average thickness. If the sheet is removed, the felt will rest on the most protruding batt fibres, which means that it is the sheet thickness, which existed at these contact points, which will be evaluated from the thickness difference calculation.

Today, there is a general agreement that, except in very slow machines, the reabsorption of water to the paper during the expansion stage of the press nip is negligible. It is proposed that the rewetting process be divided into three stages:

•   Internal rewetting in the expanding part of the press nip
•   External rewetting while the paper is in contact with the felt after the press nip.
•   Separation rewetting; the water left by the felt on the paper web surface at separation of felt from paper.
    The internal rewetting is normally negligible. The external rewetting is minimized by separation of felt and paper as soon as possible after the press nip. The separation rewetting is minimized by a smooth felt surface, so that the smallest possible water volume is available for division through "film splitting" (analogous to the splitting of printing ink between print surface and paper surface) at separation. The felt structure should further be so dense that the felt holds "its own" water and does not leave parts of it on the paper surface. In Section 12.3 it was pointed out that the felt generates an uneven pressure on the paper surface. Since both uneven pressure distribution and separation rewetting are surface effects generated by an uneven felt surface, they are difficult to evaluate separately.

12.4.4 Web Transfer

The basic rule for the movement of the paper web through a press section is that it always follows the smoothest surface. After a single-felted press nip, it will therefore always follow the smooth roll and not the more uneven felt surface. In double-felted press nips, special measures must be taken so that the web follows the desired felt. A special felt guide roll is then placed below the felt, which the web should not follow; see examples in *Figure 12.15* and *Figure 12.16*. If the felt is lifted from the underlying roll, air has access to the back of the felt, and the web is transferred to the other felt surface. However, it is not certain that the web follows a felt, which is curved along a roll surface, since the centrifugal force strives to throw the web off. A press roll must therefore be equipped with a suction zone if the web is to be carried safely to the next press nip. Such press suction rolls are used in *Figure 12.15* and *Figure 12.16*.

12.5 The Design of a Press Section

The main task of the press section is to increase the dry solids content of the paper web before it enters the dryer section, both to increase the strength of the web, and thereby the runnability, and to save drying energy. However, it is not possible to strive for maximum dry solids content without observing the influence of the pressing on certain paper properties. Pressing to higher dry solids content normally gives a higher sheet density and thus a lower bending stiffness. This limits the optimum press dry content especially for carton board products. Felts have a negative effect on paper surface smoothness, but at the same time a positive effect on paper surface density. These parameters effect print quality. For both running and paper property reasons, the design of the press section must be adapted to the actual product concerned. Some examples of the design of press sections for liner board and newsprint are shown below.

12.5.1 Liner Board

The press section of the liner board machine is characterized primarily by a demand for high outgoing dry solids content, which is important for the production economy. This was the main reason why extended press nips were first applied on liner board machines. A typical rebuilt press section on a liner board machine is shown in *Figure 12.15*, where an extended press nip has been added after a conventional roll press. The press section starts with a one double-felted and one single-felted roll press nip. The third press nip is a double-felted, extended press nip of the shoe-press type. Compared with a conventional third press nip, the shoe-press gives ca eight per cent higher dry solids content. This permits a considerable production increase. The density of the product also increases which means a certain increase in the compression strength in the cross-direction, an important property for liner board. In a modern liner board machine, the press section would consist of two double felted extended press nips.

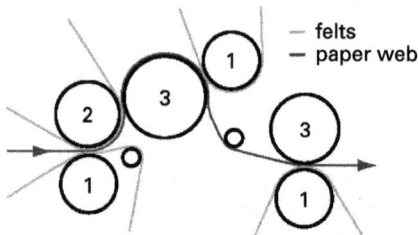

Figure 12.15. Press section with extended 3rd press nip.
1 = venta nip or blind-drilled roll; 2 = suction press roll; 3 = smooth roll; 4 = roll with internal press shoe.

12.5.2 Newsprint

A press section for a traditional newsprint paper machine is shown in *Figure 12.16*. The press section begins with a double-felted press nip followed by two single-felted nips. Later a fourth press nip was added. The fourth nip gives a slightly higher dry solids content which means a lower energy requirement in the dryer section, but also an improved runnability in the drying section through a greater web strength. In the fourth press nip, the upper side of the web contacts a smooth roll surface, which is favourable for the equilibration of the surface properties of the two web sides, since the bottom side of the web has contacted a smooth roll surface in the two previous press nips.

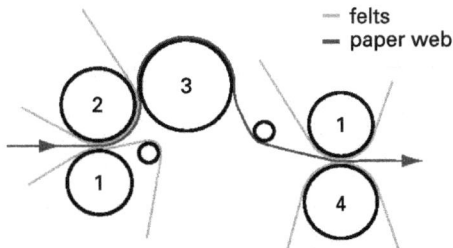

Figure 12.16. Four-nip press section for newsprint.
1 = venta nip or blind-drilled roll; 2 = suction press roll; 3 = smooth roll.

In a modern press section for newsprint, two double felted, extended press nips are applied, see *Figure 12.17*. This means a closed web transfer along the entire press and drying sections, with correspondingly high runnability. As mentioned in Section 12.4.3, Rewetting, double felting in the last press nip will give separation rewetting on both web sides. Furthermore, the web will remain on a press felt for some distance (and time) after the last press nip, which will gen-

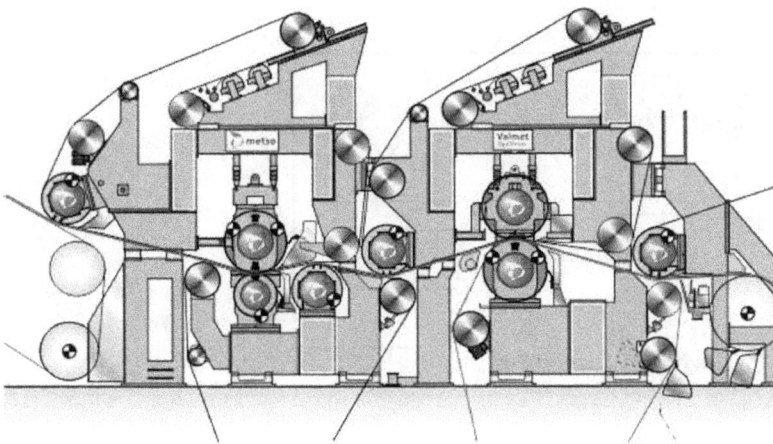

Figure 12.17. Modern newsprint press section, with two extended press nips (Metso).

erate a considerable amount of external rewetting. To increase web dryness, the bottom felt in the second press nip is therefore sometimes replaced by a press belt.

Figure 12.18 and Figure 12.19 show the web two-sidedness, expressed as top side value divided by bottom side value, for surface roughness and oil absorption respectively, using different press section setups.

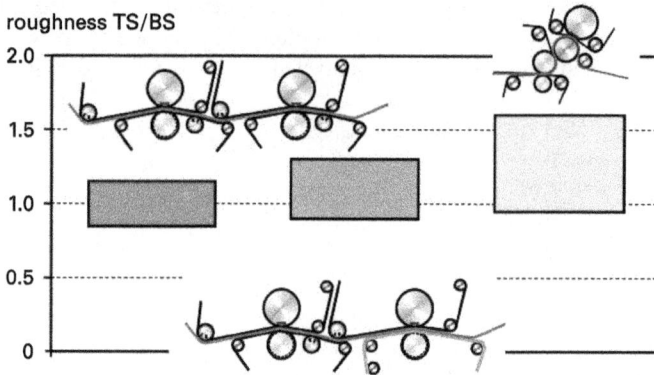

Figure 12.18. Surface roughness two-sidedness, Top Side / Bottom Side, for three press section conditions (Metso); Left: Two double felted press nips; Centre: Bottom felt in second press nip of left case replaced by press band; Right: One double felted nip + two single felted nips.

Web surface roughness will reflect the degree of imprint by felt surfaces. Therefore the bottom surface will become smoother when the last bottom felt is replaced by a press band (centre) or when a smooth roll surface is applied on the bottom side in the last two press nips (right), *Figure 12.18*. Correspondingly, the web surfaces with lower roughness will show a lower surface density and thus a higher oil absorption, *Figure 12.19*.

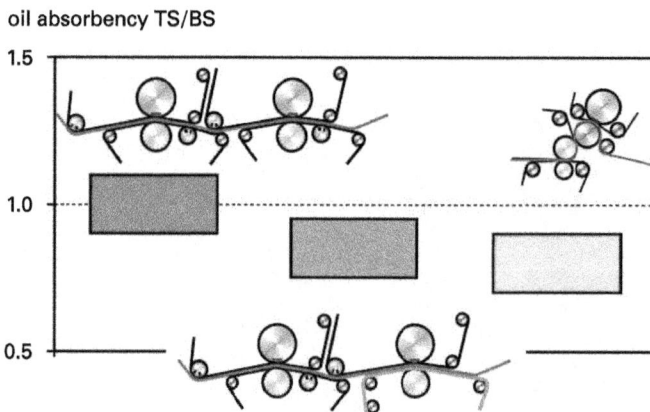

Figure 12.19. Oil absorbancy two-sidedness, Top Side / Bottom Side, for three press section conditions (Metso) Left: Two double felted press nips . Centre: Bottom felt in second press nip of left case replaced by press band. Right: One double felted nip + two single felted nips.

12.5.3 Single Nip Press Section

Single nip press sections have been introduced, see *Figure 12.20*, initially on a fine paper machine producing copy paper. Outgoing web dryness was high, but some problems regarding paper surface roughness was observed. This is understandable, since only one felt is in contact with each web surface during the whole pressing operation. This has enforced press felt surface developments, and further work is also made on improving felt life time.

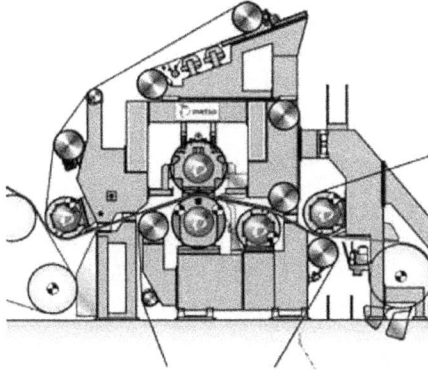

Figure 12.20. Single nip press section for copy paper (Metso).

13 Drying of Paper

Stig Stenström
Department of Chemical Engineering, Lund University

232

13.1 Introduction

Drying of paper is the final step in the water removal and web consolidation process where the dry matter content is increased from 35–45 % to 93–95 %. Thus 1.2–1.8 kg water per kg paper is removed by evaporation in the dryer section representing less than 1 % of the total water in the stock flow to the headbox. Evaporation of water requires supply of large amounts of energy as well as effective removal of the water vapour from the web. The energy supplied with the condensing steam is transferred to the humid air in the hood and efficient recovery of this part is necessary for low production costs. The paper dryer should be designed to facilitate high capacity, high productivity, good energy efficiency and producing paper with the specified quality parameters.

The largest part of the produced paper is dried on a multi-cylinder paper dryer where the surfaces of the web are alternating between contact with a hot cast iron cylinder and transport in hot humid air between the cylinders. This process is repeated with different boundary conditions from 30 to over 100 times until the moisture content has been reduced down to the desired level. The principle of using cast iron cylinders for drying of paper was first established around 1820 when the cylinders were heated by burning coal on the inside of the cylinders. The first paper machine in Sweden with this design was installed in 1832 at the Klippan paper mill.

Today the principle is the same but the cylinders are normally heated with condensing steam even if gas and radiation heated cylinders also exist. The capacity of the paper machine has doubled many times over by increasing the machine speed up to 2000 m/min, increasing the width up to 10 m requiring more advanced control of process parameters. A number of technical components have been added such as single tier designs and vacuum rolls and the complexity has increased by including on-line coating and calender stations.

13.2 Overview of Different Paper Drying Processes

13.2.1 Paper Drying Techniques used in Sweden

The multi-cylinder dryer is by far the most common dryer and is used for all kinds of paper except for tissue. Newsprint is dried on modern multi-cylinder dryers while coated papers are predried on a multi-cylinder design, followed by coating and drying the coated surface with infrared, air foil dryers and finally a small number of cylinders. Some board qualities require a smooth glazed surface which could be accomplished with a MG-Yankee cylinder in the middle of the dryer section. Tissue is dried on one large Yankee-cylinder equipped with an impingement hood and pulp is normally dried with an airborne web design or in a flash-dryer. A summary of the different kinds of paper and pulp dryers used is shown in *Table 13.1*.

Table 13.1. Different types of paper and pulp dryers used in Sweden.

	Multi-cylinder	IR	Air-foil	Yankee		Air-borne web	Flash-dryer
				MG	**Tissue**		
Product	All kinds of paper	Coated papers	Coated papers, Sack paper	Cardboard	Tissue	Pulp	Pulp
Energy supply	Steam	Electricity, natural gas or propane	Steam, natural gas or propane	Steam	Steam, natural gas or propane	Steam	Natural gas, propane or oil
Number in operation in Sweden 2002	85	20 (3 gas-heated)	20	10	14	27	6

The cylinder dryer is heated with condensing steam and the heat is transferred by contact between a hot metal surface and the paper. IR-dryers require higher temperatures and are heated either with electricity or gas and the heat is transferred by radiation between a hot surface and the paper. Impingement hoods (such as airfoils or hoods used for the drying of tissue) as well as the flash dryer operate at high temperatures which require natural gas or propane as the energy source and the heat is transferred by convection between the air or the gas stream and the paper.

The number of dryers in operation in the Swedish pulp and paper mills in 2002 is shown in *Table 13.1*.

13.2.2 Some Definitions

Before coating stations and surface sizing the dry matter content is constant throughout the dryer section and by using moisture ratio the amount of water removed between various positions can be directly calculated as the difference of moisture ratio values. Depending on the purpose the moisture content in the web can be given by the following variables:

$$\% \text{ dry matter content } (DMC) = \frac{kg \ dry \ matter}{kg \ dry \ matter + kg \ water} 100 \tag{13.1}$$

$$\% \text{ moisture content } (MC) = \frac{kg \ water}{kg \ dry \ matter + kg \ water} 100 = 100 - DMC \tag{13.2}$$

$$\% \text{ moisture ratio } (MR) = \frac{kg \ water}{kg \ dry \ matter} = \frac{MC}{100 - DMC} \tag{13.3}$$

Example 13.1	Estimate the amount of water removed in the dryer part of a paper machine producing newsprint with a basis weight of 45 g/m2 running at 1100 m/min. The machine width is 8.2 m, the dry matter content after the press section is 40 % and the final dry matter content is 94 %. According to equations (2.2) and (2.3) the result is: $$\text{Water removed} = 8.2\frac{1100}{60}0.045\left(\frac{100-40}{40}-\frac{100-94}{94}\right)$$ $$= 9.71 \text{ kg water/s} = 35.0 \text{ t/h}$$

13.3 Multi-Cylinder Paper Dryer

13.3.1 Overall Principles

The principle for the multi-cylinder dryer is repeated contact between the paper and a hot cast iron cylinder followed by a free draw between two cylinders, the principle is shown in *Figure 13.1*. The multi-cylinder dryer is built of three main components:

- a steam and condensate system for the energy supply
- a number of cast iron cylinders for the energy transfer from the steam to the paper
- a hood covering the dryer to evacuate the evaporated water vapours and facilitate energy recovery from the humid air

Traditionally the cylinders have been arranged in an upper row and a lower row of cylinders but today machines for newsprint and printing papers are built with only one row of steam heated cylinders. Also designs where the cylinders are placed above each other in what is called a ''stack'' are in operation. In the dryer the web is supported by a dryer fabric, traditionally one for the upper row of cylinders and one for the lower row of cylinders (double tier design), but today a number of different designs are in operation. This will be discussed in more detail in Section 13.3.4. The cyclic process is repeated (dryers with up to 170 cylinders have been built) until the web has reached the dry matter content of 93–95 %.

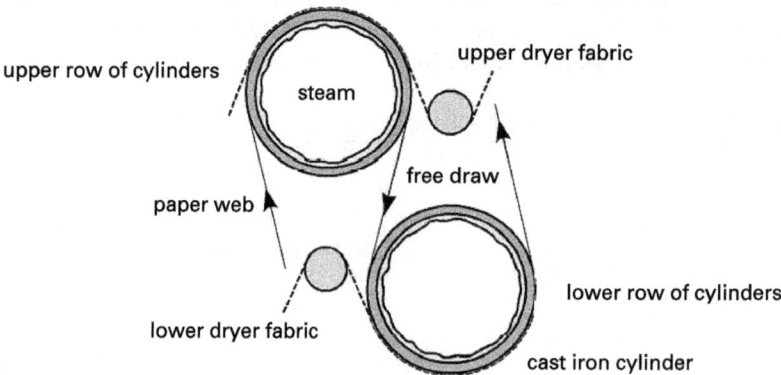

Figure 13.1. Main components in the multi-cylinder dryer.

The steam inside the cast iron cylinders condenses at the saturation temperature forming a condensate layer on the inside of the cylinder. The latent heat is transported by convection and conduction from the condensate surface through the condensate layer to the inside surface of the cylinder. The condensate layer with a thickness of a few mm represents an important resistance for heat transfer and efficient condensate removal is vital for an efficient operation of the dryer. This will be treated in more detail in Section 13.3.2.

The heat is transported by conduction through the the cylinder and transferred to the paper web. The energy is used for heating up and evaporation of water from the paper web. As the paper leaves one cylinder it enters what is called the free draw where water evaporates and cooling of the web is taking place. In the free draw heat is also transferred due to convective heat transfer between the web and the surrounding hot humid air but this part is normally much lower than the heat transferred from the cylinder and the process in the free draw is close to adiabatic.

13.3.2 Web Temperatures and Evaporation Rates

During drying on and between the cylinders different conditions for heat and mass transfer occur in different positions. The classical description by Nissan and Hansen for these phenomena in a double tier design has been to divide the processes in four different phases as shown in *Figure 13.2*. The corresponding web temperatures and evaporation rates are shown in *Figure 13.3* and *Figure 13.4*.

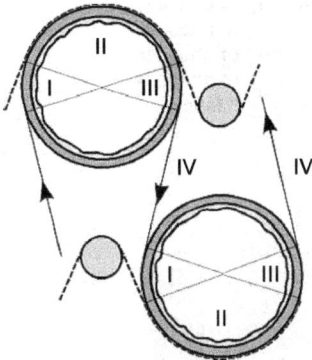

Figure 13.2. Classical four phases in the multi-cylinder dryer.

The pases can be described by:

- Phase I The web is brought into contact with the hot cylinder and heat transfer takes place to the lower side of the web rising the web temperature. Evaporation takes place from the upper side of the web at a low rate due to the low web temperature after the free draw.
- Phase II Heat transfer takes place to the lower side rising the web temperature. The upper side is covered by the dryer fabric resulting in an additional resistance for mass transfer. The increased paper temperature results in an increased drying rate.

- Phase III The fabric is no longer in contact with the web resulting in less resistance for mass transfer and an increased drying rate. The temperature of the contact side continues to increase while the temperature of the open side starts to decrease.
- Phase IV The web enters the free draw between the cylinders and evaporation takes place from both sides of the web resulting in high drying rates. The web is cooled by evaporation and receives some energy by convective heat transfer from the surrounding hot humid air but the total result is a decreased paper temperature.

The multi-cylinder dryer is a contact dryer and thus the conditions for wet bulb equilibrium are not applicable and consequently the web does not reach the wet bulb temperature and the corresponding constant drying rate. This is shown in *Figure 13.4*. The conditions for heat and mass transfer are different in each phase and thus the drying rate varies during the different phases. The drying rate is a result of *machine parameters* such as cylinder diameter and length of the free draw and *operating parameters* such as steam pressures in the different steam groups and moisture content in the hood. Some typical operating data for the cylinder dryer are given in *Table 13.2*.

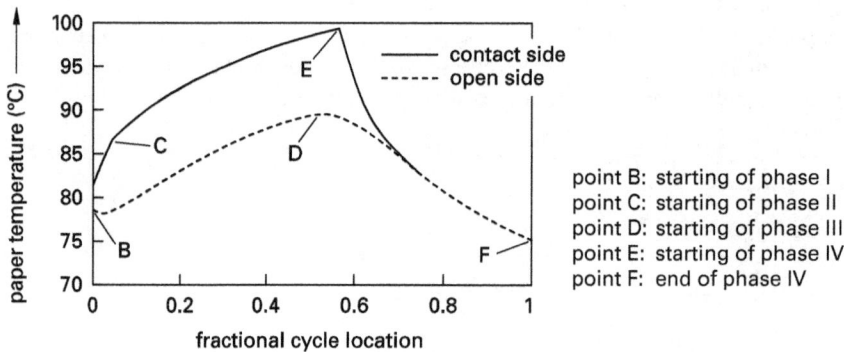

Figure 13.3. Local web temperatures during the four phases of cylinder drying.

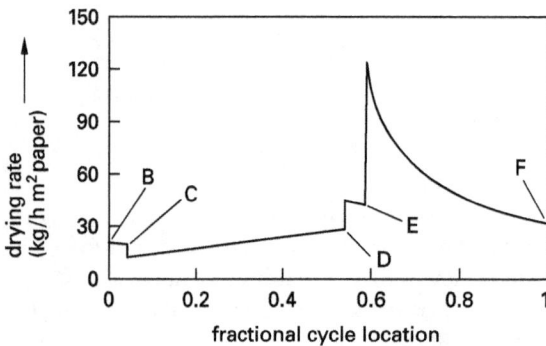

Figure 13.4. Local evaporation rates during the four phases of cylinder drying.

Table 13.2. Operating data for the multi-cylinder paper dryer.

Drying rate[1]	Cylinder surface temperature	Web temperature[2]	Mechanical con-tact pressure	Total resi-dence time	Specific energy use
kg/m²h	°C	°C	kPa	s	kJ/kg water
10–30	70–150	< 100	2–5	10–100	2800–5000

[1]Based on the cylinder surface area.
[2]At low moisture content the web surfaces could rise above 100 °C.

The drying rate can be calculated either based on the total paper surface area in the dryer or the total cylinder surface area. Using the cylinder surface area is the standard method when calculating the Tappi drying rate as will be shown in Section 13.5.1.

13.3.3 Condensate System

Overall principles. The necessary steam for heating the drying cylinders is supplied either directly from the power boilers or from the backpressure turbine. To control the energy transfer and thus the drying rates in different parts in the dryer, the cylinders are divided in a number of steam groups with different steam pressures. The number of groups range between 4 and 20, one system with 9 groups is shown *Figure 13.5.*

Most of the steam supplied condenses in the cylinders but 10–15 % of the steam is used for removal of the condensate and goes through the cylinder. This steam is called blow-through steam and is normally recirculated to a steam group with a lower pressure. The steam groups are divided in upper groups and lower groups supplying steam to a number of either upper or lower cylinders. In this way it is possible to control the evaporation from both sides of the web and control shrinkage and quality parameters such as curl. The steam leaving the final group in the system has to be condensed by cooling water and the total amount of condensate is returned to the power boilers for steam production.

The amount of energy transfer and thus the drying rate is controlled with the steam pressure; a high pressure results in a high saturation temperature, a high temperature difference for the heat transfer and a large heat transfer rate. A low pressure results in a low saturation temperature, a low temperature difference for the heat transfer and a low heat transfer rate. Normally the steam pressure is gradually increased from the wet to the dry end of the machine and range from subatmospheric (< 100 kPa) in the wet end to 500 kPa in the dry end of the dryer.

Two different steam supply systems exist:

- a cascade system where the blow-through steam is led to the next group see *Figure 13.5*
- a thermocompressor system where high pressure steam is used to compress and
- recirculate the blow-through steam to the same steam group.

The advantage with the cascade system is that it is a simple and reliable design which requires steam of lower pressure. It can be arranged and controlled in different ways. The advantage with the thermocompressor design is that it allows a wider control range and recirculation of steam reduces the amount which is led to the condensor.

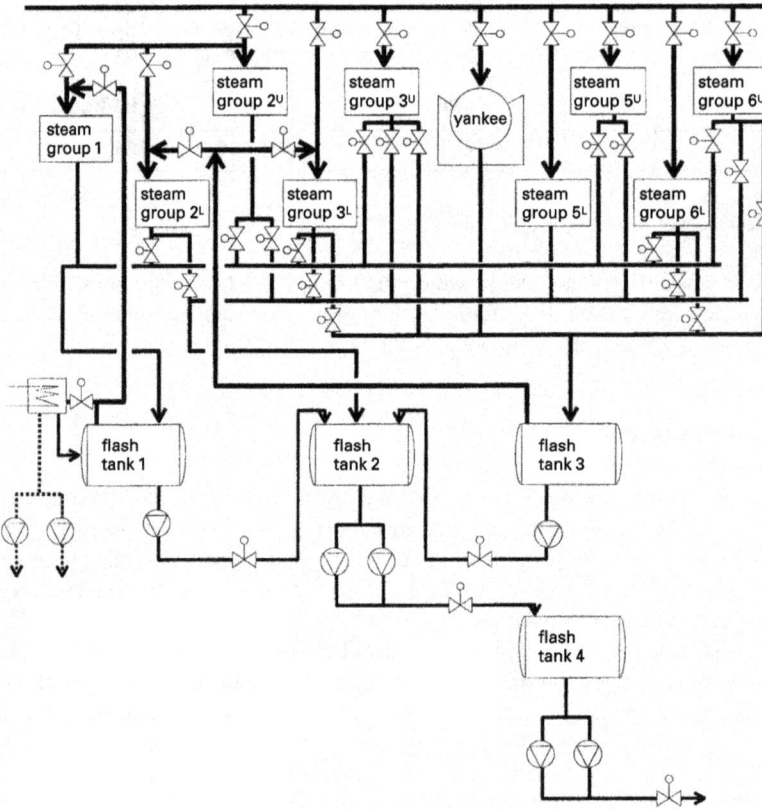

Figure 13.5. Layout of a cascade condensate system for a cardboard dryer.

Condensate behaviour and condensate heat transfer coefficients. The steam condenses on the inner cylinder surface forming a layer of condensate. The behaviour of the condensate layer varies with the cylinder speed, the cylinder diameter, the layer thickness and the design of the inner surface. The behaviour of the condensate is determined by a force balance for the condensate, that is gravitational, frictional and centrifugal forces. For a smooth inner surface the behaviour is normally classified in three different modes as shown in *Figure 13.6* below.

At low cylinder speeds the steam flows as a thin film on the inner cylinder surface to the bottom of the cylinder where it accumulates as a puddle. As the speed is increased the friction at the wall tends to drag the condensate along the inner cylinder wall but at the top position gravity is too strong and some of the condensate falls down into the puddle, This regime is called the cascading regime and the heat transfer coefficients remain high but high powers are needed for driving the cylinder. Increasing the speed further results in higher frictional forces, higher condensate velocities and higher centrifugal forces on the condensate which stabilize the condensate layer as a thin film along the inside surface of the cylinder. This behaviour is called the rimming case and a first estimate of the rimming speed can be obtained by setting the centrifugal acceleration equal to the gravitational acceleration. In practice the rimming speed is also influence by other variables such as layer thickness and physical data and tends to be higher than the value calculated with the simple balance between gravity and centrifugal force.

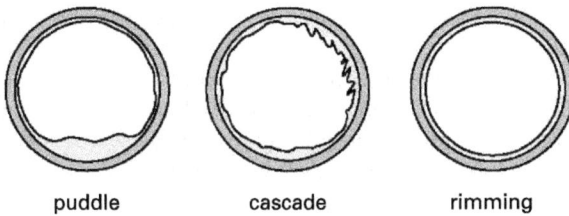

| puddle | cascade | rimming |

Figure 13.6. Condensate behaviour in drying cylinders.

The heat transfer coefficient for the condensate layer is determined by the thickness and the flow behaviour (laminar–turbulent) of this layer. At low speeds where the condensate flows along the inside wall to a puddle in the bottom the coefficients can be calculated from film condensation correlations. The process is characterised by high heat transfer coefficients through the condensate, typical values are in the range 2000–4000 W/m² °C.

Above the rimming speed the condensate is stabilised by the centrifugal force and as a first estimate can be considered as a laminar film of condensate. In this case the heat transfer coefficient can be calculated as heat conduction through a layer of water according to:

$$h_{cond} = \frac{k_{cond}}{l_{cond}} \qquad (13.4)$$

Typical condensate heat transfer coefficients for the rimming case are in the range 500–2000 W/m² °C

Example 13.2	Estimate the condensate heat transfer coefficient in a 1.5 m drying cylinder with a condensate thickness of 2 mm. The cylinder is running at 1200 m/min. The centrifugal acceleration on the condensate is calculated as:
	$$a_{cond} = \frac{v_{cyl}^2}{R_{cyl}} = \frac{20^2}{0.75} = 533 \text{ m/s}^2$$
	The centrifugal acceleration on the condensate can is 533 m/s² or 54 g. The speed is well above the rimming speed and the layer can be assumed to be laminar. The thermal conductivity for water (taken at 120 °C) is 0.68 W/m °C. The result for the condensate heat transfer coefficient based on equation (3.1) is
	$$h_{cond} = \frac{0.68}{0.002} = 340 \text{ W/m}^2{}^\circ\text{C}$$
	This is a rather low condensate heat transfer coefficient representing a major resistance for heat transfer which shows the importance of effective condensate removal especially at high cylinder speeds.

As shown in Example 13.2, the condensate heat transfer coefficients are low at high cylinder velocities even if the thickness of the condensate layer is maintained low. The most common method to increase the condensate heat transfer coefficient is to install so-called "spoiler bars" on the inside of the cylinders. With this method a number of rectangular steelbars (typical height 20 mm) are fastened on the inside of cylinder inducing turbulence in the condensate layer by forcing the condensate to flow over these bars.

Condensate Drainage. For proper operation of the dryer it is important that the condensate and the non-condensable gases are evacuated from the cylinders. By applying a differential pressure over the cylinder the excess steam (blow-through steam) entrains the condensate with a design called siphon. The siphon is in principle a tube extending close to the inner cylinder surface where high steam velocities break up and entrain the condensate and flows as a two-phase mixture in the tube out of the cylinder. Three different designs are used:

Scoop	If the condensate is in the puddle mode it can be emptied with a scoop which goes through the puddle during each rotation and forces the condensate to flow from the periphery to the centre of the cylinder
Stationary siphon	The siphon is stationary and the siphon head is positioned close to the inner surface for efficient condensate removal. The dynamic pressure of the condensate layer can be recovered and the pressure differential is low with this design. The mechanical design should be careful due to the low clearance between the cylinder and the siphon head. steam and condensate steam **Figure 13.7.** The deltasint stationary siphon (Deublin Italy).
Rotating siphon	The siphon is rotating with the cylinder and the blow-through steam is used to entrain the condensate into the siphon. The centrifugal forces on the steam and the condensate has to be overcome in this design resulting in high pressure differentials at high cylinder speeds.

The steam and the condensate are transported as a two-phase mixture in the siphon and the total pressure drop is given by the sum of the frictional, gravitational, accelerational and centrifugal components. The gravitational component is due to elevation changes of the steam and the condensate. It changes sign during the rotation and can normally be neglected. The accelerational component is due to the velocity increase arising from density changes as the pressure decreases. Except at very low pressures or very high pressure drops this component can also normally be neglected. The frictional component is due to friction at the siphon inlet and the two-phase flow in the siphon tube and the centrifugal component (only rotating siphons) due to centrifugal forces on the steam and the condensate given by the density of the two-phase mixture and the centrifugal acceleration. Normally these are the main components in the total pressure drop.

The pressure drop for a rotating siphon for different amounts of blow-through steam is shown in *Figure 13.8*.

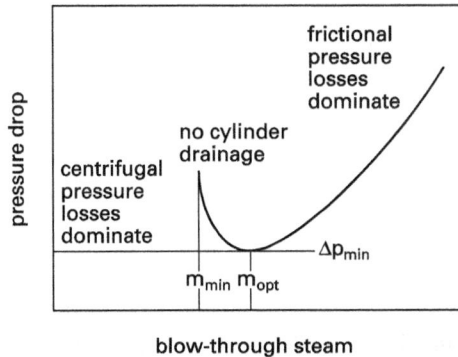

Figure 13.8. Pressure drop curve for a rotating siphon.

Increasing the amount of blow-through steam increases the velocities of the steam and the condensate in the siphon tube and thus also the frictional losses. The reduced density results in a low centrifugal pressure drop. Decreasing the amount of blow-through steam decreases the velocities resulting in low frictional losses but the increased density (due to more liquid in the flow) results in a rapidly increasing centrifugal pressure drop. Below a certain amount of blow-through steam the entrainment forces on the condensate are not large enough to overcome the centrifugal forces and cylinder drainage stops and the cylinder fills up with condensate. Thus the siphon design should be performed with great care in order to ensure cylinder drainage without excessive use of blow-through steam.

13.3.4 Dryer Fabric

The paper web is supported by a dryer fabric and the main functions of the fabric are the following:

- establish a good contact for the heat transfer between the cylinder and the web
- support the weak wet web throughout dryer section so that the number of web breaks are limited
- facilitate pocket ventilation
- control paper properties such as shrinkage and cockling
- facilitate water vapour transport through the fabric
- act as driving element for the cylinders

The dryer fabric is a weaved open structure made from synthetic fibres with appropriate surface, permeability and mechanical properties. Some examples of different designs are shown in *Figure 13.9.*

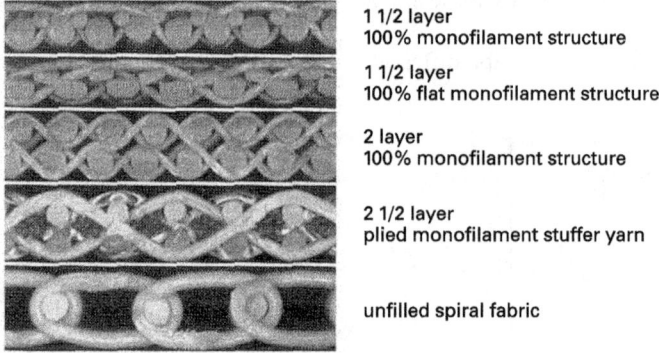

1 1/2 layer
100% monofilament structure

1 1/2 layer
100% flat monofilament structure

2 layer
100% monofilament structure

2 1/2 layer
plied monofilament stuffer yarn

unfilled spiral fabric

Figure 13.9. Different dryer fabric structures (by courtesy from Albany Finland).

Traditionally one fabric was used for the upper row of cylinders and one for the lower row of cylinders, In this design the web travels without support in the free draw and as the speed was increased for newsprint machines the number of web breaks became too large and machines with a single tier were built. Today modern newsprint and printing paper machines utilizes non-heated vacuum rolls. The principles for the different designs are shown in *Figure 13.10, Figure 13.11* and *Figure 13.12.*

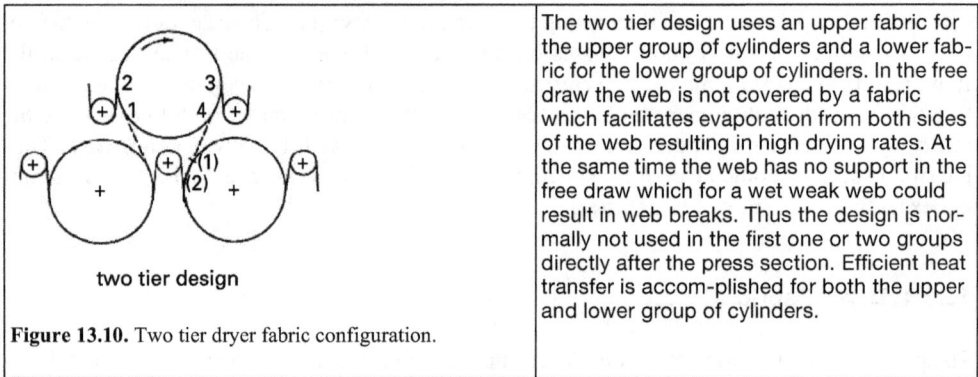

two tier design

Figure 13.10. Two tier dryer fabric configuration.

The two tier design uses an upper fabric for the upper group of cylinders and a lower fabric for the lower group of cylinders. In the free draw the web is not covered by a fabric which facilitates evaporation from both sides of the web resulting in high drying rates. At the same time the web has no support in the free draw which for a wet weak web could result in web breaks. Thus the design is normally not used in the first one or two groups directly after the press section. Efficient heat transfer is accom-plished for both the upper and lower group of cylinders.

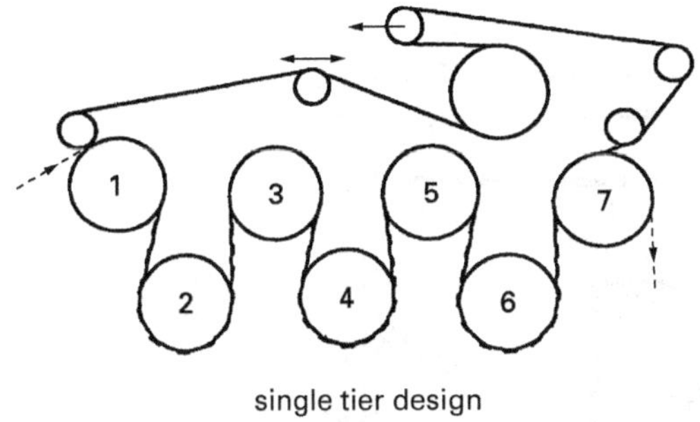

single tier design

The single tier design uses the same fabric for the upper and the lower group of cylinders, thus the fabric follows and supports the web in the free draw. In the free draw most of the evaporation occurs from one side of the web reducing the drying rate. In the lower group the fabric lies between the hot cylinder and the web drastically reducing the heat transfer from these cylinders. This design was originally developed for newsprint machines to avoid web breaks at higher machine speeds and is today used in the first one or two groups after the press section.

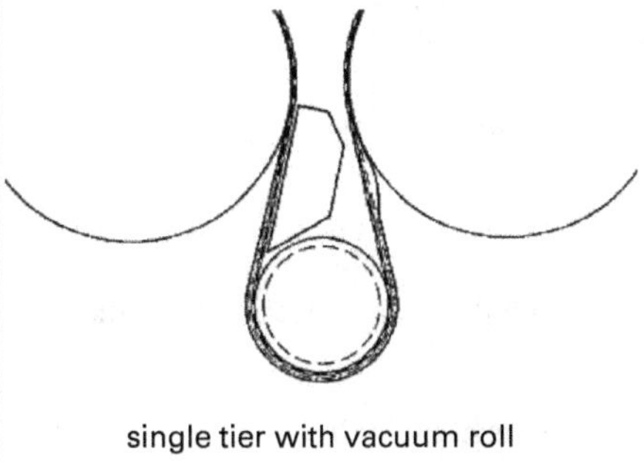

Figure 13.11. Single tier dryer fabric configuration.

single tier with vacuum roll

In the vacuum roll design the lower group of cylinders has been replaced with a vacuum roll with the pupose to transport the paper to the next upper cylinder. The vacuum roll is not heated and the vacumm used does not result in any significant amounts of air flow through the web. Heat is supplied to only one side and most of the evaporation occurs from this side of the web. Several newsprint and fineprint dryers are today designed in this way and the onesidedness of the paper corrected for in a calender after the dryer section.

Figure 13.12. Single tier with vacuum roll dryer fabric figuration.

13.3.5 Dryer Hood

The dryer hood is designed to remove the evaporated water, establish constant drying conditions across the width of the machine and recover as much of the energy in the humid air as possible. The principle is shown in *Figure 13.13*.

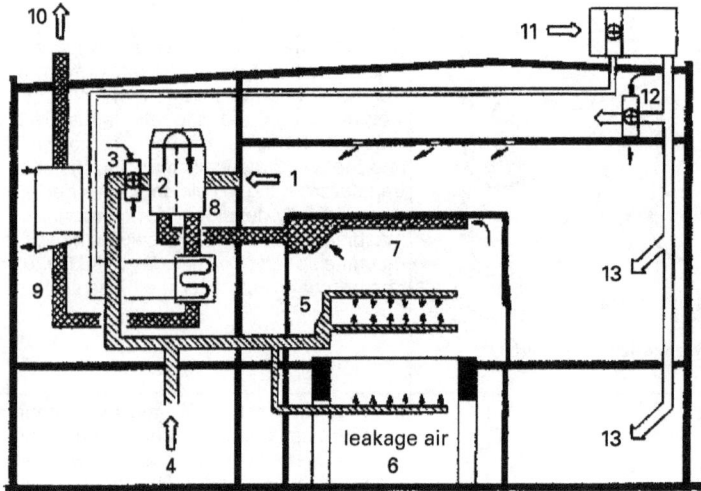

Figure 13.13. The dryer hood and the air distribution system.

The process will be described by following the air through the system according to the numbers in *Figure 13.13*.

1. Inlet air to the hood is taken from below the roof of the machine hall.
2. The air is preheated with the outgoing humid air from the hood.
3. Further preheating is performed in a steam/air heat exchanger.
4. Air from the vacuum pumps is mixed with the inlet air.
5. The hot dry air is supplied as pocket ventilation between and below the cylinders to remove the evaporated water and ensure a uniform drying.
6. Leakage air entering the hood in the cellar.
7. The humid air with a lower density than dry air accumulates at the top of the hood.
8. The energy content in the humid air is recovered, initially by preheating the ingoing air. The outgoing air from the first heat exchanger is further cooled in a water/air heat exchanger and the water used for heating of the air supplied to the machine hall.
9. The outgoing air from the second heat exchanger enters into a scrubber where the humid air is brought into direct counter-contact with cool water. The air is cooled and part of the water vapour is condensed heating up the water.
10. The cooled saturated air leaves the building through the chimney.
11. Outdoor air used for ventilation of the machine hall.
12. Preheating and ventilation of the machine hall.

A mass balance for the water vapour determines the parameters in the hood according to:

$$G_{in}X_{in} + G_{leak}X_{leak} + m_{evap} = (G_{in} + G_{leak})X_{out} \qquad (13.5)$$

where G is the air flow-rate given as kg dry air/s, X is the water content given as kg water/kg dry air and m_{evap} the amount of evaporated water in the hood given as kg water/s.

Example 13.3	Estimate the amount of dry air needed in the hood for the dryer described in Example 13.1.
	Assume that the amount of leakage air is 25 % of the inlet air, the moisture content of the inlet air is 0.02 kg water/kg dry air, the humidity of the outlet air is 0.13 kg water/kg dry air and the leakage air moisture content is 0.015 kg water/kg dry air. According to Example 13.1 and Equation (3) the inlet air flow-rate can be calculated as:
	$$G_{in}0.02 + 0.25G_{in}0.015 + 9.71 = (G_{in} + 0.25G_{in})0.13$$ $$G_{in} = 70.0 \text{ kg/s} = 252 \text{ t/h} = 300\,000 \text{ m}^3/\text{h}$$
	Thus very high air flow-rates are needed and correct design of fans and air ducts are of utmost importance for high energy efficiency and good control of the evaporation in different parts of the dryer.

The process changes in Example 13.3 can be described in a Mollier diagram as is shown in *Figure 13.14* (the air from the vacuum pumps is not included).

In the heat exchangers the inlet air is heated with a constant moisture content up to a temperature of 120 °C. Before the air is blown into the dryer it is mixed with the leakage air reducing the moisture content and the temperature. In the dryer the moisture content increases and the temperature is slightly decreased. The outgoing air is cooled with a constant moisture content in the first heat exchangers and in the scrubber it is cooled down along the saturation curve causing condensation of water vapour and recovery of the latent heat of evaporation.

The humidity level in the hood affects both the drying rate and the possibilities for energy recovery. Low humidities increases the drying rate (see Equation (12) below) but also results in less efficient energy recovery while higher humidities decreases the drying rate and leads to more efficient energy recovery. The operating point must thus find the optimum between these two parameters.

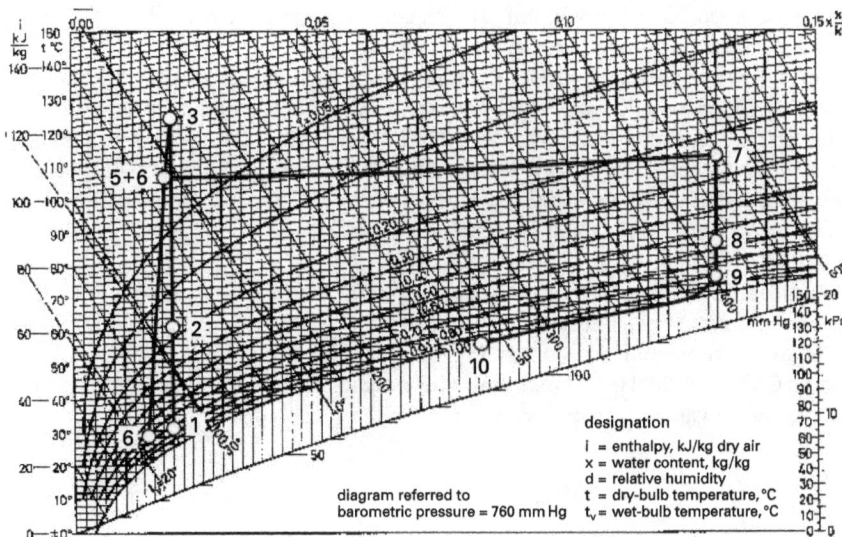

Figure 13.14. Mollier chart for the air in the dryer hood.

Example 13.4	Estimate the amount of energy that can be recovered based on the data used in Example 13.3. The energy recovered is the air flow-rate multiplied with the difference in the enthalpy of the air from the hood and the scrubber according to: $$Q = G_{\text{out}}(H_7 - H_{10})$$ $$Q = 87.5(438 - 263) = 15\,300 \text{ kW}$$ By condensing the water vapour large amounts of energy can be recovered. However as shown in Figure 13.14 this requires cooling the air down to a temperature of 48 °C which limits the use of this energy

13.4 Mass Transfer Inside the Web

13.4.1 Web Structure and some Definitions

Wet paper is a heterogeneous material consisting of a three-phases; liquid water, a gas mixture of air and water vapour and a solid phase built of a porous cellulose fibres. The proportions of these phases are not constant but vary as drying proceeds, free water in the inter-fibre pores (pores *between* the fibres) evaporates, the intra-fibre pores (pores *inside* the fibres) closes and the porous structure shrinks and changes shape as the paper dries. Some fundamental understanding of the porous structure and where the water is in this structure is necessary in order to understand the drying process and the development of quality parameters.

The total web thickness typically ranges from 75 µm to 600 µm built of fibres with a thickness in the range between 15 µm and 30 µm. The size of the inter-fibre pores is of the magnitude 1 to 10 µm while the intra-fibre pores are much smaller.

The definitions of water content appropriate for drying of paper are given in *Table 13.3*.

The ability of the fibres to hold water can be measured with centrifugation (Water Retention Value) or solute exclusion techniques (Fibre Saturation Point). The most applied on the industrial level is the WRV-value. Normally it is assumed that this results in the removal of the water in the inter-fibre pores and that the remaining water is held by the fibres in the intra-fibre pores. The values vary depending on pulp and treatment but are typically in the range 1–2 g/g dry matter.

The critical moisture content (CMC) is the moisture ratio where the evaporation rate drops due to some internal mass transfer resistance. Normally the CMC occurs at a higher moisture content than the hygroscopic moisture content (HMC) but in some cases the reverse could also occur. Below the CMC and the HMC the drying rate is reduced both due to moisture diffusion inside the material and a vapour pressure reduction from the sorption isotherm.

Table 13.3. Water content values applicable for drying of paper.

Water retention value	WRV	The amount of water held by the pulp as measured after centrifugation at 3000 g during 15 minutes
Fibre saturation point	FSP	The amount of water held by the pulp as measured by solute exclusion methods
Critical moisture content	CMC	The critical moisture content is the moisture content when the surface of the material no longer behaves as a wet surface. Thus evaporation will take place either on a limited part of the surface area or inside the material and consequently the drying rate will decrease. The CMC is a function of both the basis weight and the evaporation rate.
Hygroscopic moisture content	HMC	The hygroscopic moisture content is the moisture content where the vapour pressure of the water in the material is lower than the vapour pressure of pure water. The material then enters the hygroscopic region and the sorption isotherm of the material gives the relation between the relative humidity and the HMC.

13.4.2 Distribution of Water and Water Transport Mechanisms

Comparing the WRV-values for the pulp mixture used and the water content after the press section gives a good indication of how the evaporation of water from the web will take place in the dryer section. At moisture ratios of 3–4 kg water/kg dry matter both the fibres and most of the inter-fibre pores are saturated with water. Drying takes place both from the free water between the fibres and from the water held by the fibres. Below the WRV-value all of the water in the inter-fibre pores is removed and below this point the remaining water in the fibres is removed, the principle behaviour is shown in *Figure 13.15*.

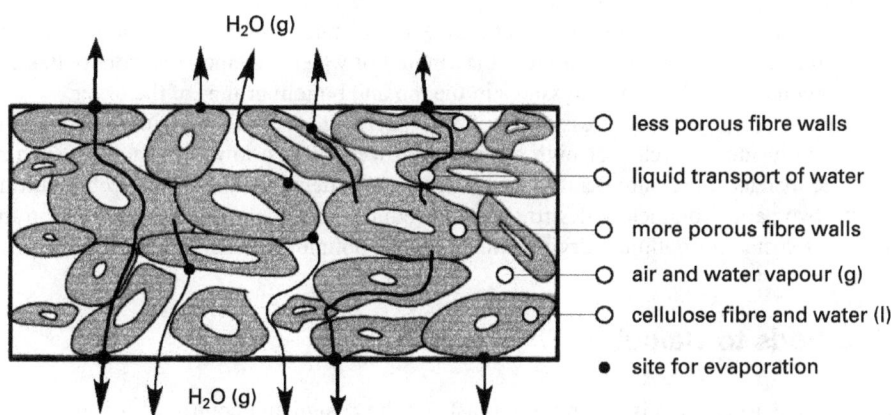

H$_2$O (g)

○ less porous fibre walls

○ liquid transport of water

○ more porous fibre walls

○ air and water vapour (g)

○ cellulose fibre and water (l)

● site for evaporation

H$_2$O (g)

Figure 13.15. Distribution of water and water transport *below* the WRV.

Below the WRV the amount of free water in the inter-fibre pores for most pulps is limited and the water is transported either as liquid due to capillary forces in the fibres and between the contact points of the fibres or as vapour due to diffusion in the inter-fibre pores. In the drying section the evaporation will thus take place from the surfaces of the fibres resulting in a moisture gradient in the web. As the water is removed the intra-fibre pores close a process which increases the resistance to mass transfer inside the fibres. Finally the sheet enters the hygroscopic region and the drying process is also affected by the sorption isotherm for the fibres used. When the sheet is dry the fibre wall is essentially non-porous.

Molecular transport (diffusion) of water vapour is governed by Fick's law:

$$J_A = -D_{eff} \frac{dC_A}{dz} \qquad (13.6)$$

where J_A is the molar flux of water vapour per unit area (mol/m^{2}s), D_{eff} is the effective diffusivity for the porous web (m^2/s) CA is the concentration of water vapour (mol/m^3). The effective diffusivity for a dry sheet can be measured by measuring the weight loss from a water reservoir under controlled conditions. Given a concentration difference over the paper web the transport of water vapour through the paper can then be calculated using Equation (6).

13.4.3 Web Shrinkage and Paper Quality

As water is removed during the consolidation process the fibres collapse and the fibre network changes shape and shrinks in the machine, in the cross and in the thickness directions. Cross direction shrinkage on the cylinder will be resisted by the felt but in the free draw between the cylinders almost free shrinkage can occur. This CD-shrinkage is typically a few percent but uneven over the width of the machine causing large variations in mechanical properties such as strength and stiffness. The paper shrinkage in the thickness direction is closely related to the distribution of moisture affecting the tendency of the paper to curl, that is how the paper is changing its shape during exposure to temperature and moisture. The most common method to avoid curl in the dryer is to control that the same amount of water is evaporated from both sides of the web by controlling the steam pressures in the top and bottom groups of the dryer.

Application of too high steam pressures in the dryer rises the web temperature and thus the vapour pressure inside the web. For high basis weight webs with a low permeability, such as cardboard, the increased pressure inside the web may cause delamination of the web (splitting the web into two parts) completely destroying the product. Thus the process parameters in the dryer must be selected for optimum drying rate as well as optimum quality parameters.

13.5 Methods to Calculate Drying Rate

Calculation of the drying rate is an important task for the design and control of the dryer. The process can be modelled quite well by setting up *the geometry* for the process combined with a systematic analysis of *the physical phenomena* occurring in the dryer. Depending on the accuracy and the time available this can be performed with different methods, from simple and rapid methods with a rough accuracy to more advanced and time consuming computer based methods but also much better accuracy. In this paragraph three levels of calculation methods for the drying rate will be presented.

13.5.1 Tappi Drying Rate

Tappi publishes charts for the drying rate as a function of the steam temperature inside the cylinders for different paper grades. One example for newsprint with machine speeds above 915 m/min is shown in *Figure 13.16*. It should be noticed that the drying rate is based on the total cylinder surface, not on the paper surface where evaporation takes place. By summing up the contributions from all cylinders in the steam groups in the dryer the total drying rate can be quite easily calculated. The limitation with these diagrams is that they are based on compilations from a number of machines with different designs and the scatter in the data is substantial. The best use of these diagrams is to calculate the capacity increase from a given base case for a given increase in steam pressure in one or several of the steam groups.

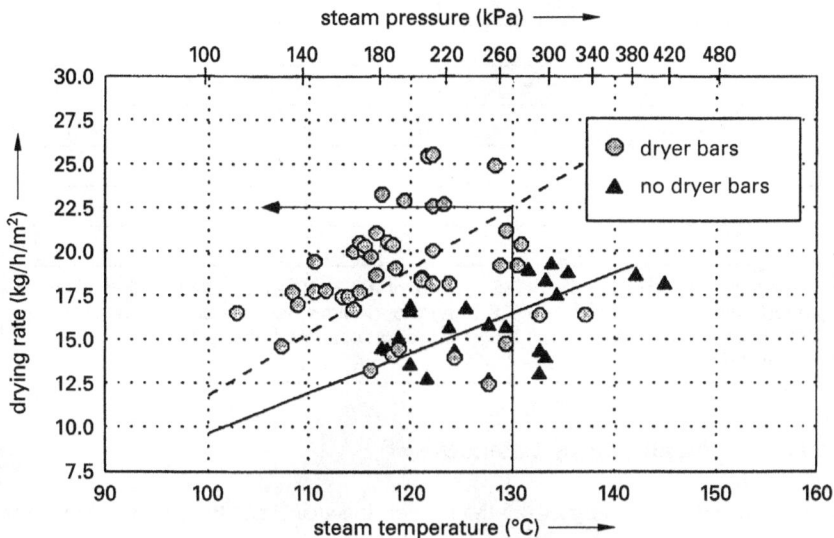

Figure 13.16. Tappi chart for paper machine drying rate for newsprint (machine speed > 915 m/min).

Example 13.5	The dryer on a newsprint machine consists of 51 cylinders with a diameter of 1.5 m. The cylinders are equipped with spoiler bars and divided in four groups as given in the table below. Estimate the capacity increase in the dryer if 4 more cylinders are installed in group 4. The drying rate per cylinder surface in each group can be read from Figure 13.16. The results are:

Group	Number of cylinders	Steam pressure, kPa
1	7	110
2	12	140
3	22	185
4	10	220

Group	Tappi drying rate	Evaporation per meter width	
		before	after
	kg/m^{2}h	kg/mh	kg/mh
1	12.6	415	415
2	15.3	865	865
3	18.5	1917	1917
4	20.3	956	1339
Total		4153	4536

The capacity increase is 4536/4153 = 1.092 or 9.2 %. The increase in cylinder area is proportional to 55/51 = 1.078 or 7.8 % and for dryers where the drying rate does not differ much between the groups the area proportional method can be used for a quick estimate of the capacity increase.

13.5.2 Method Based on an Energy Balance

The balance between the supplied energy from the condensing steam and the energy needed for the evaporation of water determines the operating point for the drying process. Assuming a constant cylinder temperature around the cylinder and in the web, the equations describing this process can be quite easily solved by a graphical procedure. Typically two-thirds of the cylinder surface is in contact with the web, thus the areas for the condensate layer and the paper contact are not the same. However the energy supplied to the web on the cylinder is partly used for evaporation in the draw (from one or both sides of the web), thus as a first estimate this difference in areas can be neglected.

The principle for the energy transfer between the steam and the paper is shown in *Figure 13.17.*

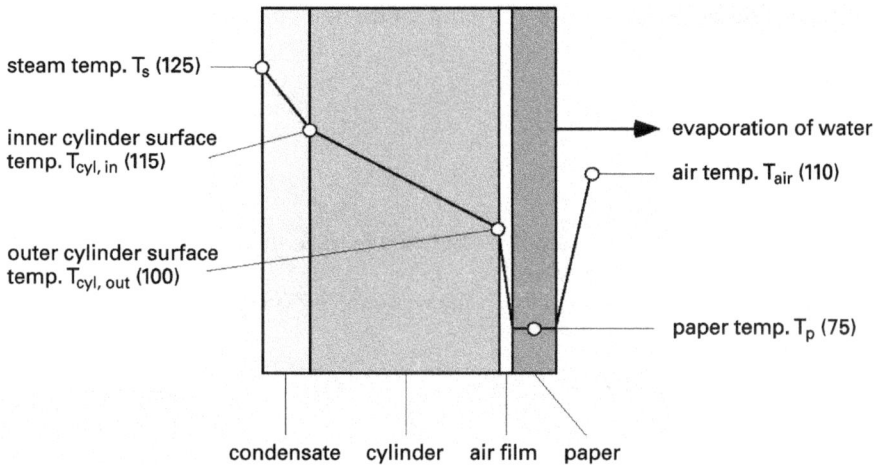

Figure 13.17. Temperature profile from steam to paper (typical figures in brackets).

The heat flux from the steam to the paper is transferred through the condensate layer, the cylinder and the air layer to the paper. The heat flux trough the condensate layer is calculated from a condensate heat transfer coefficient:

$$q_s = h_{cond}(T_s - T_{cyl,in})$$ (13.7)

The magnitude for the condensate heat transfer coefficient ranges between 500 and 4000 W/m^2 °C, see Section 13.3.3. Heat transfer through the cylinder is given by Fourier's law of heat conduction:

$$q_s = \frac{k_{cyl}}{l_{cyl}}(T_{cyl,in} - T_{cyl,out})$$ (13.8)

Typical data for a dryer cylinder are 50 W/m °C for the thermal conductivity and 25 mm for the thickness of the cylinder.

The heat transfer from the cylinder to the paper is calculated with a contact heat transfer coefficient which accounts for the resistance (the air film and scale on the cylinder) between the cylinder and the paper:

$$q_s = h_{cont}(T_{cyl,out} - T_p)$$ (13.9)

Normally this coefficient is considered to be a function of the web moisture content. Typical values are 1000–1500 W/m^2 °C at high moisture contents and 200–500 W/m^2 °C at low moisture contents. Assuming a constant heat flux through the cylinder the cylinder temperature can be eliminated from Equations (13.7–13.9). The result for the heat flux is:

$$q_s = \frac{(T_s - T_p)}{\left(\dfrac{1}{h_{cond}} + \dfrac{l_{cyl}}{k_{cyl}} + \dfrac{1}{h_{cont}}\right)}$$ (13.10)

For a given steam pressure and thus steam temperature the heat flux in Equation (13.10) is a linear function of only the paper temperature.

Convective heat transfer between the air and the web supplies energy to the web according to:

$$q_{air} = h_{air}(T_{air} - T_p)$$ (13.11)

The convective heat transfer coefficient between the surface and the air can be calculated from standard Nusselt-Reynolds-Prandtl correlations and is typically in the range 20–70 W/m^2 °C and normally this amount of energy is much lower than the energy transferred from the steam. In the following calculations it will be neglected.

Example 13.6	Calculate the heat flux in a dryer operating with a steam saturation temperature of 125 °C, a condensate heat transfer coefficient of 2000 W/m^2 °C, a cylinder thickness of 25 mm, a contact $$q_s = \frac{125 - 75)}{\left(\dfrac{1}{2000} + \dfrac{0.025}{50} + \dfrac{1}{1000}\right)} = 25000 \text{ W/m}^2$$ coefficient of 1000 W/m^2 °C and a paper temperature of 75 °C. According to Equation (10) the heat flux is given by:

Assuming no moisture gradients in the sheet and moisture ratios above the HMC (no vapour pressure decrease from the sorption isotherm) the rate of mass transfer from a wet paper surface can be calculated with Stefan's equation according to:

$$m_{evap} = \frac{k_g M_w P}{R(T_p + 273)} ln\left[\frac{P - p_{air}}{P - p_p}\right]$$ (13.12)

The mass transfer coefficient k_g given as m/s is normally calculated from the heat transfer coefficient based on the analogy between heat and mass transfer according to:

$$k_g = \frac{h_{air}}{\rho c_p}$$ (13.13)

Example 13.7	Estimate the rate of evaporation from a wet paper surface with a temperature of 75 °C. Assume that the heat transfer coefficient between the paper surface and the air is 25 W/m2 °C, the surrounding air moisture content is 0.1 kg water/kg dry air and the air temperature is 110 °C.

The mass transfer coefficient is calculated from Equation (13):

$$k_g = \frac{25}{0.91 \cdot 1100} = 0.025 \text{ m/s}$$

The water vapour pressure in the air is calculated from Equation (14):

$$p_{air} = \frac{0.1}{\frac{18}{29} + 0.1} \cdot 100000 = 13900 \text{ Pa}$$

The water vapour pressure at the paper surface is calculated from Equation (15):

$$p_p = 133.322 e^{\left(18.3036 - \frac{3816.44}{75+227.02}\right)} = 38600 \text{ Pa}$$

Finally the rate of evaporation can be calculated from Equation (12):

$$m_{evap} = \frac{0.025 \cdot 0.018 \cdot 10^5}{8.314(75+273)} ln\left[\frac{10^5 - 13900}{10^5 - 38600}\right] = 0.0053 \text{ kg/m}^2\text{s} = 18.9 \text{ kgm}^2\text{h}$$

The density and the specific heat should be based on data for the humid air. The total pressure P is normally atmospheric and the partial pressure of water vapour in the air pair can be calculated from the air moisture ratio:

$$p_{air} = \frac{X}{\frac{18}{29} + X} P \tag{13.14}$$

Above the HMC the vapour pressure at the paper surface is given by the vapour pressure for pure water. It is only a function of the paper temperature and can be calculated with the following equation:

$$p_p = 133.322 e^{\left(18.3036 - \frac{3816.44}{T_p + 227.02}\right)} \tag{13.15}$$

Thus the amount of evaporation from Equation (13.12) is also only a function of the paper temperature.

The energy required for the evaporation of water from the web is the mass flux calculated with Equation (13.12) multiplied with the latent heat of evaporation. The result is:

$$q_{evap} = \frac{k_g M_w P \Delta H}{R(T_p + 273)} \ln\left[\frac{P - p_{air}}{P - p_p}\right] \tag{13.16}$$

Thus the energy transferred from the steam as given by Equation (13.10) must equal the energy required for evaporation as given by Equation (13.16). Both these equations can be plotted in a graph as function of the paper temperature and the intersection of the curves represents the operating point (given as paper temperature or heat flux) for the studied section in the dryer. The following example explains the technique in more detail.

In spite of the simplified assumptions of a constant paper temperature and lumping the cylinder and the free draw sections into one process, the described procedure gives more accurate estimates of the drying rates than the method based on Tappi drying rates. The method provides possibilities to investigate not only the influence of the steam pressure but also other variables such as the heat and mass transfer coefficients and the moisture content in the hood.

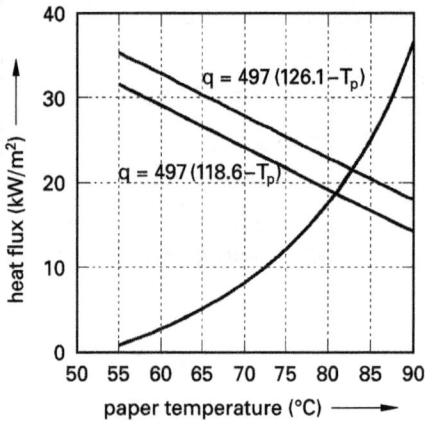

Figure 13.18. Heat flux for different paper temperatures. The operating point is given by the intersection of the lines and occurs at a paper temperature of 81.0 °C corresponding to a heat flux of 18.6 kW/m². This represents a drying rate of 0.0080 kg/m²s or 28.9 kg/m²h.

Example 13.8a	Estimate the drying rate for a steam group in a cylinder dryer operating with the following parameters:

Steam pressure   190 kPa
Condensate heat transfer coefficient   2500 W/m² °C
Cylinder thickness   25 mm
Contact coefficient   900 W/m² °C
Air moisture content   0.1 kg water/kg dry air
Air heat transfer coefficient   25 W/m² °C

At 190 kPa the steam saturation temperature is 118.6 °C and inserting the given values in Equation (10) gives the following equation for the heat flux from the steam to the paper:

$$q_s = \frac{118.6 - T_p}{\left(\dfrac{1}{2000} + \dfrac{0.025}{50} + \dfrac{1}{900}\right)} = 497(118.6 - T_p)$$

This equation can be plotted in a diagram as a function of the paper temperature. Based on the given data and the results in Example 13.7, Equation (16) takes the following form:

$$q_{evap} = \frac{0.025 \cdot 0.018 \cdot 10^5 \cdot 2320 \cdot 10^3}{8.314(T_p + 273)} ln\left[\frac{10^5 - 13900}{10^5 - P_p}\right] =$$

$$= \frac{1.26 \cdot 10^7}{T_p + 273} ln\left[\frac{86100}{10^5 - P_p}\right]$$

The latent heat of evaporation is only a weak function of the temperature but was taken as constant at 75 °C in these calculations. The result can be plotted in the same diagram as the previous equation, the result is shown in *Figure 13.18* below.

Example 13.8b	Estimate the drying rate for the following change in the dryer: The steam pressure is increased to 240 kPa. At 240 kPa the steam saturation temperature is 126.1 °C which gives the following equation for the heat flux:

$$q_s = 497(126.1 - T_p).$$

This Equation is plotted in Figure 13.18. The intersection between the equations now occurs at 82.8 °C corresponding to a heat flux of 21.6 kW/m². This represents a drying rate of 0.0093 kg/m² s or 33.5 kg/m² h an increase of 16 %.

Example 13.8c	Estimate the drying rate for the following change in the dryer: The air moisture content in the hood is increased to 0.2 kg water/kg dry air In this case the water vapour pressure in the air is calculated as 24400 Pa. Plotting the new curve for the energy for evaporation results in the new graph shown in *Figure 13.19*.

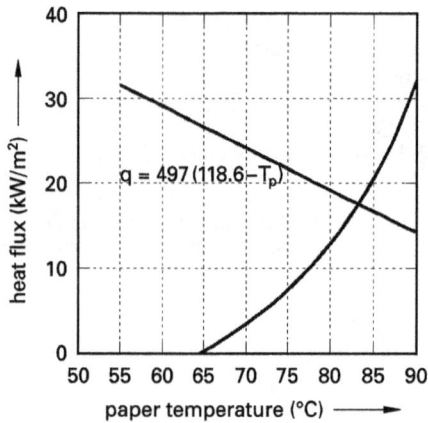

Figure 13.19. Heat flux for different paper temperatures.
Notice that no evaporation occurs at paper temperatures below 65 °C, the dewpoint of the air. The intersection between the equations now occurs at 83.3 °C corresponding to a heat flux of 17.5 kW/m². This represents a drying rate of 0.0075 kg/m²s or 27.2 kg/m²h a decrease of 6 %. Notice that the paper temperature increases with an increase in the hood moisture content.

13.5.3 Computer Simulation Tools

In order to obtain accurate results for the local web temperature, the drying rate, and moisture gradients in the web, the exact geometry of the dryer must be specified and the equations for the unsteady heat and mass transfer processes in the web solved with a step by step procedure from the press section to the pope. This is well suited for computer based simulation programs. Often the programs start with the Nissan description of dividing the process in four different phases. However with the development of new dryer designs, such as single felted dryers and vacuum rolls, this description is not sufficient to describe and model the modern paper dryers. One example of different computational phases is shown in *Figure 13.20* below where 17 different phases were used to model the paper dryer. Still today many other combinations exist which can not be modelled using that description.

The heat and mass transfer phenomena inside the web could be treated in a number of ways depending on the goals with the program. If only the local drying rate should be calculated it could be sufficient to calculate temperature gradients but neglect moisture gradients. If on the other hand web shrinkage is to be modelled moisture gradients must be included and depending on basis weight and drying rate calculation of the temperature gradients could also be necessary.

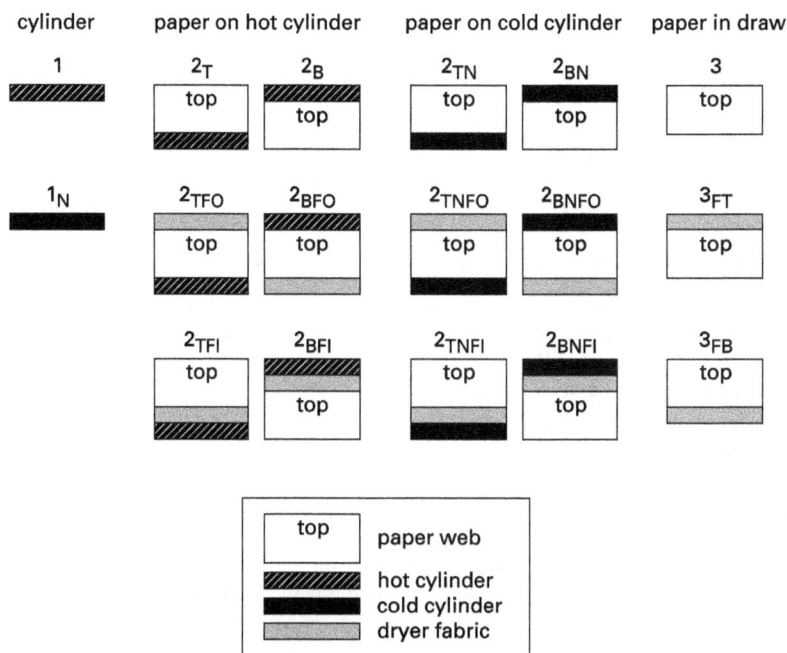

Figure 13.20. Different computational phases used in simulation programs.

13.6 Energy Use

The dryer section is the largest single consumer of steam on an integrated paper mill. The average thermal energy use in the dryer section of the paper machine at an ingoing moisture content of 40 % is 6000 MJ/ton paper. This corresponds to 4000 kJ/kg water evaporated or 1.7 times the latent heat of evaporation. Based on a total paper production of 11 million tons (year 2000) the total thermal energy use for drying of paper in Sweden would be 18 TWh. Further the electrical energy use in the paper machine (not only the dryer) can be estimated at 8 TWh. The total figures are very high and efficient energy use in the dryer is of high importance for low production costs.

Condensing steam supplies the largest amount of energy but electricity, natural gas or propane are also used. In Sweden the supply of natural gas pipeline extends from Malmö to Gothenburg and is thus limited to the mills at Hyltebruk and Mölndal.

The minimum specific energy use (energy use per kg water evaporated) includes the latent heat of evaporation, heating up the web from the press section to the reel, heating up the leakage air to the hood and some heat losses. An estimate for a modern machine is given in *Table 13.4* below.

Table 13.4. Specific energy use for a modern multi-cylinder paper dryer.

Energy use	Specific energy kJ/kg water evaporated
Latent heat of evaporation including heat of sorption	2410
Heating the web and the remaining water	50
Heating up leakage air	125
Heat losses	150
Total	2735

Some of the paper produced needs to be repulped and redried which explains part of the difference between 4000 for the existing and 2735 kJ/kg for the modern machine but also indicates further possibilities for energy savings in the dryer.

Knowing the specific energy use for the process the steam consumption in the dryer can be estimated quite easily, one example is shown below.

Example 13.9	Estimate the amount of steam needed in the dryer for the paper machine in Example 13.1. Assume that low pressure steam is available at 400 kPa and that the specific energy use for the dryer is 3400 kJ/kg water removed. According to Example 13.1, the amount of water removed is 9.71 kg/s. Thus the thermal energy to the dryer and the steam consumption can be calculated as: $$\text{Thermal energy required} = 9.71 \cdot 3400 = 33000 \text{ kW}$$ $$\text{Thermal energy required} = \frac{33000}{2134} = 15.5 \text{ kg/s} = 55.7 \text{ t/h}$$

The most efficient way to reduce the amounts of thermal energy in the dryer is to reduce the amount of water that is to be evaporated in the dryer by increasing the dryness of the web from the press section. A good role of thumb is shown in the example below. Some techniques which can be applied are using shoe-press technology, using heated centre rolls and preheating the web with steam boxes before the press section.

Most of the energy supplied with the steam will appear in the hot humid air leaving the hood, thus efficient energy recovery is essential for an energy efficient paper mill. As shown in examples 3-3 and 6-1 the energy that could be recovered was 15 300 kW as compared with the supplied amount 33 000 kW. Thus the percentage of the energy that could be recovered was 46 %. On modern machines this percentage could increase to 50–55 % during the cold period of the year.

Example 13.10	Estimate the reduction in steam consumption in the dryer if the dry matter content after the press section is increased from 40 to 41 %. Use the data in Example 13.1. According to Example 13.1, the amount of water removed at 40 % dryness is 9.71 kg/s. Increasing the dryness to 41 % results in the following change:
	$$\text{Water removed} = 8.2 \cdot \frac{1100}{60} \cdot 0.045 \cdot \left(\frac{100-41}{41} - \frac{100-94}{94} \right) = 9.30 \text{ kg water/s}$$
	The amount of water to be removed is decreased to 9.30 kg/s or by 4 %. Either the steam consumption can be reduced with this value or the machine speed increased with the same amount. Rule of thumb! 1 % increase in dry matter content after the press section at 40 % will require 4 % less steam to the dryer. Alternatively the capacity could be increased by the same amount for the same steam flow to the dryer.

13.7 Drying of Tissue, Coated Papers and Pulp

13.7.1 Drying of Tissue

General. Tissue products (kitchenpaper, towells, toilet paper, napkins and tablecloths etc.) have a basis weight between 15–50 g/m^2 and a porous elastic structure with a good capacity to absorb water. Two different drying techniques are used, the Yankee-cylinder and the through air drying technique (TAD). The Yankee-cylinder has been the traditional method but during the last 10 years a number of TAD-dryers have been built. So far there are 8 TAD-dryers in operation in Europe but none in Sweden.

The Yankee-dryer. On the Yankee-dryer the paper is dried on one large cast iron cylinder with a diameter of 5–7 m. It is heated with steam on the inside and hot air blown at high velocities on the outside of the web. The principle is shown in *Figure 13.21* below.

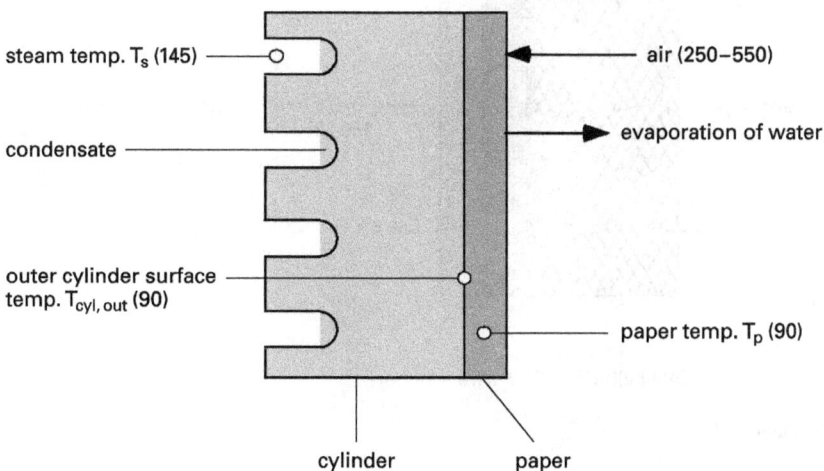

Figure 13.21. Principle for a Yankee-dryer.

After forming the tissue web in one or more layers it is transferred to a wire which presses the web against the Yankee-cylinder so that it sticks to the hot surface resulting in good contact between the cylinder and the web. Normally a dry matter content of 35–45 % is achieved after pressing the paper against the cylinder. The energy from the condensing steam is transferred by conduction through the cylinder and in order to achieve high heat transfer rates it is equipped with grooves to reduce the resistance for heat transfer through the cylinder. Due to the centrifugal force the condensate accumulates in the bottom of the grooves and is taken out with a number of small tubes (''straws'') and collected in siphon tubes similar to the ones used in multicylinder dryers.

The drier is equipped with two hoods on the outside blowing hot air with a temperature of 250–550 °C at velocities of 75–150 m/s on the web resulting in very high convective heat transfer coefficients between the air and the web. Water can evaporate only to the hood side and the water vapour is taken out with the hood exhaust gases.

The wed is dried to its final moisture content on the cylinder after ¾ of a revolution and is removed by cutting it from the cylinder with a steel blade normally called a "doctor blade" which produces a creped elastic sheet.

By summing up the contributions from the conduction through the cylinder and convection from the hood the total energy transfer and the drying rate can be estimated in the same way as was shown for multi-cylinder dryers in Section 13.5.2. Typically 50 % of the energy for evaporation is supplied from the steam and 50 % from the hoods.

Due to the efficient energy transfer on the Yankee-dryer the drying rates are much higher than for multi-cylinder dryers, typical drying rates are in the range 100–250 kg/m^{2}h. The energy efficiency is lower, the specific energy use is around 6000 kJ/kg water evaporated. Further substantial amounts of electrical energy is needed for the operation of the fans in the hoods.

The TAD-dryer. In the TAD-dryer the web is dried on a porous TAD-cylinder by sucking hot air through the web. The principle is shown in *Figure 13.22*.

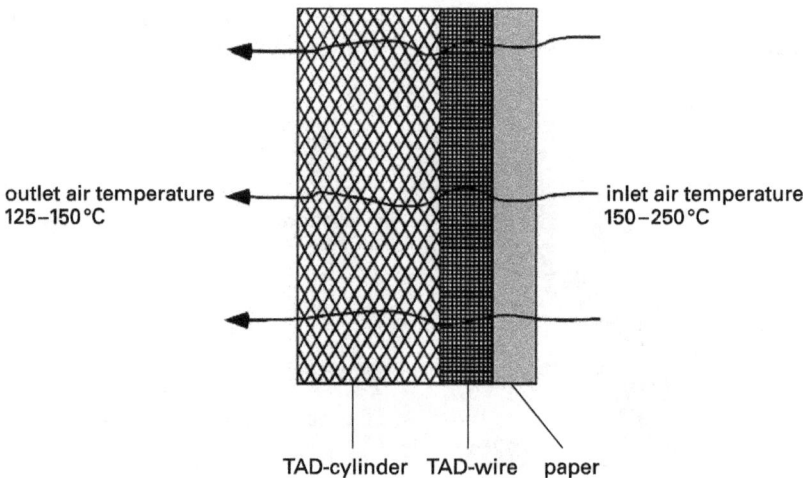

outlet air temperature
125–150 °C

inlet air temperature
150–250 °C

TAD-cylinder   TAD-wire   paper

Figure 13.22. Principle for a TAD-dryer.

In the TAD-process the web is transferred to the dryer at a dry matter content of typically 25 %. Thus the amount of evaporated water per m² product is roughly the double as compared with the Yankee-dryer. By creating an underpressure inside the TAD-cylinder, hot air is drawn through the web, the TAD-wire and the porous TAD-cylinder which dries the web. Designs exist with one or two TAD-cylinders as well as combinations with a Yankee-cylinder. Natural gas or propane is normally used as the thermal energy source for heating the air.

In the TAD-dryer the energy transfer between the hot air and the porous web is very efficient and very high drying rates can be obtained, typical rates are in the range 150–500 kg/m²h. The energy efficiency is lower, the specific energy use is around 8000 kJ/kg evaporated water. Also large amounts of electrical energy are needed for the operation of the fans.

The advantages with the TAD-process are:

- a paper with a higher capacity for water uptake per kg paper is produced
- the paper is softer and has a higher bulk (lower density)

13.7.2 Drying of Coated Papers

General. In the coating process a pigment and a binder are applied on the surface of the paper to improve its surface finish and printability. After the coating station the surface is wet and sticky and bringing the surface in contact with a hot cylinder surface will destroy the coated surface and deposit part of the coating on the cylinder. The normal technique to dry a coated surface is to initially use non-contact dryers such as radiation dryers and convection dryers (sometimes called air-foils) and then perform the final drying on traditional cylinder dryers. One example of a typical installation is shown in *Figure 13.23*.

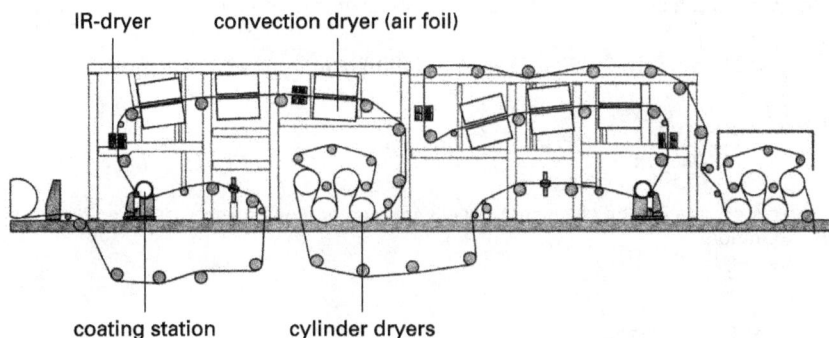

Figure 13.23. Drying of coated papers (by courtesy from Solaronics-IRT).

One typical feature for IR-dryers is that they can be turned on and off in sections over the width of the machine. This is normally referred to as profiling and makes it possible to correct for variations in moisture content over the width of the machine. The coating and the drying process can be performed either directly after the dryer on the paper machine or in subsequent step on an off-machine coater.

IR-Dryers. In infrared dryers energy transfer occurs by radiation from a hot surface to the paper web. The radiation can be absorbed in the wet coating layer, in the base paper, be reflected from the web or be transmitted through the web. Two fundamental radiation laws are important for understanding the behaviour of IR-dryers. Stefan-Boltzmann's law gives the total amount of energy emitted from a surface at a given temperature and emissivity:

$$q_{emit} = \varepsilon\sigma(T_{surf} + 273)^4 \qquad (13.17)$$

The emissivity ε is 1 for a black body but for all other materials it is lower than one. For paper it varies strongly with the wavelength of the radiation. The constant σ has the value $5.67 \cdot 10^{-8}$ W/m^2 K^4 nd the amount of radiation from a surface is a very strong function of the surface temperature. The wavelength at which maximum emission occurs is calculated with Wien's displacement law:

$$\lambda_{max} = \frac{2898 \cdot 10^{-6}}{(T_{surf} + 273)} \qquad (13.18)$$

The temperature of the sun is 6000 K and results in a maximum wavelength of about 0.5 μm which lies in the visible light. For IR-dryers the maximum wavelength is in the near infrared spectrum of 1–5 μm.

Example 13.11	Calculate the total amount of radiation and the wavelength at which maximum emission occurs for a ceramic gas-heated IR dryer heated to a temperature of 950 °C. Assume that the emissivity of the surface is 0.9. Calculate the drying rate in an IR-dryer assuming that all of the emitted radiation from the unit is absorbed in the paper and can be used for evaporation of water.
	According to equations (17) and Equation (18), the amount of emitted radiation and the wavelength at which maximum radiation occurs are given by:
	$$q_{emit} = 0.9 \cdot 5.67 \cdot 10^{-8}(950 + 273)^4 = 114000 \text{ W/m}^2$$
	$$\lambda_{max} = \frac{2898 \cdot 10^{-6}}{(950 + 273)} = 2.4 \cdot 10^{-6}\text{m} = 2.4 \text{ μ/m}$$
	The drying rate can be calculated directly from an energy balance assuming that the evaporation occurs at 75 °C:
	$$114000 = m_{evap} \cdot 2320 \cdot 10^3$$ $$m_{evap} = 0.491 \text{ kg/m}^2\text{s} = 177 \text{ kg/m}^2\text{h}$$
	The example shows that very high heat fluxes and consequently high drying rates can be achieved using this technology.

Two types of IR-dryers are used:

- electrical heated units where a tungsten filament in a quartz tube is heated to temperatures of 1900 –2500 °C.
- gas heated units where natural gas or propane is burnt on the surface of a porous ceramic or metal fibre material. The surface temperature is normally in the range 850–1050 °C.

Due to the limited gas distribution network in Sweden the majority of installations are based on electrical heated units, the principle is shown in *Figure 13.24*. The tungsten lamps are heated to very high temperatures which results in short wave lengths and some of the radiation is transmitted through the paper. This radiation is normally absorbed by a co-radiator on the other side of the paper which absorbs and reemitts this part. The lamp glass requires cooling with air and the amount of cooling determines the radiation efficiency (the amount of radiation relative to the electrical energy supplied to the unit) and normally this figure is around 60 %. The outgoing hot air temperature is in the range 100–150 °C which somewhat limits the use of this energy source.

The specific energy use for both the electric and the gas heated units is normally in the range 5000–6000 kJ/kg water evaporated. The gas-fired IR-dryers operate at a lower temperature and emits radiation more in accordance with the absorption spectra of coated papers. Different arguments are raised for using electrical or gas-heated IR from a paper quality point of view.

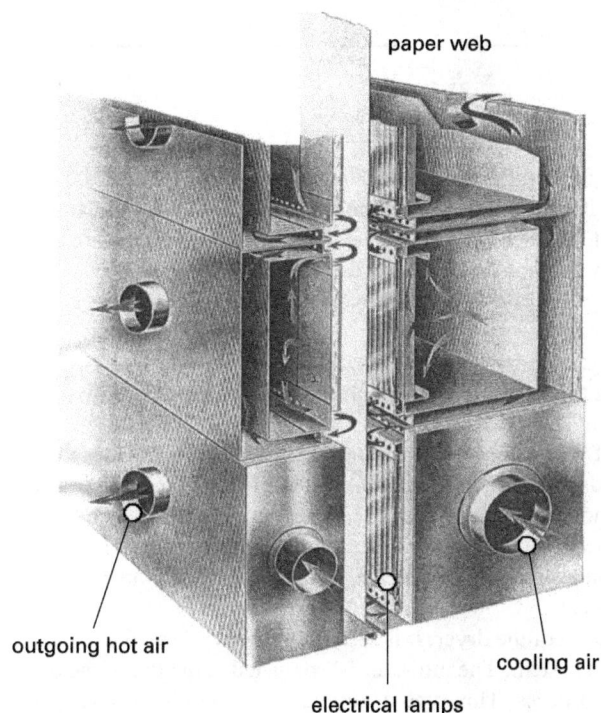

Figure 13.24. Electrical heated IR-dryer (by courtesy from Solaronics-IRT).

Convection dryers. The IR-dryer is normally followed by a convection dryer where hot air is blown with high velocities on to the surface of the web. One possibility to achieve a high energy efficiency of the dryer for coated papers is to use the outgoing hot air from the electrical or gas heated IR-dryers in the convective dryers reducing the amount of steam or gas needed for heating these dryers.

Example 13.12	Calculate the amount of energy transfer in a convective dryer using a hot gas with a temperature of 130 °C and a coated paper web with a temperature of 75 °C if the heat transfer coefficient has been estimated to be 120 W/m² °C. Calculate the drying rate in the same dryer assuming that all of the transferred energy is used for evaporation of water. The amount of energy transfer is given by: $$q_{transfer} = 120 \cdot (130 - 75) = 6600 \text{ W/m}^2$$ The drying rate can be calculated directly from an energy balance: $$6600 = m_{evap} \cdot 2320 \cdot 10^3$$ $$m_{evap} = 0.0028 \text{ kg/m}^2\text{s} = 10.2 \text{ kg/m}^2\text{h}$$ The drying rate for convection dryers is considerably lower as compared with IR-dryers.

13.7.3 Drying of Pulp

General. Drying of pulp is necessary for two main reasons. Firstly in a wet form it will be damaged by biological and chemical activity. Secondly, market pulp is transported globally and it is therefore necessary to remove the water content to a low level. Two different designs are available for drying of pulp; the airborne web dryer and the flash dryer.

Airborne Web Dryer. The principle for the airborne web dryer is to dry a continous web by blowing hot air with high velocities on the web. The hot air is heated with steam to a temperature of 120–160 °C and blown from nozzles in a top and bottom row of blow-boxes. The blow boxes are arranged in a deck allowing the web to pass through a number of rows of blow-boxes in the dryer. The principle is shown in *Figure 13.25*.

The pulp web has a basis weight of between 600 and 1400 g/m² and today dryers are in operation with a capacity of up to 2500 t/day. The moist web enters at the top of the dryer at a web moisture content of typically 50 % and the web then makes several horizontal passes back and forth before it leaves the dryer at approximately 10 % moisture. It is then transferred to a cutter or winder. The drying rate depends on several factors such as, type of pulp, impingement heat transfer coefficient and the air conditions.

The dry supply air enters at the bottom of the dryer. It is recirculated and reheated with steam several times in the internal circulation system. The moist air is exhausted at the top of the dryer which gives a counter current drying process. This type of drying results in high capacity and good energy efficiency.

The top blow boxes have round holes for perpendicular impingement in a configuration for maximum heat transfer. For the bottom blow box, aerodynamic forces are used to give a stabilised transport of the web a few millimetres above the blow box deck. It will also have a self-adjusting effect on the web. If a disturbance occurs which will increase the floating height and thereby try to remove the web from the normal position over the bottom blow box, a force will act to hold it back. On the other hand, if the floating height is decreased, a force will act to reset the floating height. The resulting balanced forces keep the web floating in a stabilised position over the blow boxes and the pulp web travels flat and flutter free through the dryer.

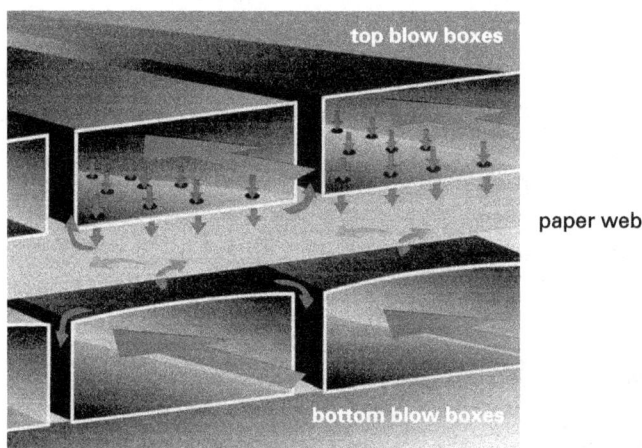

Figure 13.25. Principle for the airborne web dryer (by courtesy of ABB Industries).

Flash Dryer. The principle for the flash dryer is to disintegrate the web after the press section and dry the dispersed fibres in a hot air stream. The hot air is normally heated with natural gas or propane but designs using oil or steam are also in operation. After drying the pulp is pressed into a bale. The principle is shown in *Figure 13.26.*

In the first stage the recirculated air from the second stage is heated to between 225 and 350 °C with natural gas and mixed with the disintegrated pulp. The hot air stream transports and dries the pulp in the first stage. In the cyclone the humid air leaves the dryer and the partially dried pulp is fed to the second stage in the dryer. In the second stage the air temperature is limited to between 160 and 200 °C which makes it possible to heat the air with steam. In the second stage the pulp is dried to its final moisture content, the air is recirculated to the first stage and the pulp sent to a cooling cyclone before it is pressed to bales. For a modern flash dryer the specific energy use is around 3000 kJ/kg evaporated water. The flash dryer is mainly used for drying of mechanical pulps such as TMP and CTMP and units with capacities of 500 t/day have been built.

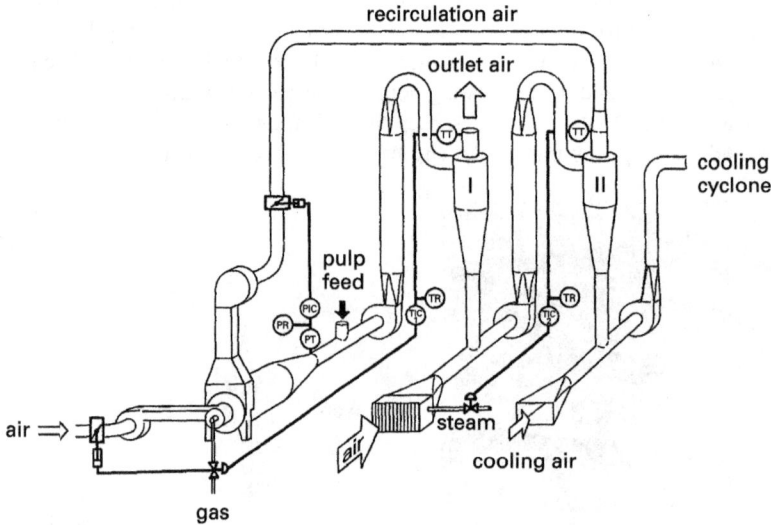

Figure 13.26. Principle for the flash dryer.

13.8 New Developments

The principle for the multi-cylinder paper dryer has with some modifications remained the same for a long time but during the last 20 years a number of new designs have been proposed. Some of them have come to a commercial stage and some are still on a pilot stage.

13.8.1 Designs in Commercial Operation

Condebelt. Completely new ideas for drying of paper were developed by Prof. Jukka Lehtinen from Finland in the Condebelt process. Two ideas were different from the conventional multi-cylinder design. Firstly steel belts instead of cast iron cylinders were used as the energy transfer material from the steam to the web. Secondly removing most of the air when the web enters the dryer eliminates the diffusional resistance of water vapour in the air inside and outside the web. The principles for the design are shown in *Figure 13.27*.

Before the web enters the dryer steam is drawn through the web which evacuates most of the air in the inter fibre pores in the web. The web is dried between two steel belts with a thickness of 1 mm, thus the distance for conduction through the belt is much smaller as compared with the cylinder design. The upper side of the web is in contact with the upper belt which is heated with steam at temperatures of 110–170 °C corresponding to steam pressures of up to 800 kPa. This results in very good contact between the belt and the web and the contact resistance is practically eliminated. The lower side of the web is in contact with a fine wire which is followed by a coarse wire and the lower belt which is cooled with water. The purpose with the fine wire is to avoid markings on the web and the purpose of the coarse wire is to act as a reservoir for the condensate.

Since the gas essentially consists of only water vapour the transport of water vapour is pressure driven instead of being controlled by diffusion and differences in concentration. This is one reason for the high drying rates observed with Condebelt dryers, figures in the range of 100–200 kg/m²h have been reported.

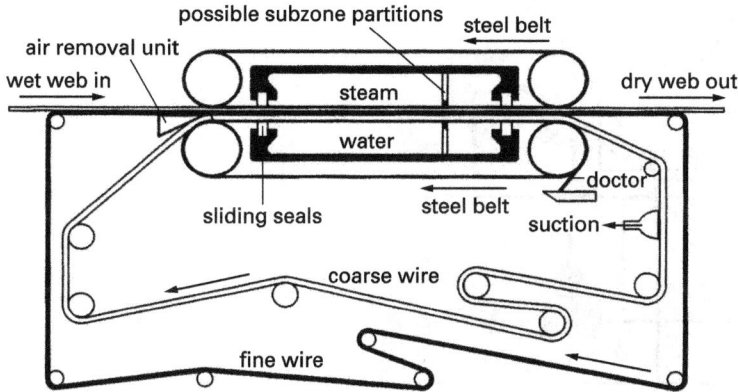

Figure 13.27. Principle for the Condebelt dryer (by courtsey of Metso).

The temperature of the outgoing cooling water can be controlled between 60 and 90 °C. A high outgoing cooling water temperature makes energy recovery much more attractive from a mill perspective compared with the traditional scrubber design where the temperature is limited to 50 °C or below.

During the drying process the web is held firmly at the high pressure between the upper belt and the fine wire. This reduces the web shrinkage to very low values and the tensile and compression strengths will increase dramatically, especially in the CD-direction. The Scott bond value will increase and the web will densify remarkably. One possibility to implement the increased strength is to use lower basis weights or less expensive raw materials for the products dried with the Condebelt process.

Two commercial machines have been built since 1996. The first is a board machine in Finland with a maximum speed of 240 m/min and the second machine is a testliner machine in South Korea with a maximum speed of 700 m/min.

Impingement Hood. On single tier machines or on single tier machines with vacuum rolls the energy transfer and thus drying capacity on the lower row of cylinders is very limited. One possibility to overcome this drawback is to install impingement hoods on the traditional lower cylinder or on a new large diameter vacuum roll. The principle for the design is shown in *Figure 13.28*.

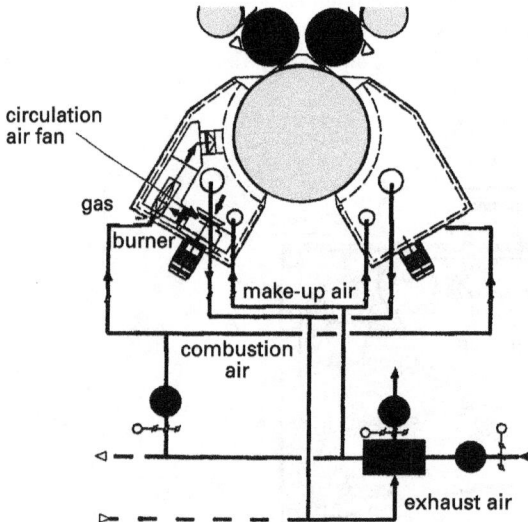

Figure 13.28. Impingement hood (by courtsey of Metso).

Hot air with a temperature of 250–400 °C is blown with high velocities (typically 70–150 m/s) on the web resulting in very high convective heat transfer coefficients resulting in drying rates between 100 and 160 kg/m2h. The impingement hood competes with a rebuild of the press section for increased capacity. The cast iron cylinder has a large thermal mass which results in slow control during capacity and quality changes on the machine while the impingement hood has less thermal mass enabling more rapid quality changes on the machine. Also it could be designed in sections enabling cross machine control. The dryer is primarily designed for high speed printing paper machines and the first unit was installed in Germany during 1999.

Gas Heated Cylinder. Gas-heated cylinders are available using natural gas or propane for firing an IR-burner inside the cast iron cylinder. Energy transfer from the cylinder to the web then occurs in the same way as for the steam heated cylinder. High heat fluxes and thus also high drying rates can be achieved with this design. Roughly 10 units have been installed world-wide, so far none in Sweden

13.8.2 Designs in Pilot Stage

Impulse Drying. Mechanical dewatering in the press section is very energy efficient and if this amount could be increased the total energy efficiency for the paper machine would also be increased. As shown in example 6-2 increasing the dewatering a few % in the press section results in large amounts of water removal and thus a more energy efficient production or the possibility to increase the capacity on the machine.

One of the most studied techniques to achieve increased dewatering in the press section is called impulse drying. Impulse drying was originally proposed by Douglas Wahren in the late 1970's and since then has been extensively studied in laboratory as well as in a pilot scale in USA, Canada, and Sweden. The idea came up when Douglas after heating up a hammer with a gas burner hit a wet paper and noticed the efficient water removal and drying of the spot of the paper which was hit by the hammer. The industrial design today is based on using a shoe-press and a heated counter roll, the principle is shown in *Figure 13.29.*

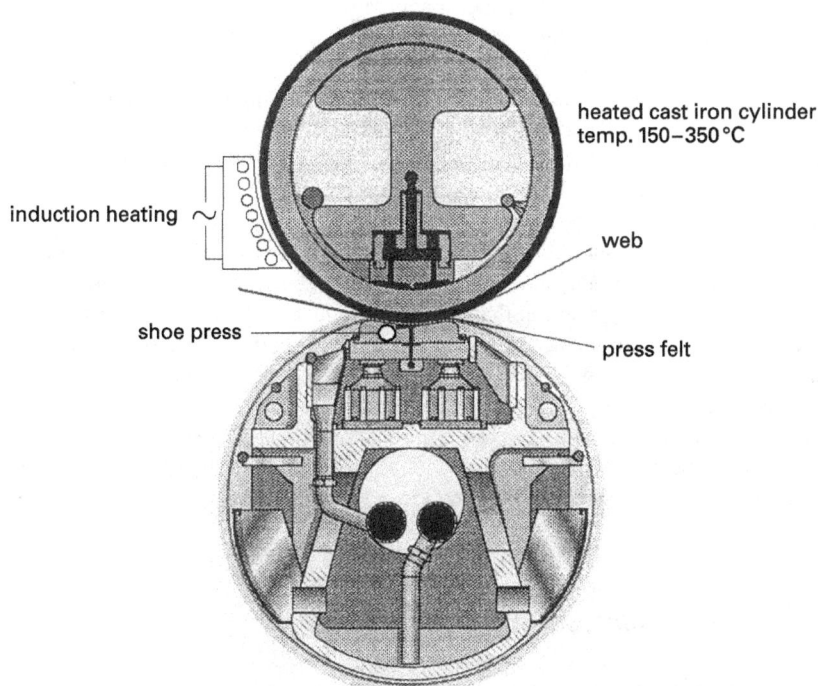

Figure 13.29. Principle for impulse drying (by courtesy of Metso).

In the press nip high pressures of 1–5 MPa and high temperatures of 150–350 °C are used resulting in efficient water removal and high heat fluxes from the hot cylinder to the web. Water removal occurs by a wet pressing mechanism and by evaporation and dry matter contents of up to 70 % have been achieved on pilot machines. Considering that the heat transfer rates are in the range 0.5–3 MW/m^2 (two orders of magnitude larger than for the multi-cylinder dryer) and that the area for heat transfer is much smaller (5–10 % of the multi-cylinder area) the overall amounts of energy transfer per area paper is something like 5–10 times higher than for a traditional cylinder.

High temperatures of the web results in high vapour pressures inside the web as it leaves the press nip. If the z-direction strength in the web is not enough to withstand the increased pressure inside the web it will split into two or more layers, a phenomenon called delamination. The high pressures used in the press nip will result in less bulk for the paper and the surface in contact with the heated cylinder will have a more smooth surface. Considering the high heat fluxes needed, the electrically powered induction heaters are the most probable solution for heating the cylinder but high capacity gas fired hoods or a gas-fired IR emitter could also be possible.

The total energy use for the paper machine will be determined by the water removal rates in the press and drying sections combined with the energy use for theses processes. If more water can be removed by using less energy in the press section the total energy use for the process will also be lower. The electrical energy is more expensive than steam produced in the recovery or bark boilers and the total energy cost for the paper machine will be based on the total amounts of electricity and steam used in the pressing and drying sections.

Drying in Superheated Steam. By superheating saturated steam it can be used for drying purposes. In a pure steam atmosphere the product will attain the boiling point of water, thus at atmospheric pressure drying of a material will take place at 100 °C. Only steam is present in the gas and no diffusional mass transport resistances exist. The superheated steam will be cooled from the superheated temperature close to the saturation temperature. Examples in Sweden include drying of biomass materials and peat.

For paper the superheated steam should be blown at high velocities at the web in an impingement dryer and in pilot scale drying rates of up to 100 kg/m^{2}h have been measured. The big advantage with drying in superheated steam is that the evaporated water will mix with the pure steam with excellent possibilities for energy recovery. By condensing the excess steam the energy could be used in several positions in the mill and the remaining part recirculated to the superheater and back to the dryer.

One of the problems to overcome is to avoid mixing the superheated steam with air as this will reduce the drying rate and make energy recovery more difficult. Thus the dryer will have to be operated at a slight overpressure and the inlet of the web to the dryer designed with great care.

13.9 Experimental Evaluation of Paper Dryers

13.9.1 Cylinder Surface Temperature

The functioning of condensate drainage can be determined by measuring the cylinder surface temperature. A correct temperature indicates a good operation of the system while a decrease in the temperature indicates that the heat transfer from the steam to the condensate is not efficient. The best way to determine the cylinder surface temperature is to use a contact thermometer such as the Swema SWT instrument. With this instrument a correct value is obtained within 10 seconds. IR instruments are not suitable for smooth cylinder surfaces since they are dependant on the emissivity of the surface and radiation reflected from other objects.

13.9.2 Hood Moisture Content

The hood moisture content determines the rate of evaporation and the energy recovery efficiency. Some modern hoods are equipped with instruments (such as a dew point meter) for monitoring the dew point and use this value for control of the air flow rate. If not available the best way to measure the moisture content of the hood is to measure the dry and the wet bulb temperature of the air and read the moisture content from a Mollier chart. The dry bulb temperature is the normal temperature measured with an ordinary thermocouple or thermistor. The wet bulb temperature is the temperature measured with a thermocouple equipped with a wet surface from which water can evaporate to the surrounding air. The wet surface is most easily obtained by drawing a short piece of a shoelace (from ice-hockey skates etc.) over the thermocouple and dipping it in water. Inserting it into a position in the hood where the air velocity is above 2 m/s quickly results in reading a lower temperature than the dry temperature, the wet bulb temperature. As the shoe-lace dries out the temperature rises again to the dry temperature and in order to avoid dipping the thermocouple in water for each measurement it could be equipped with a small water bottle from where the shoe-lace is continuously fed with water.

13.9.3 Web Moisture Content

The web moisture content gives an indication of the operation of the dryer in different sections. Instruments based on using microwaves as well as IR are available but he most reliable data are obtained with an instrument measuring the backscatter of γ-radiation. This instrument measures the total amount of mass in front of the instrument and since the dry matter content can be obtained from existing instruments before the reel the moisture content in the web is easily calculated. The instrument senses solid matter (such as the cylinder) roughly within 0.5 m so measurements should be performed in the free draw or the background accounted for by calibration. The instrument is sliding in good contact with the web but this does not normally cause any problems with web breaks and measurements have been reported on newsprint machines directly after the press section.

13.9.4 Web Temperature

The web temperature is an indication of the operation of the steam and condensate system as well the drying behaviour of the web. It is most easily measured with IR-pyrometers or possibly a contact thermometer.

13.10 Nomenclature

C	concentration	mol/m³
c_p	specific heat	J/kg °C
D	mass diffusivity m²/s	m²/s
h	heat transfer coefficient	W/m² °C
G	air flow rate	kg dry air/s
H	enthalpy	J/kg
ΔH	latent heat of evaporation	J/kg
J	molar flux	mol/m²s
k	thermal conductivity	W/m°C
k_g	mass transfer coefficient	m/s
l	length	M
m	mass flowrate	kg/s
M	molecular weight	kg/mol
p	partial pressure	Pa
P	total pressure	Pa
q	heat flux	W/m²
Q	power	W
R	gas constant	J/mol K
T	temperature	°C
X	air moisture content	kg water/kg dry air
z	length coordinate	m
ρ	density	kg/m³
ε	emissivity	–
λ	wavelength	M
σ	Stefan Boltzmann's constant $5.67 \cdot 10^{-8}$	W/m² K⁴

Subscripts

air	air		leak	leakage
cond	condensate		max	maximum
cont	contact		out	outlet, outside
conv	convective		p	paper
cyl	cylinder		s	steam
eff	effective		surf	surface
evap	evaporated		w	water
in	inlet, inside			

13.11 Literature for Further Reading

Karlsson, M. (Ed.) (2000) *Drying. Papermaking Science and Technology, Vol 9, Part 2,* Fapet Oy.

Stenström, S. (1996) Mathematical Modelling and Numerical Techniques for Multi-Cylinder Paper Dryers. In: Mujumdarm, A. S., and Turner, I. W. (Eds.), *Mathematical Modelling and Numerical Techniques in Drying Technology*, pp. 613–661.

Wilhelmsson, B. (1995) *An experimental and theoretical study of multi-cylinder paper drying (PhD Thesis).* Lund University.

A good overview of recent work in the paper drying area can be found at
www.chemeng.lth.se/research.jsp

14 Sizing

Tom Lindström
Innventia AB

14.1 Internal Sizing of Paper and Paperboard

One important property of paper and paperboard products is their ability to resist wetting and penetration by liquids such as water, hot and cold drinks, fruit juices, various inks, oils, greases or organic solvents. A product is said to be "sized" if it possesses this ability to withstand wetting or liquid penetration. Historically, paper was sized to prevent ink feathering, so one could write on paper. It is important to distinguish between surface sizing and internal sizing. Surface sizing operations consist of the application of modified starches and polymer dispersions to the surfaces of already formed paper or board at an appropriate location at the dry end of the machine. Although polymer dispersions may be hydrophobic, or internal sizing agents may be added in size presses, the purpose of surface sizing is to control the surface porosity for printing, decrease surface fuzz, improve surface finish and improve surface picking resistance or the internal strength of paper.

Internal sizing is a method to reduce the rate of liquid penetration into the paper structure by treating the stock with hydrophobic substances, which make the dried sheet water repellent.

In paper the penetration can take place either:

- as liquid penetration in the capillary system of the sheet
- as surface diffusion on the fibres
- as diffusion through the fibres (when swollen)
- as vapour phase transport

Paper and board are internally sized either to control the wetting and liquid penetration in the final product or to control size press or coating pick up in paper/board manufacture. When paper is wetted in a size press, the fibre-fibre bonds are weakened, and this may lead to paper breaks, if the rate of liquid penetration is not appropriately controlled through internal sizing.

Internal sizing methods cannot prevent vapour to penetrate the structure. In order restrict vapour diffusion a physical barrier must be created. Which of these processes dominate depends, among other things, on the properties of the liquid, on the network structure, on the prevailing pressure situation, on the time available and on the hydrophobic properties of the paper.

The liquid penetration is not static but dynamic and must be treated as such. This means, for instance, that these different transport mechanisms can be transformed into each other without sharp limits. The changes occurring in the network during the transport process are also very important. For instance, when a paper sheet becomes wetted, residual stresses in the paper can be released when hydrogen bonds are being destroyed and the structure may become more absorbent. Water absorption in the fibres makes them swell and change shape, which leads to changes in the network structure.

The use of a hydrophobic agent leads to a considerable reduction in the liquid penetration provided external pressure is not applied. The main penetration mechanism for sized papers is most often diffusion in fibres/swelling of fibres.

14.1.1 The Theory of Liquid Penetration in a Capillary System

If a small drop of liquid is placed on a smooth surface, it will under certain conditions adopt an angle to the solid body as indicated in *Figure 14.1*.

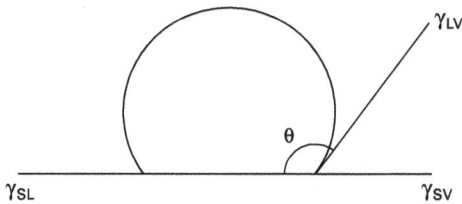

Figure 14.1. Force components around a liquid droplet on a flat surface.

The equilibrium is dependent on three surface tensions, as is evident in the figure:
γ_{VS} = surface tension in the contact surface solid body – vapor
γ_{SL} = surface tension in the contact surface solid body – liquid
γ_{LV} = surface tension in the contact surface liquid – vapor
Between these surface tensions and the contact angle, the following relationship exists:

$$\gamma_{VS} - \gamma_{SL} = \gamma_{LV} \cos\theta \tag{14.1}$$

This equation, which is called Young's equation, is only valid for an absolutely plane surface, which is never found in nature and which certainly is not true of paper. In practice, the contact angle is therefore not a well-defined quantity but is dependent on how it is measured. One speaks of contact angle hysteresis and different values are obtained depending on the direction from which the equilibrium is approached when the drop is applied. One speaks of "advancing" angle θ_a and "receding" angle θ_r, defined as in *Figure 14.2*.

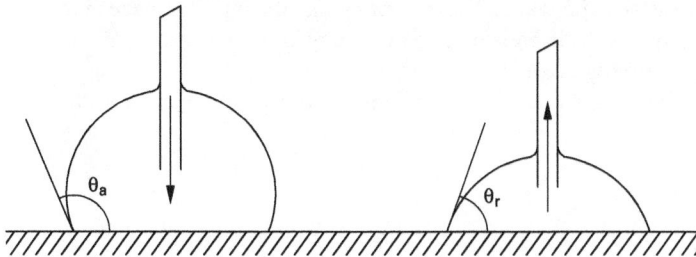

Figure 14.2. Contact angle hysteresis and definition of "advancing" (θ_a) and "receding" (θ_r) contact angle.

Since paper surfaces are not smooth, it is also important to note that the roughness has an important effect on the contact angle. Through the introduction of a roughness factor r, defined as the ratio between the real and apparent surface areas of the body concerned, an apparent contact angle can be defined as the angle between the liquid and the envelope of the surface. The following expression is obtained:

$$\cos \theta_{obs} = \cos \theta \, r \qquad\qquad (14.2)$$

During direct measurement of the contact angle, it is θ_{obs} that is measured. It can be seen in Equation (14.2) that if the real contact angle is above 90°, an increase in roughness gives an increase in the apparent contact angle, whereas if the real contact angle is lower than 90°, the opposite is true (*Figure 14.3*).

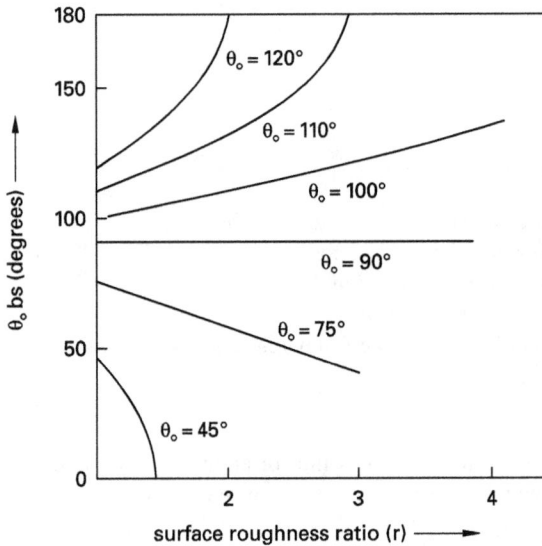

Figure 14.3. The observed contact angle θ_{obs} as a function of the roughness factor.

The quantity of water, which penetrates into a paper, is directly dependent on the forces of attraction between the paper surface and water. These forces make the liquid surface in a capillary assume a curved shape, a pressure difference being created over the surface. That this is the case is shown by the following simple consideration:

Consider a water drop on a surface and the forces by which it is affected, *Figure 14.4*.

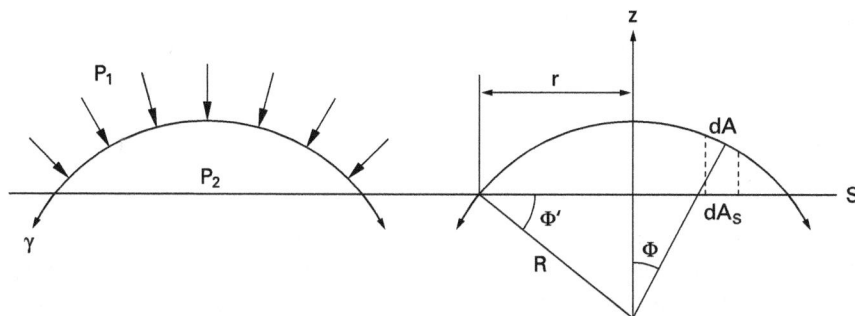

Figure 14. 4. Force components for a drop on a flat surface.

The pressures P_1 and P_2 act perpendicular to the surface, If gravity is neglected, P_1, P_2 and γ (the surface tension of the liquid) must balance each other. Observe a surface element dA, the projection of which on the plane S is dA_S. The force perpendicular to dA is $(P_1 - P_2)\,dA$ and the force component along the z-axis is given by:

$$(P_1 - P_2)\,dA \cos\theta = (P_1 - P_2)dA_S \tag{14.3}$$

The sum over the whole surface is:

$$(P_1 - P_2)\,A_s = (P_1 - P_2)\,\pi r_d^2 \tag{14.4}$$

On the periphery of the base surface, the surface tension exerts a force gdL on each length unit dL. The component of this force in the z-direction is $g \cos q'dL$ and it is summed over the whole periphery $2\,pr_d^2/r$. The system is symmetrical about the z-axis and the force components cancel each other:

$$(P1 - P2)\,\pi r_d^2 = 2\pi r^2 \gamma / r \tag{14.5}$$

or

$$P_1 - P_2 = \frac{2\gamma}{r} \tag{14.6}$$

which can be generalized to:

$$P_1 - P_2 = \Delta P = \gamma_L \left(\frac{1}{r_1} + \frac{1}{r_2}\right) \tag{14.7}$$

where P_1 and P_2 respectively are the pressures on the concave and convex side of the meniscus, ΔP is the pressure difference over the meniscus or the capillary pressure and γ_L is the surface tension of the liquid. R_1 and r_2 are the principal radii of the curved liquid surface.

If the liquid exists in a cylindrical capillary with a constant radius r and the attraction forces are such that a defined contact angle θ exists between the liquid and the capillary wall, the liquid surface becomes part of a sphere and $r_1 = r_2 = r$. The capillary pressure P_C is then given by:

$$P_C = \frac{2\gamma_L \cos\theta}{r} \tag{14.8}$$

which is often called the Young-Laplace equation.

The capillary pressure is counteracted by the pressure drop due to the flow resistance, P_F, which can be calculated according to the Hagen-Poiseuille law:

$$P_F = \frac{8\eta v l}{r^2} \tag{14.9}$$

where l is the penetration length, $v = dl/dt$ is the rate of penetration and η is the viscosity of the liquid. The pressure equilibrium at the liquid front is thus:

$$P_E + P_C = P_F \tag{14.10}$$

where P_E is the external pressure. Substitution of the expression for P_C and P_F in Equation (14.8) and Equation (14.9) respectively yields:

$$P_E = \frac{2\gamma\cos\theta}{r} = \frac{8\eta l(dl/dt)}{r^2} \tag{14.11}$$

The solution to this differential equation shows that the penetration distance is proportional to the square root of the time, $\sqrt{t}$.

$$1 = \sqrt{\frac{2r_\gamma\cos\theta + P_E r^2}{4\eta}}\sqrt{t} \tag{14.12}$$

This equation is often called the Washburn equation. The pressure term is often wrongly omitted from the equation. The equation is strictly valid only for a single capillary or several identical parallel capillaries not connected with each other. The pores in paper consist, however, of a complicated system of connected cavities and channels of different sizes. As a consequence, liquid transfer will take place from large to small cavities due to differences in capillary pressure. It follows that the network structure is also of importance for the liquid penetration. It is very important to realize that the equation relates only to capillary penetration and does not take into account other processes for water transport (see below).

Equation (14.12) shows that in order to obtain a paper with low capillary sorption, it is necessary:

- to create a surface with low surface energy
- to create a structure with narrow pores, i.e. small r

(Note: however, that r increases with time due to the expansion of the fibre network)

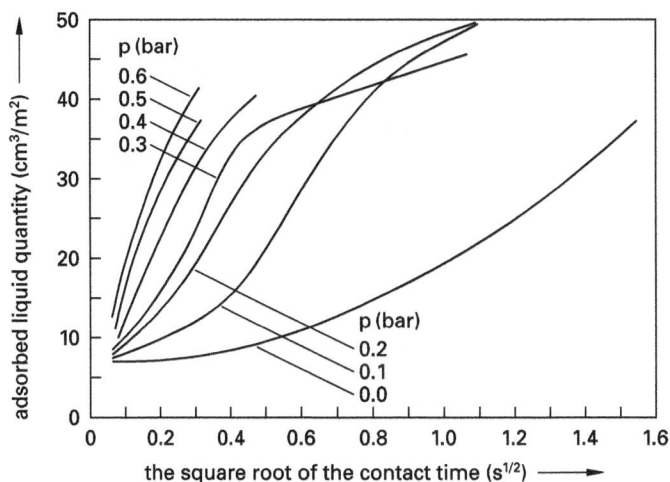

Figure 14.5. Pressure penetrations.

If an external pressure exists, which is often the case in surface sizing/coating operations, the importance of surface chemical effects is reduced and the capillary structure becomes increasingly more important. *Figure 14.5* shows the effect of pressure penetration on liquid sorption.

In this context, it is important to note that the surface energy of the paper can be changed during contact with liquid, due to a redistribution of surface active substances in the boundary layer and also due to chemical hydrolysis reactions during liquid contact.

The classic penetration model, represented by the Washburn Equation (14.12), shows that the liquid penetration should be a direct function of the square root of time. This means that when the sorption is plotted as a function of the square root of time, straight lines should be obtained. *Figure 14.5* shows that this is usually not the case for low external pressures and short times. The reason can be related at least to the following:

- The dynamic character of the capillary pressure
- Expansion of the fibre network
- Liquid transport mediated by the vapor phase
- External pressure

In the classical model, the contact angle is assumed to be independent of time and rate of penetration. Molecular processes in the wetting zone and in advance of the liquid front, however, probably affect the dynamic capillary pressure.

When liquid penetrates into a fibre network, the fibre network expands, initially because of the breaking of fibre bonds and fibre relaxation. In a later stage, liquid is sorbed into the fibre with a consequent change in the form of the fibre. The driving potential for capillary transport is the capillary pressure. According to the Young-Laplace (Equation (14.8)), the capillary pressure is negative when the contact angle between the liquid state and the solid state exceeds 90° and the capillary pressure then counteracts penetration into the pore system. Water transport in the vapor phase is, however, independent of the surface tension forces between the solid state and

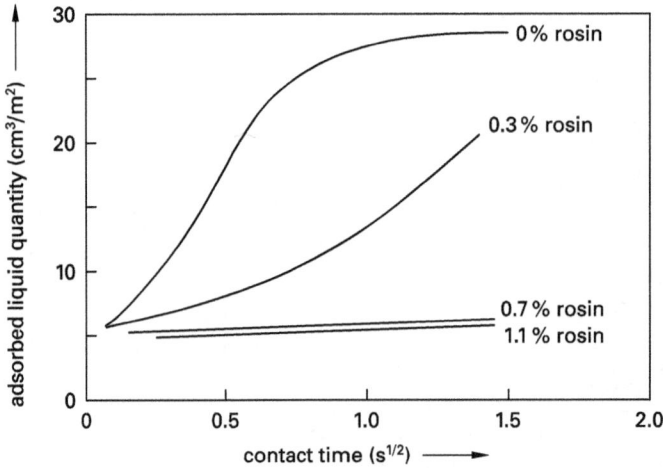

Figure 14.6. Liquid penetration in hydrophobic paper (no external pressure).

the liquid. For hydrophobic papers, vapor phase water transport can therefore be an important mechanism. Other possible effects are surface diffusion through the fibres and osmotic effects.

If a paper is sized, the liquid transport is affected drastically, as is illustrated in *Figure 14.6*. Hardly any liquid penetration takes place at all when the amount of hydrophobic sizing agent is high. With 0.7 % added rosin size, (full hydrophobicity) has been obtained, the capillary pressure has become negative and the only transport mechanism is diffusion.

14.1.2 Water Sorption in Paper Structures

From the aforementioned discussion it is obvious that there are two well-defined paths of penetration of water through the thickness of paper. The penetration through the pores of the paper i.e. inter-fibre penetration as described by the Washburn equation (Equation (14.12)), driven by the capillary pressure and the penetration through the fibres, i.e. intra-fibre penetration, resulting from diffusion of moisture or water through the fibres. The extent to which each of these paths of transfer of water is measured depends also on the type of size tester used.

Hence, in an un-sized or partly sized paper, the liquid transfer is determined by variables such as the viscosity of the liquid, the liquid pressure, the typical pore size of the paper and the time of contact as described by the Washburn equation. A physical modification, which will decrease the pore size, will retard the rate of wetting and increase the penetration resistance. Decreasing the paper porosity and increasing the apparent density may achieve this. In un-sized paper, both inter- and intra-fibre penetration are so fast, that it is almost impossible to distinguish between the two mechanisms.

In a fully sized paper, liquid transudation is solely determined by the intra-fibre water transfer, which is *not* described by the Washburn equation although attempts have been made to describe the simultaneous swelling and capillary sorption (Hoyland et al 1977).

For papers, where intra-fibre diffusion is controlling the penetration there will be no effect of the viscosity of the aqueous media, if the large solute controlling the viscosity can not penetrate the cell wall of fibres. Water transport will here be determined by the extent of fibre-fibre bonding. A more well-beaten stock should therefore, in theory, be better to promote surface for water diffusion in the structure. In practice, this is seldom the case, because fibre swelling opens up void spaces between the fibres, which can provide liquid transport by capillary suction. It is therefore of general advantage to have a structure where the pore size is smaller.

Most sizing agents are provided as emulsions and dispersions and these can by definition not spread or migrate in the areas, where fibres are bonded and bonds are therefore un-sized. When these bonds are broken, when the fibres start to swell, water will penetrate these areas due to capillary action. Some fibre structures, for instance mechanical fibre based papers, are inherently difficult to size because residual stresses in the fibre structure are released under the action of moisture opening up fibre-fibre bonds, which will provide high surface energy surfaces mediating water transport. Wet strengthening may help the prevention of fibre swelling in highly sized papers and represents one strategy to restrict water sorption in such structures.

14.2 Aluminium and its Chemistry in Aqueous Solutions

Aluminium salts have been an almost ubiquitous ingredient in papermaking chemistry since the beginning of the 19th century, when Moritz Illig invented rosin sizing. Long before synthetic polymers were used as retention agents, papermakers alum, $Al_2(SO_4)_3 \times 14H_2O$, was used as the only coagulant available to retain fines, fillers and other papermaking adjuvants. Apart from papermakers alum, sodium aluminate, $NaAl(OH)_4$, and various so called polyaluminium salts are used in acid and neutral papermaking systems. Aluminium salts can be used as coagulants, because, when hydrolysed, they form various cationic species, with an affinity to anionic surfaces. Aluminium chemistry is highly complex, mostly because slow and often irreversible reactions, have often hampered the determination of appropriate equilibrium constants.

The chemical activity of aluminium salts is a result of the high formal charge, +3, and the small radius of the Al-ion. In order to reduce the high charge density, the Al-ion forms a typical Lewis acid (electron pair acceptor), strongly coordinating with Lewis bases (ligands) such as the oxygen and fluorine (Baes and Mesmer 1976). The aluminium ion forms complexes with H_2O, OH^-, $H_2PO_4^-$, SO_4^{2-}, $R\text{-}COO^-$ but not with Cl^-, NO_3^-, ClO_4^-, see *Figure 14.7*.

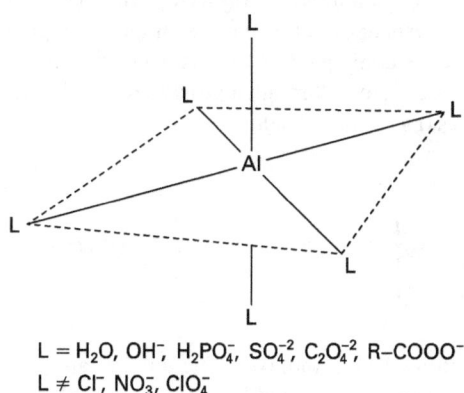

$L = H_2O$, OH^-, $H_2PO_4^-$, SO_4^{-2}, $C_2O_4^{-2}$, $R\text{-}COOO^-$
$L \neq Cl^-$, NO_3^-, ClO_4^-

Figure 14.7. General complexing behaviour of the aluminium cation in dilute solution. L may be H_2O, OH^-, $H_2PO_4^-$, SO_4^{2-}, $R\text{-}COO^-$ but not Cl^-, NO_3^-, ClO_4^-.

The aluminium cation complexes with six ligands to form an octahedral structure. It should be recognized that there are in fact two coordinate spheres, one octahedral internal and one external outer sphere coordination shell. The ligand equilibrium is, of course, dependent on a number of factors such as type of ligand, solution concentrations etc. In dilute solutions, fluoride and the bidentate ligands (oxalate, malonate and tartrate) form inner-sphere complexes, whereas sulfate, formate and acetate mainly form outer-sphere complexes. The stability of the aluminium ligand bond is controlled by the ability of the ligand to emit electrons, by the receptivity for the Al-ion and by the stability of the ligand to occupy more than one coordination site with the formation of a stable Al-structure.

This also implies that water molecules form relatively strong bonds with the aluminium ion, forming the $Al(H_2O)_6^{3+}$ structure in aqueous solution. The $Al(H_2O)_6^{3+}$ ion acts as a weak Bronsted acid and hydrolyzes as $Al(H_2O)_6^{3+} \Leftrightarrow Al(H_2O)_{6-x}(OH)_x^{3-x} + xH^+$.

Finally the aluminium ion may also hybridize into its tetrahedral form, when forming aluminate ions, $Al(OH)_4^-$, under aqueous alkaline conditions.

14.2.1 Al-Hydrolysis

When an aluminium salt of a noncomplex forming anion is diluted in water, the single trivalent ion, $Al(H_2O)_6^{3+}$, is only obtained at acidic pH-values. The aluminium salts are said to undergo hydrolysis, which formally be described by a series of hydrolysis reactions:

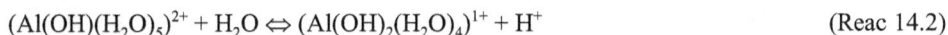

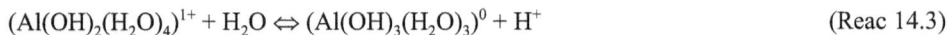

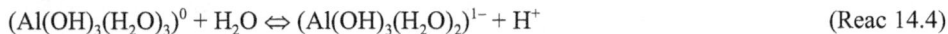

$$Al(H_2O)_6^{3+} + H_2O \Leftrightarrow (Al(OH)(H_2O)_5)^{2+} + H^+ \qquad \text{(Reac 14.1)}$$

$$(Al(OH)(H_2O)_5)^{2+} + H_2O \Leftrightarrow (Al(OH)_2(H_2O)_4)^{1+} + H^+ \qquad \text{(Reac 14.2)}$$

$$(Al(OH)_2(H_2O)_4)^{1+} + H_2O \Leftrightarrow (Al(OH)_3(H_2O)_3)^0 + H^+ \qquad \text{(Reac 14.3)}$$

$$(Al(OH)_3(H_2O)_3)^0 + H_2O \Leftrightarrow (Al(OH)_3(H_2O)_2)^{1-} + H^+ \qquad \text{(Reac 14.4)}$$

In aqueous solutions, the Al-ion yields significant quantities of $Al(OH)_2^+$, $Al(OH)_3$, and $Al(OH)_4^-$. More important, however, is that Al-hydrolysis does not take place simply through a series of mononuclear processes, but that the hydrolysis complex $Al(OH)^{2+}$ is coupled to polynuclear compounds. These polynuclear compounds are formed through a process referred to as an olation process (hydroxyl bridge formation process) as outlined below:

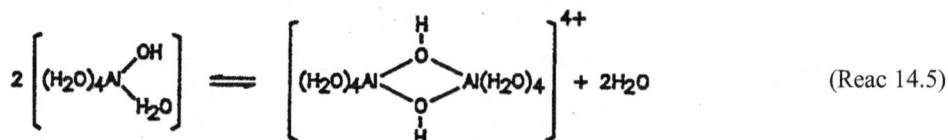

$$\qquad \text{(Reac 14.5)}$$

An extension of the olation process leads to multinuclear complexes. The number and formulae of these are uncertain since their formation is strongly affected by counterions, Al-concentration, temperature and ageing conditions.

14.2.2 Aluminium Hydroxide, Al(OH)$_3$

The solubility in the aluminium system has been studied mostly with respect to Gibbsite, (Al(OH)$_3$ (s)), The solubility constants in this system is very dependent on the purity in the system. A synthetic, clean Gibbsite, gives a solubility product, which may be only one fourth of that for natural Gibbsite. If the Gibbsite is newly formed and amorphous its solubility product is high. With ageing, the crystallinity increases and the solubility product increases. *Figure 14.8.* shows the concentration of aluminium and its hydrolysis products in equilibrium with Gibbsite.

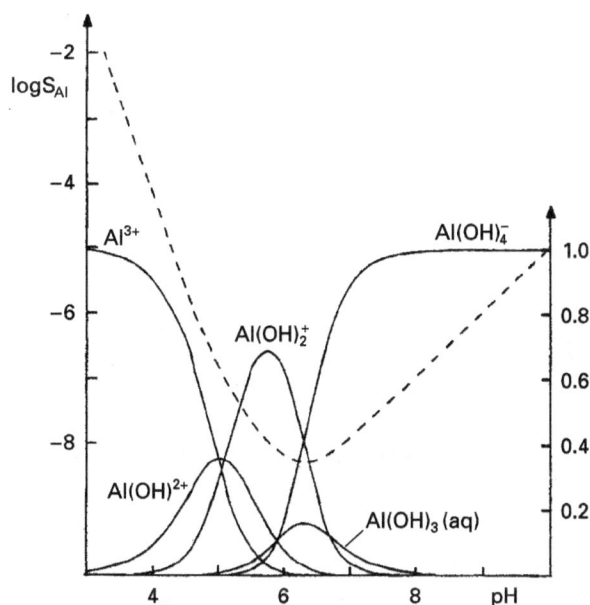

Figure 14.8. Diagram showing the aqueous solubility of a 10^{-2} M (Al (III) solution in equilibrium with crystalline Gibbsite (dotted curve). The solid curves give the corresponding distributions among Al-hydrolysis products (Öhman and Wågberg 1997).

The narrow span from Al (H$_2$O)$_6^{3+}$ to Al (OH)$_4^-$ is usually explained in terms of the concomitant and gradual decrease in coordination number from 6 to 4. As *Figure 14.8.* is based on thermodynamically stable Gibbsite, it is not relevant for freshly formed amorphous precipitates, which have a much higher solubility. Under such circumstances various polynuclear species are formed and the complexity of the system increases to a large extent. Secondly the system will be very sensitive to the nature of the counterion. If the counterions are Cl$^-$, or NO$_3^-$, that is no ligand forming anions, polynuclear ions such as Al$_{13}$O$_4$(OH)$_{24}^{7+}$ ("keggin polymer") is formed as a distinct species, whereas in the sulfate system insoluble cationic polynuclear species will be formed (see "Papermakers alum"). Aluminium hydroxide precipitates have both basic and acidic groups and participate in the following ion-equilibrium:

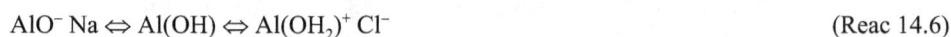

$$AlO^- \, Na \Leftrightarrow Al(OH) \Leftrightarrow Al(OH_2)^+ \, Cl^- \qquad \text{(Reac 14.6)}$$

Hence the precipitate is basically amphoteric in nature and posses both cationic and anionic sites. *Figure 14.9a* shows the distribution of active surface sites vs. pH and *Figure 14.9b* shows the zeta potential of Al-precipitates at various NaCl concentrations. The isoelectric point of the precipitate is located at pH = 8.2.

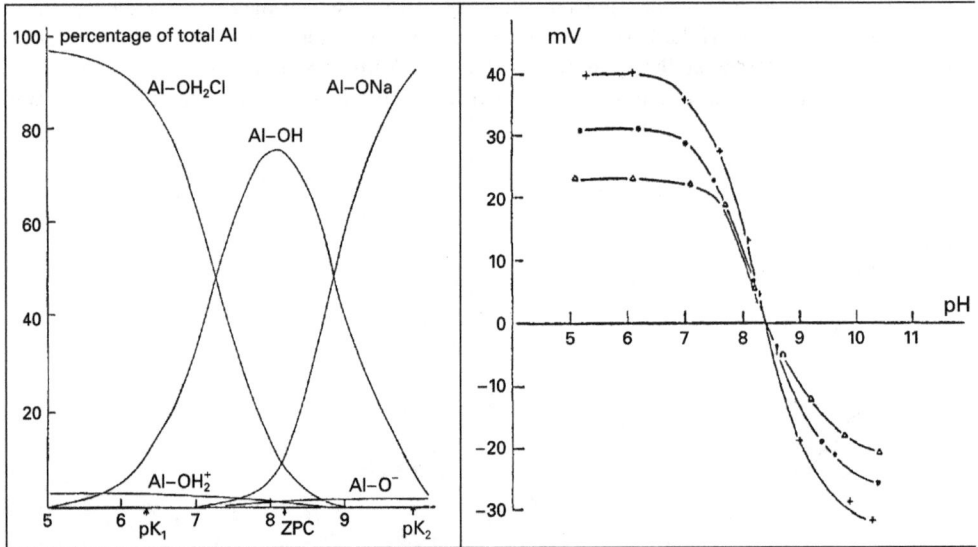

Figure 14.9. a) Distribution of active surface sites vs. pH in solutions with no NaCl. b) Zeta potential of Al precipitates in NaCl electrolyte solutions. Total Al concentration is 1 mM. (NaCl) = 0 mM (+); (NaCl) = 10 mM (o); (NaCl) = 100 mM (∇). (Bottero and Fiessinger 1989).

14.2.3 Papermakers Alum (Al$_2$(SO$_4$)$_3$ x 14 H$_2$O)

As papermaker's alum contains the sulfate ion, the hydrolysis process is distinctly different from aluminium chloride and aluminium nitrate hydrolysis products. For many years there have been conflicting evidence as to the nature of these hydrolysis products. Early studies (Hayden and Rubin, 1974) indicated the presence of five aluminium species, Al^{3+}, $Al(OH)^{2+}$, $Al_8(OH)_{20}^{4+}$, $Al(OH)_3$ and $Al(OH)_4^-$ in the absence of coordination anions. These studies were extended (Arnson, 1982) to the sulfate system, maintaining the existence of $Al_8(OH)_{20}^{4+}$ or $Al_8(OH)_{10}(SO_4)_5^{4+}$ (Matijevic and Stryker, 1966).

Meanwhile it has long been known that precipitates from aluminium sulfate always have the composition 2.5 OH/Al (Brosset et al 1954), later confirmed by Linke and Reynolds (1963), who found the average composition of the floc to be $Al(OH)_{2.44}(SO_4)_{0.28}$ at 70 % neutralisation. Later investigations showed that the relationship between the degree of neutralisation of aluminium sulfate and the composition of the alumina flock was established up to 77 % neutralisation as shown in *Figure 14.10*.

From a material balance calculation the composition was found to be $Al(OH)_{2.3}(SO_4)_{0.28}$ in good agreement with the cited investigations. Al^{27}-NMR have also failed to recognize any $Al_8(OH)_{20}^{4+}$ or $Al_{13}O_4(OH)_{24}^{7+}$ species over a wide range of alum concentrations and degrees of neutralisation and the composition of the precipitate was confirmed to be Al $(OH)_{2.3}(SO_4)_{0.28}(H_2O)_{0.45}$ at 40–75 % neutralisation. XPS (X-ray photoelectron spectroscopy) indicate two Al-atoms (7:1) with different environment and show that the floc has octahedral symmetry. Accordingly, the basic structural unit of the floc is $(Al_7 (OH)_{16}(SO_4)_2(H_2O)_3)^{1+}$ (Pang 1997).

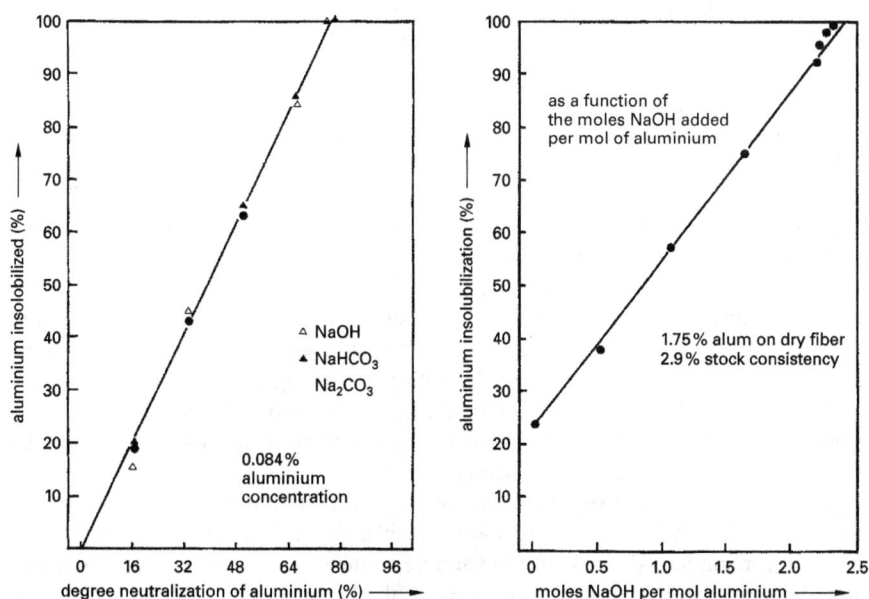

Figure 14.10. % Aluminium insolubilized as alumina floc vs. % degree of alum neutralisation in the absence of fibres (left) and in the presence of fibres (right) (Reynolds 1986).

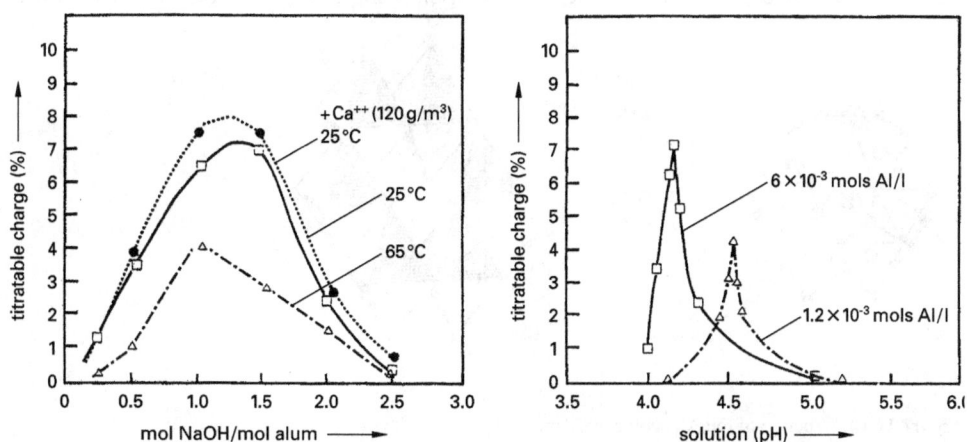

Figure 14.11. a) Effect of heat and Ca^{2+} on titratable (polyelectrolyte titrations) charge of alum versus the molar ratio of NaOH to alum. Conditions. 6×10^{-3} M Al/l. b) Titratable charge of alum as a function of pH and Al concentration (Strazdins 1989).

The basic precipitate is amphoteric in nature but possesses a titratable cationic charge up to an OH/Al ratio around 2.5. The titratable charge is given in *Figure 14.11*. vs. the OH/Al ratio (left) and vs. pH 8 (right). The titratable charge of the precipitate is not very dependent on the presence of metal ions with a charge lower than the charge of aluminium, exemplified with Ca^{2+}, but is dependent on heating of the precipitate. Heating reduces the charge because of temperature induced oxolation processes (formation of Al-O-Al bonds) in the precipitate with a concomitant decrease of pH as hydrogen ions are liberated during heating. The latter effect has important practical implications for the charge characteristics of papermaking systems.

14.2.4 Aluminium Species in Aluminium Chloride and Polyaluminium Chloride Solutions

It has long been known that if aluminium chloride or aluminium nitrate solutions are neutralised the Al_{13} "keggin polymer", see *Figure 14.12*, is formed. The Al_{13} polymer contains both Al (IV) and Al (VI) coordinated aluminium species. Basically the Al_{13} polymer is metastable and its formation is also determined by the exact experimental conditions during its formation. The soluble species present during neutralisation of aluminium chloride, and detected by Al^{27} NMR are Al $(H_2O)_6^{3+}$, $Al_{13}O_4(OH)_{24}^{7+}$ and $Al(OH)_4^-$ (e.g. Bottero and Fiessinger 1989). At a molar ratio of OH/Al = 2.0 almost 100 % Al_{13} polymer can be formed. The reason Al_{13} is not formed from aluminium sulfate is probably that the H_2O ligand in the outer coordination sphere can be partly replaced by HSO_4^- thereby breaking the symmetry.

Through special neutralisation procedures, so called polyaluminium salts can be prepared from aluminium chloride. Their compositions are variable, depending on the manufacturing procedure, but they contain the Al_{13} polymer to variable contents. The formation of tetrahedral $Al(OH)_4^-$ due to inhomogeneous conditions at the point of base introduction is essential for the formation of the Al_{13} polymer. The polyaluminum chlorides are manufactured with varying levels of basicity (OH/Al-ratio) and their formulae may be written as $Al_n (OH)_{mj} Cl_{3n-m}$. There is very little published information on their composition (Pang 1997; Crawford and Flood 1989).

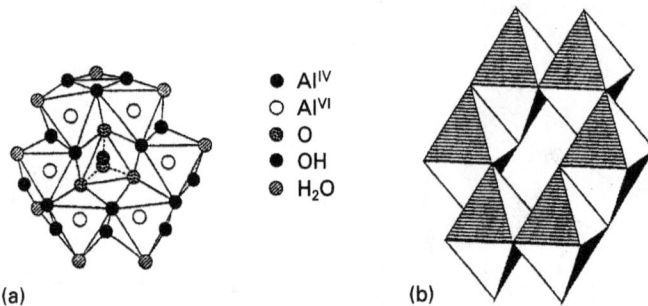

(a) (b)

- ● Al^{IV}
- ○ Al^{VI}
- ◕ O
- ● OH
- ◕ H_2O

Figure 14.12. Structure of the Al_{13} polymer (Pang 1997).

14.2.5 The Aluminium Sulfate-Cellulose System

When aluminium sulphate is contacted with cellulosic fibres, two different processes are insolubilizing the aluminium ions.

- Ion exchange in the acidic region before the aluminium precipitate is being deposited onto fibres.
- Through the deposition of the precipitate, $(Al_7 (OH)_{16}(SO_4)_2(H_2O)_3)^{1+}$.

The situation was illustrated in *Figure 14.10*. The quantity of unhydrolysed aluminium ions adsorbed by the fibres is depending on the number of charged groups in the cell wall of fibres and is thus depending on the fibre choice. Charge stoichiometry exists in the case of single cations. As a rule charge reversal of cellulose does not take place with single cations with a charge less than +3.

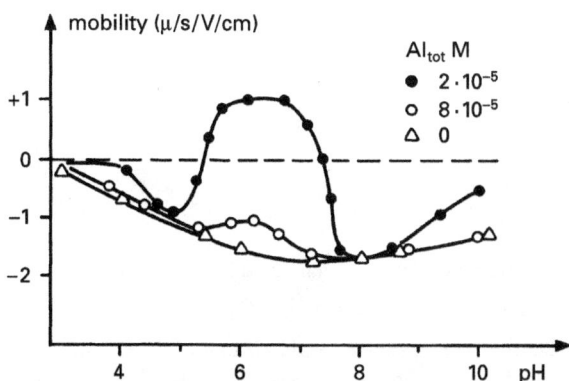

Figure 14.13. The electrophoretic mobility of a microcrystalline cellulose (sol) at different concentrations of aluminiumsulphate (Heingård et al, unpublished).

Figure 14.13 shows how the electrophoretic mobility of microcrystalline cellulose is affected by different quantities of aluminium sulfate. The charge characteristics of the cellulose powder sol are not directly reflected in the charge characteristics of the aluminium precipitate, because the local pH at the surface of the precipitate (and as well as on the surface of fibres) is affected by the surface potential. Hydrogen ions are being attracted by a negative surface and cationic surfaces by a positive surface charge attracting hydroxyl ions. The local pH and the solution (global) are related through the expression:

$$pH_{local} = pH_{solution} + e\Phi/kT \tag{14.13}$$

where e is the elementary charge of an electron and Φ the surface potential. It can be calculated that for every 25 mV surface potential the local pH will be increased one pH unit for a cationic surface. Hence, the local pH for the sol in *Figure 14.11* is much higher and there will be a shift towards higher pH-values in *Figure 14.13*. Basically no charge reversal takes place below pH = 5. The figure shows that aluminium sulfate can recharge a cellulose sol between pH 5.2–7.0 for certain concentration intervals. At higher pH-values, the precipitate is converted to anionic sols by means of aluminol ions.

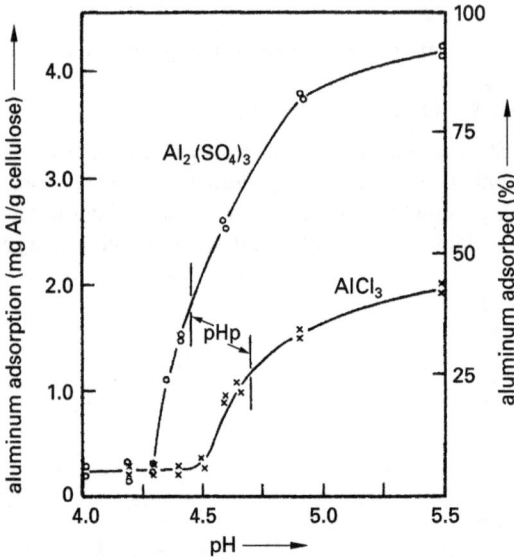

Figure 14.14. Comparison of aluminium adsorption (retention) for aluminium chloride and aluminium sulphate at different pH-values (Arnson and Stratton 1983).

The adsorption or retention of the precipitate increases with pH up to an OH/Al ratio of 3.0. The situation is illustrated in *Figure 14.14*, before decreasing at alkaline pH-values. The Al attached through an ion exchange mechanism can be said to be adsorbed, whereas the deposition of the colloidal precipitate is dependent on the shear conditions. It should also be noted that the accessible charge of the colloidal precipitate is also dependent on the state of aggregation and therefore is also affected by shear (Stratton 1986). The $AlCl_3$ system has a much larger ability to recharge the fibres, by means of the presence of the Al_{13} polymer, thereby reducing further aluminium deposition.

14.3 Rosin Sizing

Rosin is a resinous material that occurs naturally in the oleoresin in pine trees. There are three types of rosin: gum rosin, wood rosin and tall oil rosin. These types are distinguished by their source i.e. the manner in which they are derived from trees. Modern refining processes have made them essentially equivalent in terms of sizing efficiency. Gum rosin is obtained by tapping living pine trees from their oleoresin exudates. Wood rosin can also be obtained by direct extraction of aged wood stumps. Today, the primary rosin source is tall oil, which is obtained by fractional distillation of crude tall oil obtained from the kraft pulping process.

Rosin is composed of various resin acids, with minor amounts of non-acidic material (less than 10 %). Tall oil rosin can contain up to 5 % of fatty acids. The resin acids fall into two groups, the abietic acid group and the pimaric acid group, the structures of which are shown in *Figure 14.15*. They differ in the location of the double bonds and in the type of alkyl group attached to the ring.

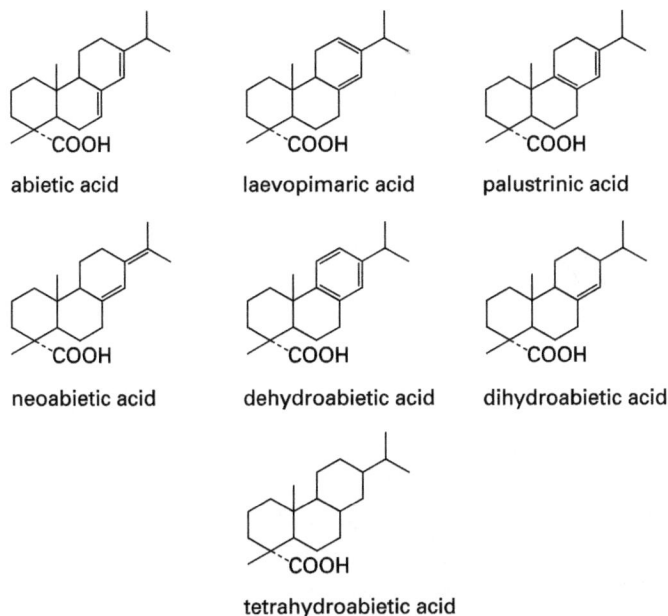

abietic acid   laevopimaric acid   palustrinic acid

neoabietic acid   dehydroabietic acid   dihydroabietic acid

tetrahydroabietic acid

Figure 14.15. The structure of rosin acids.

Commercial sizing agents are so-called fortified rosin sizes, i.e. besides the natural – COOH group in the rosin, further carboxyl groups have been introduced to increase the sizing efficiency. The fortification is achieved by means of reacting levo-pinaric acid with maleic acid anhydride or fumaric acid in a Diels-Alder reaction (see *Figure 14.16*).

The maleopimaric acid anhydride adduct is, in itself, a poor sizing agent and only a smaller fraction (around 20 %) of the rosin is fortified. There are several reasons (Strazdins 1977) why fortification improves sizing efficiency. In the case of soap rosin, the precipitated aluminium resinate particles become smaller. The greater number of carboxylic acids also facilitates the anchoring of the rosin onto the fibre surfaces via the different aluminium complexes. The maleopimaric acid anhydride adduct also has 60 % higher monolayer surface area than abietic acid indicating a horizontal orientation rather than vertical on the surface. Finally, the adduct also have an important role for the emulsification of dispersion sizes. Rosin may also be reacted with formaldehyde, in order to decrease the tendency of the rosin to crystallize.

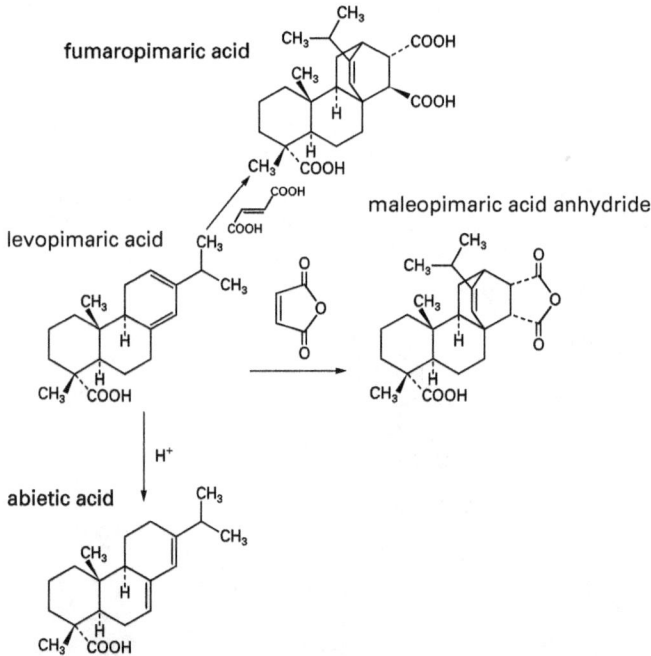

Figure 14.16. Reaction of rosin with maleic acid anhydride.

14.3.1 Types of Rosin Size

Commercial rosin sizes consist of two distinctly different types: highly neutralised (saponified size) and acid rosin particle dispersion (dispersed size):

- Pastes (Na^+-soaps containing 10–30 % free rosin acid) and liquid size (K^+-soap)
- Dispersions (H^+-form with 75–100 % free resin acid, so-called free rosin dispersions).
  There are two types of these. a) protein-stabilized (sterically stabilized) dispersions, with a particle size of 0.5–5 µm. b) Anionically or cationically (electrostatically) stabilized dispersions, with a particle size of 0.2—1 µm

In the Scandinavian countries, group 2, dispersion sizes and particularly the cationically dispersed sizes have come to dominate the market. All types of rosin sizes are used together with alum, and the role of the aluminium ion is to aid anchoring of the rosin acids onto the anionic fibre surfaces. Alum also has a distinct role for the retention of rosin size. The optimum pH-range for rosin sizing is in the range between 4.2–5.5. The quantity added depends on the furnish and the desired degree of sizing and is normally between 0.2 % to 1 %.

There is a decided mechanistic difference between soap sizes and dispersed sizes, so these sizes are discussed under separate headings.

14.3.2 Mechanistic Aspects of Rosin Sizing

In order to obtain good sizing the following conditions must be met:

- The rosin precipitate must have a low surface energy and thus give a high hydrophobicity giving a high contact angle towards water.
- The rosin precipitate must have good retention on the fibres surfaces. The size should be retained in a way to give uniform size distribution on the fibre surfaces. Hence, the size should be retained as a fine dispersed colloid rather than as aggregates. This is particularly important for soap size, where the particles are retained as solid particles with no ability to spread during drying.
- The wet rosin precipitate must be transformed to a stable low energy surface, which must remain unchanged during contact with the liquid in question.

In all sizing with rosin (hereinafter designated as Hab), the hydrophobic product is a reaction product between an aluminium salt and rosin. In this context the different resin acids are not distinguished and are simply referred to as abietic acid (Hab). When soap size is precipitated with alum the precipitate has a chemical composition corresponding approximately to a mixture of aluminium diabietate and abietic acid (Davison 1964) approximately independent of the pH (Traser and Jayme 1973) of precipitation. It must be emphasized, however, that the reaction product has not a stoichiometrically given composition. The precipitate is a highly hydrophobic substance with a contact angle around 90–105° towards water. If the rosin soap is precipitated with acid, a considerable lower contact angle, around 50–70° is obtained. The reason can be traced to the amphiphatic nature of the precipitate. When the rosin acid is being contacted with water, the molecule will turn its hydrophilic carboxyl group towards the water.

During the drying of a rosin/aluminium precipitate, the polar carboxyl groups are oriented towards the cellulose interface, whereas the nonpolar parts will become oriented towards the air phase in order to minimize the surface energy of the system. If the molecules are not sufficiently anchored to the surface of the fibres, they will turn their polar group towards water (overturning) and there will be a loss in the sizing efficiency. The role of the aluminium ions is to prevent the turning of the polar ends by anchoring them onto the fibre surface.

The reason why aluminium has favourable properties for anchoring the precipitate to the fibres can be sought in their ability to form polynuclear charged complexes with rosin and the carboxylic groups on the fibre surfaces. During drying the aluminium/rosin complex can form polymeric species via oxolation reactions (Kamutzki and Krause 1984; Subrahmenyam and Biermann 1992) and the species responsible for rosin sizing may be pictured as in *Figure 14.17*. The carboxyl groups on the fibres play a profound role for sizing and it has long been known that fibres with a low carboxyl group content are difficult to size with alum/rosin.

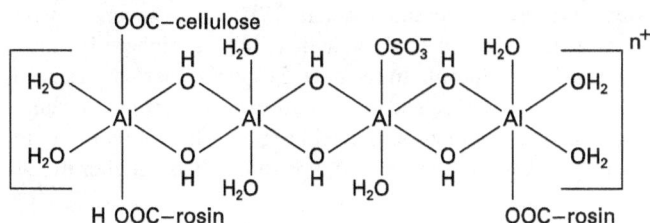

Figure 14.17. Hypothetical structure of aluminium rosin complex (Subrahmenyam and Biermann 1992).

14.3.3 Sizing with Rosin Soap (NaAb)

The degree of neutralisation varies from 75–100 % in commercial size formulations. The soap is sold either as a dry solid, a high solids thick paste (70–77 %) or as a lower solids (35–60 %) fluid size. Dilute aqueous dispersions (pH 8.0–9.5) of these saponified sizes consist of micelles containing 50–100 molecules (< 0.01 μm) and suspended particles (0.1 μm) of neutrals and unsaponified rosin acid.

During sizing, with rosin soap, the rosin is precipitated with aluminium salts in the thick stock and colloidal particles in the size range between 0.01 μm to 0.1 μm are obtained. As shown in fig. 4, these particle are cationic in the pH-interval between 4–6.5 if there is an excess of aluminium ions, and the (Al)/(Ab) molar ratio is greater than 0.5. These colloidal particles are easily retained by the anioic fibres. As shown in *Figure 14.18*, Hab is negatively charge throughout this pH-interval if common acid is used for neutralisation.

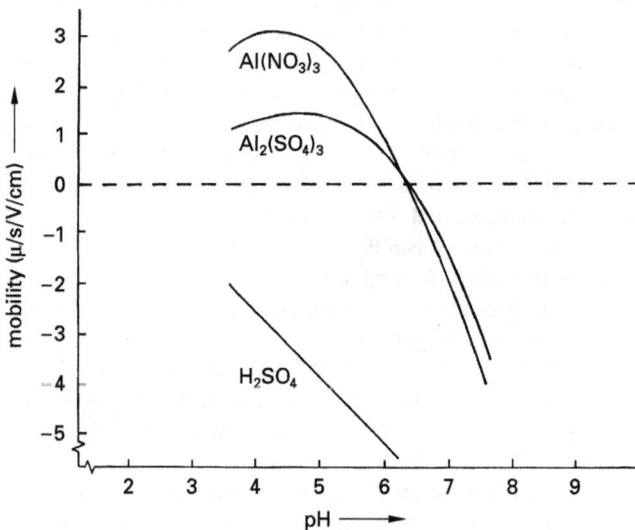

Figure 14.18. Electrophoretic mobility of rosin precipitates at different pH-values using different acids to neutralise sodium abietate (Lindström and Söremark 1977).

Aluminium nitrate forms more highly charged precipitates, but there is no advantage of using highly charged particles as fibre recharging will lead to a lower size retention. The retention mechanism for the aluminium resinate is therefore heterocoagulation. There are several reasons why it is not possible to use alum/rosin sizing at higher pH-values. Firstly, at higher pH-values aluminate ions $Al(OH)_4^-$ are formed instead of aluminium resinate. Secondly, the soap is deprotonized at high pH-values, which leads to overturning and deterioration of the formed low energy surface (Strazdins 1984). Sizing is therefore limited to the acid region. The optimum pH for soap rosin sizing is between pH 4.2 to 5.0. A higher size retention is required, the higher the pH as illustrated in *Figure 14.19*.

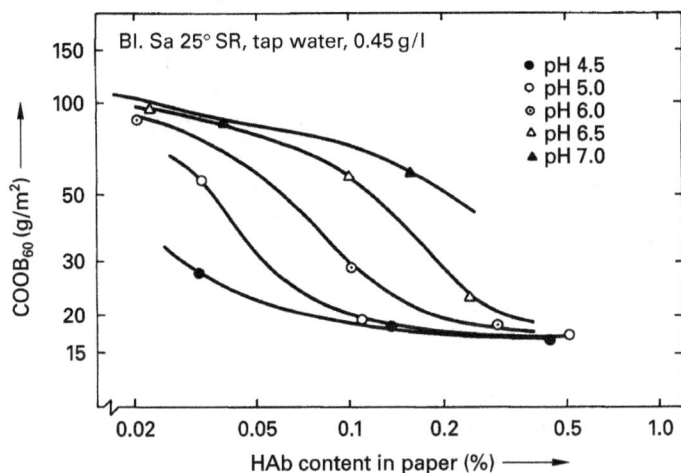

Figure 14.19. A higher quantity of rosin acid is required at a high pH-value than at a lower pH-value in order to obtain the same Cobb-value (Lindström and Söderberg 1983).

Particle size is critical in soap rosin sizing. A finely dispersed precipitate is preferred for optimal sizing. Under practical conditions agglomeration always takes place. This may give better retention but deteriorates sizing efficiency. A high electrolyte conc., a high soap conc., a high temperature and long contact time favour particle growth. Under optimal conditions a pulp may be sized with only a fraction of size required in practical applications. The rosin content in paper is therefore a poor predictor of sizing.

Electrolytes interfere with the heterocoagulation, decreasing the interaction between the cationic precipitate and the anionic fibre surfaces (Lindström and Söderberg 1983, 1984). Ca^{2+} and Mg^{2+} also interfere with sizing as they may react with the rosin, preventing the formation of aluminium resinate (Strazdins 1977, 1984). Alum is usually added after size with soap sizes, but a reversed addition is suggested in systems with a high content of divalent cations, so the rosin can react with the aluminium before the divalent cations.

Anions, which coordinate with alum, such as tartrate, oxalate and citrate, compete with Hab forming the sizing complex and are therefore detrimental to sizing. Lignosulfonate, pentosans and other highly charged substances in the stock recharge the cationic precipitate and also interferes with size retention and sizing efficiency (Lindström and Söremark 1977). Mechanical, thermomechanical and chemimechanical pulps with a high specific surface and often a high load of anionic substances are particularly difficult to size with soap rosin.

14.3.4 Sizing with Rosin Dispersions

The sizing mechanism with rosin dispersion is different from that with soap rosin. Here size retention, spreading and complex formation are the consecutive elements in sizing. The reaction between the rosin and aluminium takes mainly place in the drying section in the papermachine after spreading/surface diffusion of the dispersion droplets has taken place. Retention conditions of the rosin dispersion and the aluminium ions are therefore separate events. Commercial rosin dispersion sizes are often cationic (Eberhardt and Gast 1988), therefore they are self-attaching onto negatively charged fibres. More important, however, is the use of common retention aids to enhance the retention of both rosin and aluminium species (Davison 1988).

The addition of aluminium sulfate before the rosin dispersion (reversed addition) mode is recommended for anionic dispersed sizes (Marton 1989). Aluminium adsorption on the fibre surfaces gives cationic charge sites, where anionic rosin dispersion particles can be attached onto the fibre surfaces. The rosin dispersion is usually added to the thick stock, before the mixing pump, before dilution in the short circulation, since the collision frequency between fibres and rosin particles is higher at a higher stock consistency.

Since, the softening temperature for a rosin dispersion is around 60 °C, flowing and spreading, in principle to a monomolecular layer, can take place in the drying section, after which the rosin can react with the adsorbed colloidal aluminium species. The aluminium resinate formed in soap sizing has a sintering temperature around 120 °C and can not spread on the fibre surfaces as rosin dispersion sizes. Much less is known about the chemical composition of the reaction product, formed in dispersion sizing compared to soap sizing, but it is generally assumed that the reaction product has a composition close to that depicted in *Figure 14.17*.

Dispersed sizes do not interfere with paper strength to the same extent as rosin soap sizes, because the aluminium resinate can not redistribute during drying and will interfere with paper strength. Dispersed rosin particles contained in the fibre-fibre bond areas can not spread (Strazdins 1988), and hence, will not interfere with fibre-fibre bonding.

Filler materials with a high specific surface area are in general detrimental to both soap sizing and sizing with dispersed sizes (Marton 1986).

14.3.5 Neutral Rosin Sizing

Rosin size can not be used for neutral sizing for various reasons. Firstly, hydroxyl ions compete with the rosin acid in the complexation with the aluminium species. The hydrolysis of aluminium species may to some extent be minimized by decreasing the contact time between alum and the stock and by using alum at a low temperature. Secondly, the formation of rosin soap, will be an effective overturning agent to destroy the sizing. Thirdly, rosin reacts with calcium ions forming in calcium carbonate containing papermaking systems. Modern cationic rosin dispersions do, however, function better than anionic dispersions at higher pH-values, but this is not a solution to the problem.

In spite of these facts, there have been substantial efforts over the years to use rosin in the alkaline region and various strategies have been deviced:

- The use of rosin esters
- The use of polyaluminium compounds
- Incorporation of aluminium species in the rosin emulsion
- The use of alternative mordants to aluminium compounds

Rosin esters, such as glycerol rosinate, are in themselves no effective sizing agents because esters are difficult to anchor to cellulosic surfaces. It has, however, been found (e.g. Wang et al 1999) that esters such as glycerol diresinate stabilize the acid rosin emulsion towards dissolution of sodium rosinate. Therefore, there is an optimum amount of rosin ester for optimum sizing in such formulations. Glycerol dirosinate has been found to be effective in such formulations.

Aluminium species hydrolyse under aqueous alkaline conditions and polyaluminumchloride (PAC) is often used instead of papermakers alum. PAC is, however not a particularly effective sizing agent together with rosin and it is also necessary to use effective retention agents together with PAC. By incorporation of aluminium species in the rosin dispersion, hydrolysis can to some extent be minimized. Thirdly alternative mordants may be used (Biermann 1992; Shimada et al 1997; Wu et al 1997; Wang et al 2001). Polyallylamine has, for instance been found to be an effective mordant under alkaline conditions. It is possible to use other metal ions than aluminium ions e.g. ferric and ferrous ions (Zhuang and Biermann 1993, 1995), but these particular ions will also destroy the brightness of paper. None of these neutral rosin sizing strategies have been practically utilized to any larger extent.

14.3.6 Sizing with Stearic Acid

Instead of using rosin acids, fatty acids such as stearic (C-18) acid can be used. The sizing mechanism is very similar to rosin sizing and it functions basically in the acidic region between pH 4–5.5. The use of stearic acid is limited to specialty papers, photographic papers being the principal application. Stearic acid provides excellent sizing against aqueous solutions and its principal advantage compared to rosin sizing is that saturated fatty acids will not oxidize, providing excellent stability towards brightness/whiteness loss. Rosin is, however, more cost efficient.

14.3.7 Self-Sizing and Sizing Loss

Self-sizing is a process, primarily occurring in wood-containing papers or papers containing a high amount of extractives. The reason for self-sizing is that resins and extractives with a low softening point are redistributed on the fibre material by means of surfaces diffusion. If alum has been used during paper manufacture, the extractives may also react with the colloidal aluminium species in the paper. It is thermodynamically favourable for resin in parenchyma cells to spread to a monomolecular layer formed at the interface between air and the fibre surface, since a high-energy surface (cellulose) is being transformed into a low energy surface. The mechanism is analogous to the way in which windows become greasy through vapour phase diffusion and condensation of surfactants on to the high surface energy glass surface. Self-sizing can also depend on the reorientation of amphiphatic molecules in the boundary surface.

The opposite to self-sizing is called sizing loss. It is relatively common for pulps with a high buffer capacity (high acidic group content), which have been manufactured by alkaline cooking processes. If there are residual Na-ions adsorbed onto the acidic groups, they can saponify rosin, which has not reacted with alum, so that the sizing is being destroyed.

14.4 Cellulose Reactive Sizing Agents

Over the years, there have been great efforts to try to develop cellulose reactive sizing agents (see *Table 14.1*). The assumption in these developments have been that the covalent linkage allows permanent attachment of hydrophobic groups in a highly oriented state, which makes sizing possible at very low levels of added chemical. The main requirement of the molecule is that it should have a balance between the reactivity towards water, because of the necessity of making stable emulsions or dispersions, and its reactivity towards cellulose. These assumptions are to some extent mutually exclusive and a compromise must be sought.

Although, many different types have been tried out over the years the most important sizes used are the Alkyl Ketene Dimers (AKD) and the Alkenyl Succinic Anydrides (ASA). These sizing agents are at the opposite in terms of stability of hydrolysis and reactivity towards cellulose, where AKD´s are the least reactive species and fairly stable towards hydrolysis, whereas ASA´s are very reactive towards cellulose, but also sensitive to hydrolysis.

14.4.1 Sizing with Alkyl Ketene Dimers (AKD)

Alkyl ketene dimers were the results of direct development efforts in the 40´s. These investigations demonstrated thet the parent molecule, the diketene could derivatize hydroxyl groups and in particular those of cellulose. The strained lactone ring in ketene dimers can react both with cellulose and water forming either the β-keto ester or the β-keto acid, which spontaneously decarboxylates to the corresponding ketone as shown in *Figure 14.20*. The ketone is incapable of reacting with cellulose. The balance of reaction and hydrolysis is subtle, as with all reactive sizes, but the reaction is favoured and, hence, AKD can be used as a sizing agent under commercial papermaking conditions. The nucleophilic reaction with cellulose can also be accelerated with various so-called promoters, which will be discussed below.

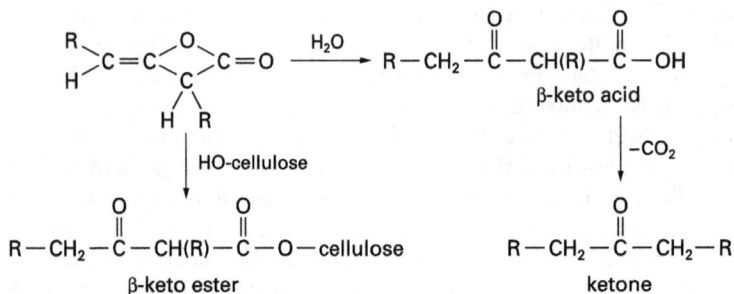

Figure 14.20. Alkyl ketene dimers can react either with cellulose forming the β-keto ester or with water forming the β-keto acid, which spontaneously decarboxylates forming the corresponding ketone.

Table 14.1. Some cellulose reactive sizing agents.

Compound	Structure	R	Cellulose Derivative	Ref
Acid cloride	O $\parallel$ $R-C-Cl$	$C_{14}-C_{18}$	O $\parallel$ $R-C-OCell$	(a)–(d)
Acid anhydride	$O \quad\quad O$ $\parallel \quad\quad \parallel$ $R-C-O-C-R$	$C_{17}H_{35}$	O $\parallel$ $R-C-OCell$	(e)–(j)
Enol ester	$O \quad\quad CH_3$ $\parallel \quad\quad \mid$ $R-C-O-C=CH_2$	$C_{17}H_{35}$	O $\parallel$ $R-C-OCell$	13
Alkyl ketene dimer	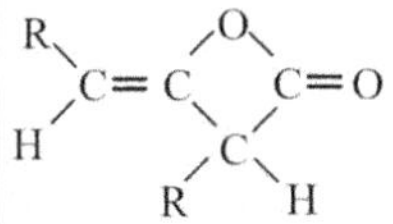 (ketene dimer ring structure)	$C_{14}-C_{18}$	$O \quad\quad O$ $\parallel \quad\quad \parallel$ $RCH_2-C-CHR-C-OCell$	(k)–(r)
Alkyl isocynate	$R-N=C=O$	$C_{17}H_{35}$	$H \quad O$ $\mid \quad \parallel$ $R-N-C-OCell$	(s)–(y)
Alkenyl succinic anhydride	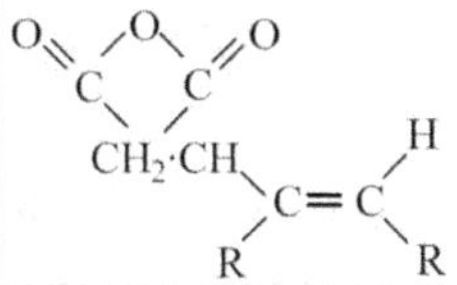 (succinic anhydride structure)	C_1-C_6	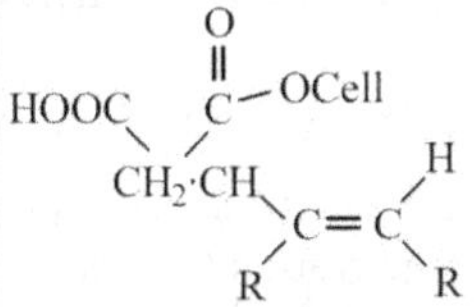 O $\parallel$ HOOC $\;C-OCell$ $CH_2\cdot CH$ $C=C$ $R \quad\quad R$	(z)
Rosin anhydride	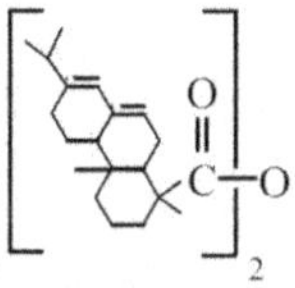 (rosin anhydride structure)		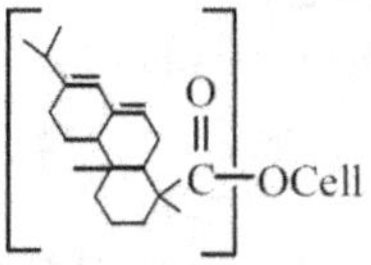 (rosin ester structure)	14

(a) German Patent 2,423,651 (1974); (b) German Patent 2,611,827 (1976); (c) German Patent 2,611,746 (1976); (d) US Patent 4,123,319 (1978); (e) British Patent 954,526 (1964) (f) Canadian Patent 770,079 (1967); (g) US Patent 3,102,064 (1963); (h) US Patent 3,409,500 (1968); (i) US Patent 3,455,330 (1969); (j) US Patent 4,207,142 (1980); (k) US Patent 2,627,477 (1953); (l) US Patent 2,785,067 (1957); (m) US Patent 2,762,270 (1956); (n) US Patent 2,856,310 (1958); (o) US Patent 2,865,743 (1958); (p) US Patent 2,961,366 (1960); (q) US Patent 2,986,488 (1961); (r) US Patent 3,483,077 (1969); (s) US Patent 3,050,437 (1962); (t) US Patent 3,589,978 (1962); (u) US Patent 3,492,081 (1970); (v)US Patent 3,310,460 (1967); (w) US Patent 3,627,631 (1971); (x) US Patent 3,499,824 (1970); (y) US Patent 3,575,796 (1971); (z) US Patent 3,821,069 (1974)

Commercial AKD′s are prepared from long fatty acids via their acid chlorides, which then dimerize to the corresponding alkyl ketene dimer.

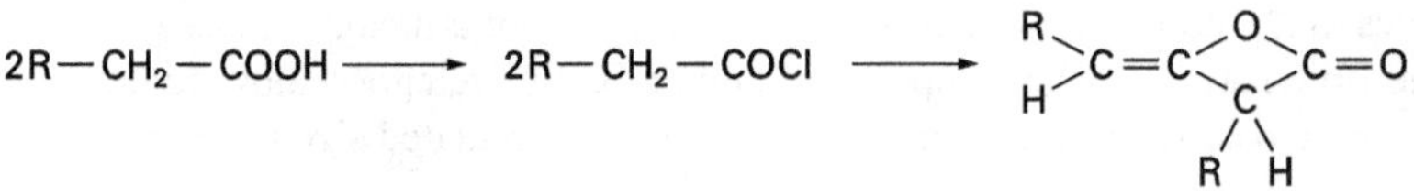

$$2R-CH_2-COOH \longrightarrow 2R-CH_2-COCl \longrightarrow$$ (alkyl ketene dimer structure)

Figure 14.21. Formation of alkyl ketene dimers from the corresponding fatty acid chlorides.

The linear saturated AKD′s are waxy substances, water insoluble solids with melting points around 50 °C, when manufactured from commercially available fatty acids, being mixtures of C-14 to C-18 fatty acids (*Figure 14.21*). C-16 is the dominating fatty acid in the most used formulations. The sizing efficiency increases with carbon chain length from C-8 and levels off at C-20.

14.4.2 Emulsification

AKD′s are dispersed by using high pressure homogenizers at elevated temperatures. Most frequently used stabilizers are cationic starches in conjunction with lignosulfonates/naphthalene sulfonic acids. Waxy maize starches with no propensity to retrogradation are the preferred choice of starch. It is important to avoid surface active substances in the dipersion formulation, because they may interfere with sizing. The dispersions are usually made slightly cationic in order for them to have a natural substantivity to negatively charged fibres, but anionic dispersions are also used in commercial practice. In order to avoid hydrolysis of the AKD, the pH is kept around 3 in the formulations.

14.4.3 Consecutive Events of AKD-Sizing

The consecutive events in AKD-sizing are:

- Retaining the AKD-size using appropriate retention strategies for the size
- Spreading/size migration to a monolayer
- Reaction of the sizing agent with the cellulosic fibres

These events are schematically illustrated in *Figure 14.22*.

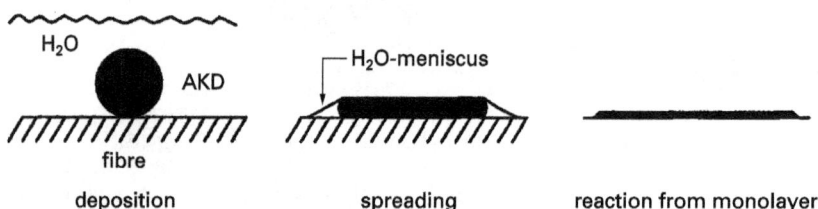

Figure 14.22. Consecutive events in sizing with AKD. AKD-deposition, spreading and reaction.

The retention mechanism is, in theory, heterocoagulation, where cationic size particles are being attached to the negatively charged fibres. This is expected to give a good distribution of the dispersion particles, but practice shows that size distribution is not critical, because of extensive spreading on the fibre surfaces. More important is that effective retention aids are used for the purpose. A high single pass retention is important, because recirculated size is being hydrolyzed in the white water.

When the size particles have been deposited on the fibres, there can be no reaction because very few molecules are in molecular contact with cellulose. Because of the high surface tension of water, no spreading can take place until the water has been removed and the size particles are in direct contact with air. Hence, an air-AKD surface must be formed before spreading can take place and this takes place during drying at a solids content exceeding 60 %. The spreading continues until a monomolecular layer has been formed. This layer then reacts with the hydroxyl groups of cellulose. This mechanism is illustrated in *Figure 14.23*, where the degree of reaction and the solids content of the paper have been plotted versus the drying time. The reaction is slow at low pH-values and, in practice, AKD can not be used except in the neutral or slightly alkaline pH-range. This will be even more obvious when the use of sizing accelerators are considered.

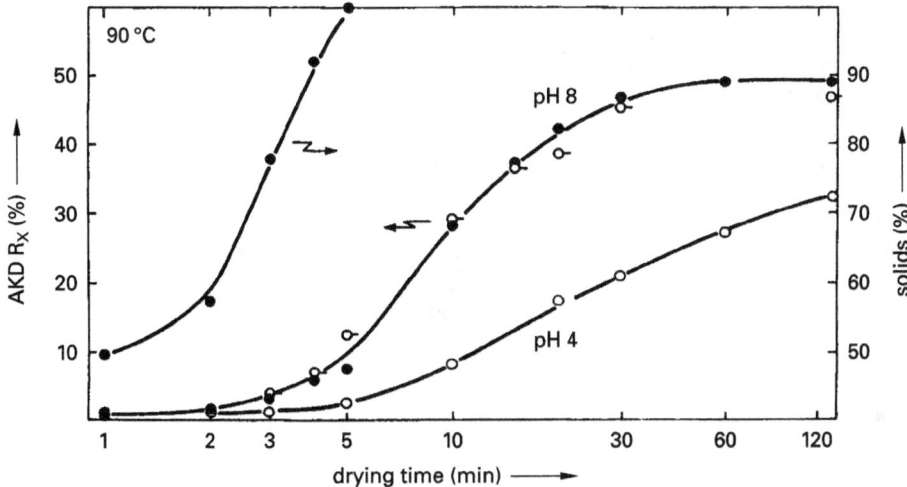

Figure 14.23. Extent of AKD-reaction and sheet solids content vs drying time (bleached sw kraft pulp (Lindström and O'Brian 1986).

AKD is especially sensitive to the presence of extractives of the fatty acid type, since these become saponified at alkaline pH-values and compete with the spreading, forming an autophobic layer blocking AKD-spreading. Poor AKD-spreading leads to poor chemical reaction with the fibre. For these reasons, groundwood, thermomechanical and chemimechanical pulps as well as chemical pulps containing high extractive contents are difficult to size with AKD. Elimination of extractives is, hence, essential to sizing with AKD. The blocking of extractives by the use of neutralised aluminium salts (PAC-type) have been suggested to aid AKD-sizing.

14.4.4 Retention of AKD

Cationic AKD-particles are retained by a heterocoagulation deposition mechanism onto the negatively charged fibres, but can also be retained by using common retention aids such as cationic starch or cationic synthetic resins of various types. This may seem contradictory, but AKD is often stabilized by cationic starch and lignosulfonates/naphthalene sulfonic acid formulations and, hence, such AKD particles are amphoteric in nature and can react with both anionic and cationic retention aids. Actually, the appropriate choice of an efficient retention aid system is crucial, because AKD will be subject to hydrolysis when circulated in the short circulation. Anionic dispersions may be equally good retained by the use of cationic retention aids (see *Figure 14.24 b*).

In the practical use of AKD, the point of addition is, of course, dependent on the wet end system and the addition order of other chemical adjuvants. As a rule of thumb it is advantageous to add AKD to the high consistency stock only a short time before dilution takes place in the short circulation. The AKD–particles are deposited at a fast rate at high stock consistency but are subsequently sheared of the fibre surfaces by the other fibres in a stirred pulp suspension. Both high shear and long contact times are known to reduce AKD retention. As fines and fillers constitute a high surface area material it is advantageous to have a high fines and filler retention. As a rule, both AKD, fines or fillers follow the general retention level and there is little selective retention of certain types of dispersed material. Anionic charged dissolved substances in the stock, such as hemicelluloses and lignin residues, are generally detrimental to size retention.

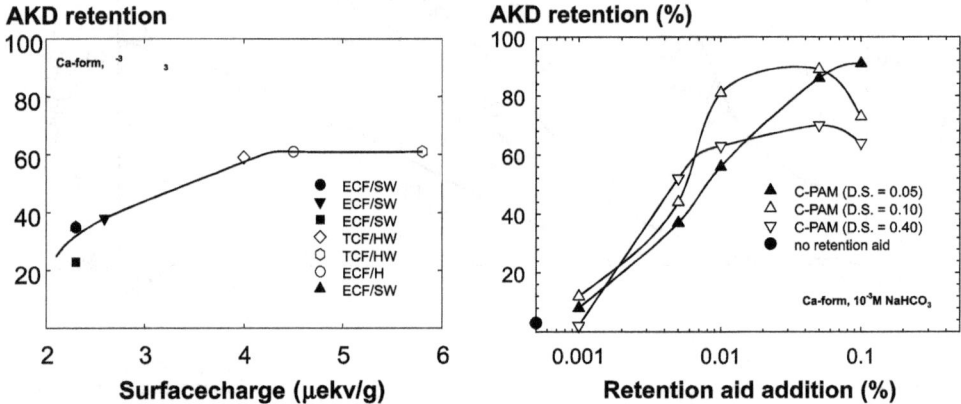

Figure 14.24. a) Retention of cationic AKD-particles onto various TCF and ECF kraft pulps using cationic PAM (Johansson and Lindström, 2003). Unpublished data.
b) Retention of anionic AKD-particles onto using cationic PAM´s with different charge densities (Johansson and Lindström, 2003). (Unpublished data).

14.4.5 Spreading/Size Migration

The distribution of AKD-size on the fibres occurs in the drying section as discussed above.
AKD readily spreads on cellulose, because the cellulose surface is a high energy surface.
The free energy of spreading, ΔGs, for AKD on a cellulose surface can be written:

$$\Delta G_s = \gamma \text{Cell/AKD} + \gamma \text{AKD} - \gamma \text{Cell} \qquad (14.14)$$

The surface free energy of AKD is 33 mN/M and the surface free energy of dry cellulose is around 45 mN/m or higher when cellulose is moist. If γ Cell/AKD is small, the conditions for spreading, $\Delta GS < 0$, are fulfilled. For an AKD particle trapped in between a fibre-fibre bond the free energy of spreading, $\Delta GS = 2\gamma$ Cell/AKD, cellulose-cellulose is a positive quantity, because it is associated with the cleavage of a high energy surface. Hence, AKD-particles trapped in between fibre-fibre bonds can not spread and can not react with cellulose. Spreading or more correct surface diffusion has been manifested by several investigators (Horn 2002; Roberts and Garner 1987; Seppänen and Tiberg 2000; Ödberg 2002). The spreading of AKD on cellulose should, however, not be associated with the common hydrodynamic phenomena of spreading (Cazabat 1989), which is a very rapid process. Instead, spreading takes place by the surface diffusion of an autophobic monolayer of AKD on cellulose (Seppänen and Tiberg 2000), which is a slower phenomenon than hydrodynamic spreading. The apparent surface diffusion coefficient of AKD on celluloise have been calculated to around 10^{-11} m^2/s at 50–80 °C (Seppänen and Tiberg 2000; Ödberg et al 2002). As AKD-sizing particles typically have the dimension of the order of a micron, such a droplet would on a cellulose surface spread within 10 sec, using this diffusion coefficient. The time of reaction is typically of the order of at least 5 minutes (comparing *Figure 14.23*), hence spreading/surface diffusion is not the rate-determining step in sizing.

14.4.6 AKD Cellulose Reaction

The ability of AKD to be oriented and immobilized on fibre surfaces is related to its molecular structure and reactivity. The four-membered lactone ring in AKD allows it to react with nucleophiles to form β-keto ester bonds with cellulose. The reaction of AKD with cellulose to form the β-keto ester is the primary mechanism by which AKD sizes paper. The covalent bond provides immobilisation and orientation of the hydrophobic tail outward, away from the fibre surface. The formation of the β-keto ester has been established in a number of publications by extraction of C-14 labelled AKD (Lindström and Söderberg, 1986), by FTIR (Ödberg et al 1987) and by solid state C-13 NMR (Bottorf 1993, 1994).

Assuming a bimolecular nucleophilic reaction between cellulose and AKD can be written (Lindström and O'Brian 1986):

$$\mathrm{d}x/\mathrm{d}t = K_0 \, [\text{Cell}] \cdot [\text{AKD}] \tag{14.15}$$

where x is the reacted amount of AKD and [Cell] and [AKD] are the molar concentrations of reactants.

If there is an excess of cellulose hydroxyls available for reaction, [Cell] >> [AKD], the equation reads:

$$\mathrm{d}x/\mathrm{d}t = K_1[\text{AKD}] = K_2 \, [\text{AKD}]_{av} = K(a\text{–}x) \tag{14.16}$$

where a is the maximum amount of reacted AKD and $[\text{AKD}]_{av}$, the available amount for reaction. Integration leads to:

$$\int \mathrm{d}x/(a\text{–}x) = \int K\mathrm{d}t, \text{ and } \ln a/(a\text{–}x) = Kt \tag{14.17}$$

So, if the quantity $\ln a/(a\text{–}x)$ is plotted versus time, straight lines should be obtained and the reaction follows what could be called a pseudo first order reaction. This is illustrated in *Figure*

14.25a, which shows such graphs for the reaction of AKD onto a bleached kraft pulp at different pH-values. It is also shown that the straight curves intersect the x-axis at a common point in time independent of pH. This is the time interval required to remove water from the sheet structure to sufficiently high solids content for spreading to occur. As expected the drying time to the solids content where spreading takes place is independent of the pH-value as the rate of drying is independent of pH for this type of pulp. If the same experiment then is performed at different pH-values, the corresponding values between the reaction rate constant K at different pH-values and temperatures can be obtained and the Arrhenius activation energies for the AKD/cellulose reaction can be calculated from the Arrhenius graphs in *Figure 14.25b*.

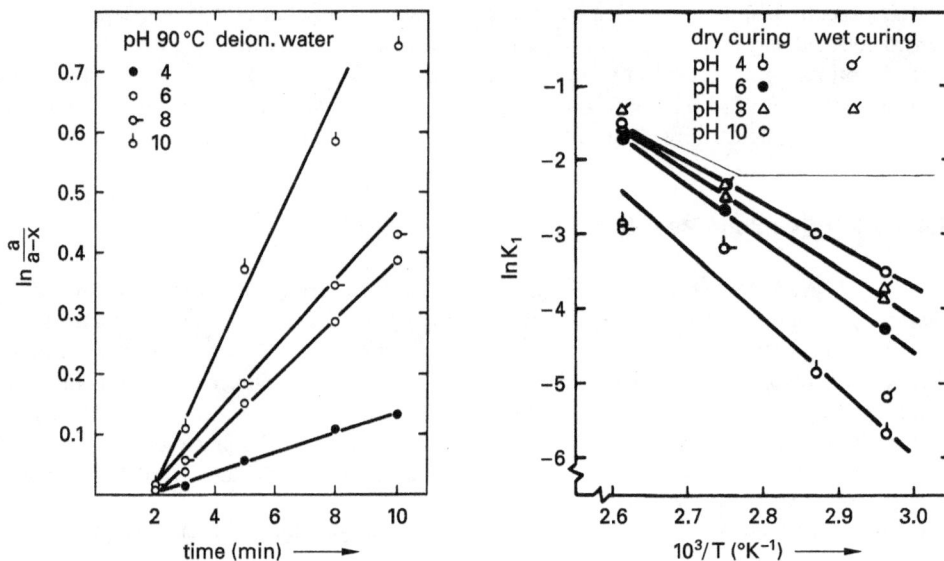

Figure 14.25. a) The quantity ln $a/(a-x)$ versus reaction time, t, where a is the maximum reacted AKD and x, the reacted amount of AKD at different pH-values (left). b) The reaction rate constant K, from equation 2 versus $1/T$, where T is the absolute temperature (Lindström and O´Brian 1986).

The so calculated activation energies for the reactrion between cellulose and AKD can be calculated from fig. 6b to be 72 KJ/mole at pH 4 to 46 KJ/mole at pH 10 (Lindström and O´Brian 1986). As expected the activation energy decreases for the nucleophilic reaction the higher the pH.

14.4.7 Sizing Accelerators

The reaction between AKD and cellulose is, however slow and sizing accelerators are invariably used in commercial operations, whereby the reaction rate easily can be increased by an order of magnitude. The most important sizing accelerations are:

- HCO_3^-
- Basic polymers with amine groups

The HCO_3^--ion has a unique ability to catalyse the reaction between AKD and cellulose. HCO_3^--ions exist in natural systems, e.g. when $CaCO_3$ is used as a filler, but it is general practice to add $NaHCO_3$ to increase the alkalinity of the stock to cellulose fibres is analysed. *Figure 14.26a* shows how the reaction of AKD by $NaHCO_3$. A further analysis of the catalysis reveals that the reaction rate is proportional also to $[HCO_3^-]$. Hence, the reaction follows the equation:

$$dx/dt = K_0 \, [\text{Cell}] \cdot [\text{AKD}] \cdot [HCO_3^-] \tag{14.18}$$

This equation clearly suggests that there is a trimolecular reaction between cellulose, AKD and HCO_3 taking place. The suggested mechanism of catalysis is given in *Figure 14.26b*. Obviously the HCO_3^--ions acts as a proton transfer agent.

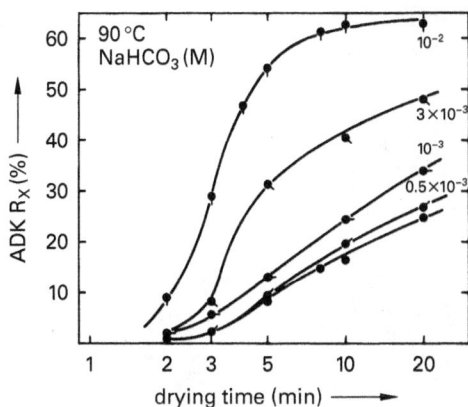

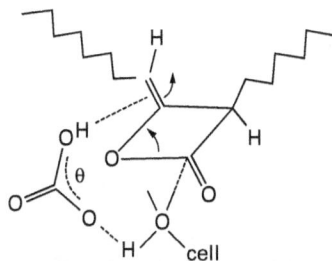

Figure 14.26. a) The amount of reacted AKD vs time of reaction at different $NaHCO_3$ concentrations in the stock. b) Mechanism of catalysis with $NaHCO_3$ (Lindström and Söderberg 1987).

Polymeric amines (e.g. PAMAM-EPI resins) having amino groups with a free electron pair are classic sizing accelerators for AKD. (Lindström and Söderberg 1987; Thorn et al 1993; Cooper et al 1995). Several different types of condensation polymers have been investigated over the years and are used in commercial contexts. The polymeric sizing accelerators are often added either to the AKD-dispersion ("rapid curing dispersion") or used separately as combined accelerator and retention aid. Moreover, the effects of simultaneously using polymers and HCO_3 are synergistic as shown in *Figure 14.27*.

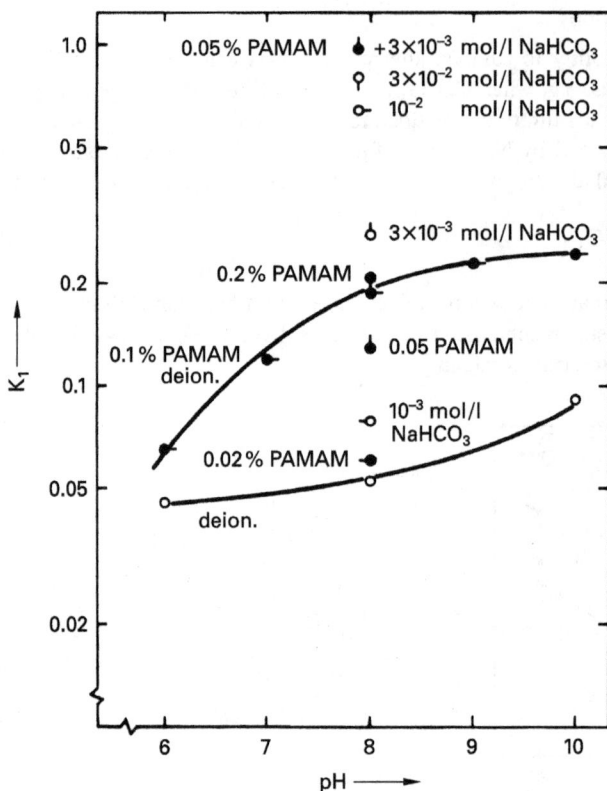

Figure 14.27. Synergistic effects between HCO_3 and PAMAM-EPI resin on the reaction rate constant, K, when simultaneously used in AKD-sizing (Lindström and Söderberg 1987).

14.4.8 AKD-Hydrolysis

AKD can also react with water forming the β-keto acid, which spontaneously decarboxylates forming the corresponding ketone. AKD is, however, stable at room temperature at acidic pH-values, allowing storage at the time scale of months. The effect of pH on hydrolysis is shown in *Figure 14.28*. The higher the pH, the faster is the hydrolysis.

The hydrolysis is slow compared to hydrolysis of ASA, but is still of significance under practical papermaking conditions (Colasurado and Thorn 1991, Marton 1990, 1991), when process water with AKD is being recirculated in the mill. High single pass retention of AKD is therefore of high importance.

PCC has also been shown to have a strong catalytic effect on AKD-hydrolysis (Jiang and Deng, 2000), which is also of importance for fugitive sizing of PCC-containing papers. Residual amounts of alkali from PCC-manufacture is most probably responsible for hydrolysis.

The hydrolysis product of AKD, the ketone, has a slight positive effect on sizing, provided there is some reacted AKD already present in the paper (Lindström and Söderberg 1986), but is in itself no sizing agent.

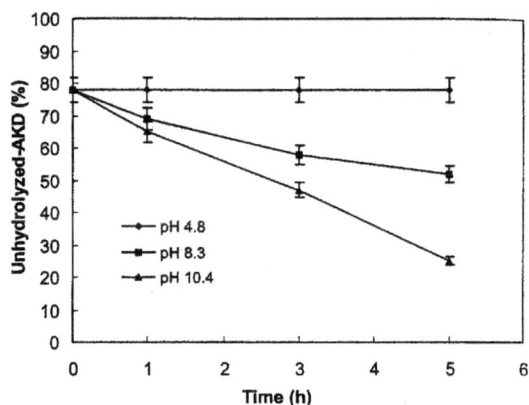

Figure 14.28. The effect of pH on hydrolysis of AKD at 50° C (Jiang and Deng 2000).

14.4.9 Amount of AKD Required for Sizing

The required amount of AKD for sizing for a given pulp depends on a number of variables and is also linked to a number of wet-end factors. Critical is the retention of the size and the extent of reaction together with the nature of the pulp furnish and the structure of the sheet.

Retention is critical, as recirculated size can be the subject of hydrolysis. The extent of reaction depends on the drying conditions together with the presence of size accelerators. The extent of reaction for AKD-sizes is dependent on the fraction of fibre surface exposed to the air phase, because it is only size on free surfaces, which can spread and potentially react with cellulosic fibres. AKD spreads all over the sheet and sizing with AKD is not dependent on size agglomeration, which is critical for instance with soap/alum sizing. The presence of surface active substances, such as extractives and certain defoamers, may be critical if they spread faster than AKD, because spreading is less efficient on hydrophobic surfaces than on hydrophilic surfaces. The extent of reaction can be quite high under ideal laboratory conditions, but under practical mill conditions it is often in the range between 15-40 %.

Figure 14.29 illustrates how sizing expressed as the $Cobb_{60}$-value depends on the amount of reacted AKD for various pulps. It is obvious that mechanical pulps with a high specific surface area requires more size than low surface area pulps, such as bleached kraft pulps. Defining the onset directly proportional to the BET surface area of the papers as shown in *Figure 14.30*. From surface of full sizing as $Cobb_{60} = 25$ g/m^2, the required amount of AKD necessary for sizing is dir balance measurements on AKD the Planar Oriented Monolayer Surface Area of AKD can be calculated to 24 Å^2. From this surface area it can be calculated from *Figure 14.29* that it is only necessary to cover 4 % of the total surface area for a given pulp in order to obtain sizing.

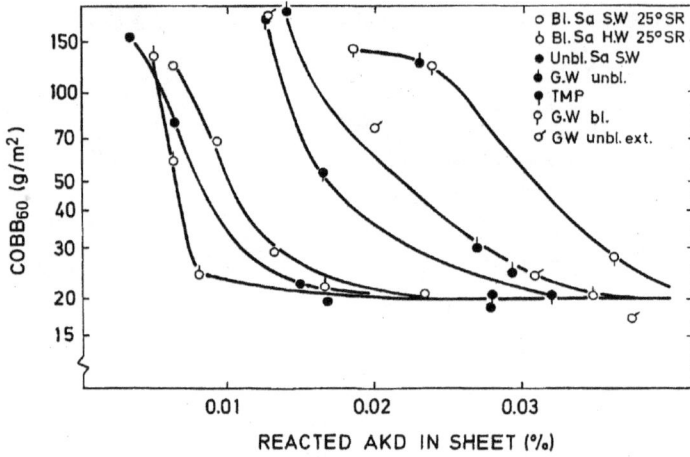

Figure 14.29. The effect of reacted amount of AKD on sizing of various types of pulps.
Abbrev: Bl. Sa. SW = Bleached softwood kraft pulp; Bl. Sa H.w.= bleached hardwood kraft pulp; Unbl. Sa.
S.W.= unbleached softwood kraft pulp; G.W. Unbl.= unbleached groundwood pulp; TMP = thermomechanical
pulp, G.W. Bl. = bleached groundwood pulp, GW unbl. Exl. = extracted unbleached groundwood pulp.

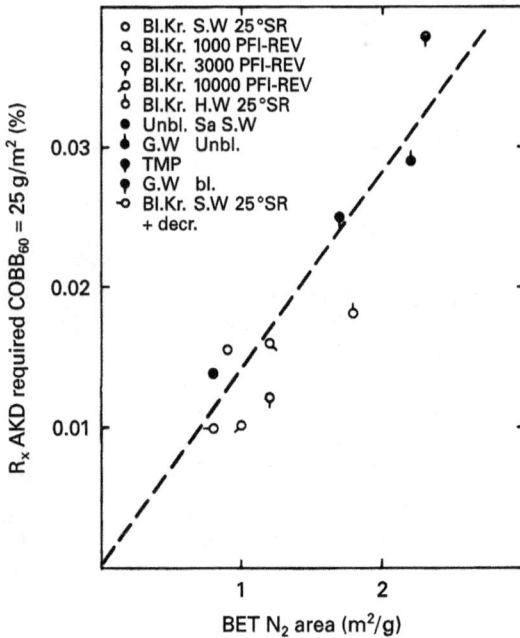

Figure 14.30. The required amount of reacted AKD, necessary to obtain $Cobb_{60} = 25$ g/m^2 for various pulps.

14.4.10 Practical Aspects and Comparisons between Different Sizing Agents

In *Table 14.2*, some major aspects between the different sizing agents have been compiled. Rosin sizing is basically restricted to acidic pH-values and so both AKD and ASA are basically restricted to neutral/alkaline papermaking, although ASA may be used at slightly acidic pH-values. Electrolytes are basically negative for all sizing agents because they interfere with retention aid use and decrease their affinity to fibre surfaces. Divalent metal ions are particularly devastating for rosin sizing because they compete with aluminium species in the complexation reaction. Ferric or ferrous ions can actually be beneficial for rosin sizing because they complex with rosin in a similar fashion as aluminium species, but can usually not be used for the purpose, because they decrease the brightness of paper. Fines/fillers have a large surface area consuming the sizing agent. Fillers can generally not be sized because rective sizes do not react with fillers and the aluminium resinate can not be anchored to the filler. The hydrolysis product of ASA can complex with Ca-ions, so it may in principle be able to to use for slack sizing of calcium carbonate.

Table 14.2. Comparisons between different sizing agents.

	Rosin	AKD	ASA
pH	4.2–5.0	7–8.5	5–8.5
Electrolytes	– –	–	–
Fines/Fillers	–	–	–
Dissolved an. substances	– –	– –	– –
Fibre-COOH	+ +	+	+
Extractives	+	– –	(+)
Hydrolysis products	irrelevant	(+)	– –
Aluminium sulfate	+ +	–	+
Stock temperature	– –	–	–
Lactic acid resistance	–	+	v

Dissolved anionic substances are in almost all cases detrimental to size retention. The charged groups on the fibres are necessary for rosin sizing, and the higher the carboxyl group content, the easier the pulp is to size with rosin. For AKD/ASA sizes, the charged groups are beneficial for retention processes. Acidic extractives may in principle be used as sizing agents in the presence of alum and non-ionic extractives contribute slightly to sizing. Extractives, of the fatty acid types are detrimental to AKD-sizing, because they interfere with retention and spreading and hence, with AKD-reaction. Extractives have in general a slight positive effect on ASA-sizing because aluminium salts are used in conjunction with ASA-sizing. The AKD-hydrolysis product has a slight positive effect on AKD-sizing, but the ASA-hydrolysis product is detrimental to ASA-sizing because the diacid is amphiphatic and will overturn in the presence of aqueous liquids in contact with the sized paper. Aluminium salts are, of course, necessary for rosin sizing, but interfere with AKD-sizing, if they contribute sufficient acidity to decrease the HCO_3- content of the water. AKD is used in conjunction with rosin sizing for liquid packaging

(Walkden 1991), but there is no simple mechanistic explanation for this synergism. Stock temperature has a strong negative effect, particularly for rosin soap sizing, but also for dispersion sizing. A higher temperature leads to aggregation of precipitated aluminiumresinates and oxolation of aluminium species making them loose some of their cationic charge characteristics. For synthetic sizes a higher stock temperature leads to a higher rate of hydrolysis of non-retained size. Neither rosin sizes nor ASA-sizes can protect paper against liquids containing strongly co-ordinating species like lactic acid

14.4.11 Fluorochemical Sizing (FC)

The sizes discussed so far are efficient in giving water resistance. If the hydrogen atoms along the carbon chain is exchanged for fluorine, surfaces can be created with so low a surface energy that they also repel oil, fats and organic solvents; the surface is said to become oleophobic.

When a drop of oil is placed on a "grease proof" paper, the oil will not penetrate because the paper has no pores and represent a physical barrier for the oil. The oil will still, however, spread on the paper, because it is a high energy surface. If oil is placed on a FC-treated paper the oil will not spread but will form a contact angle in excess of 90°.

The fluorine-based chemicals have long carbon chains with fluorine atoms instead of hydrogen atoms as in conventional sizes and a hydrophilic end, which is fixed to the fibre material one way or another. The FC chemicals can be divided into FC-chrome complexes, FC-phosphates and FC-polymers (Reynolds 1989). Typical FC phosphate structures are shown in *Figure 14.31*.

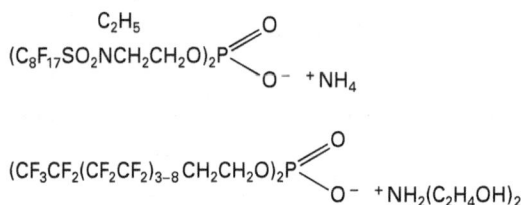

$$(C_8F_{17}SO_2NCH_2CH_2O)_2P \overset{C_2H_5}{\underset{O^-\ ^+NH_4}{\overset{O}{<}}}$$

$$(CF_3CF_2(CF_2CF_2)_{3-8}CH_2CH_2O)_2P \overset{O}{\underset{O^-\ ^+NH_2(C_2H_4OH)_2}{<}}$$

Figure 14.31. FC-phosphate structures (Reynolds 1989).

FC-phosphates may be used as internal sizing agents, but the phosphate group is not stable in the presence of multivalent ions such as aluminium species and does not give water repellence. The phosphate group is anchored to cellulose by means of hydrogen bonding and a synthetic sizing agent, such as AKD is commonly used in conjunction with FC-phosphates in order to impart both water and oil repellence.

14.5 Sizing with Alkenyl Succinic Anhydride (ASA)

The development of sizing agents based on alkenyl succinic anhydrides took place in the beginning of the 70´s and, hence, the technology is the youngest of the major sizing technologies employed today. Like AKD, they are able to undergo reaction with cellulose and is therefore characterized as a reactive sizing agent, see *Figure 14.32*.

Compared to AKD, ASA's have substantially higher cellulose reactivity, but are also more prone to hydrolysis. The higher reactivity makes them suitable to control size press pick up in fine paper manufacture, but they are less suitable for hard sizing in, for instance, liquid packaging applications.

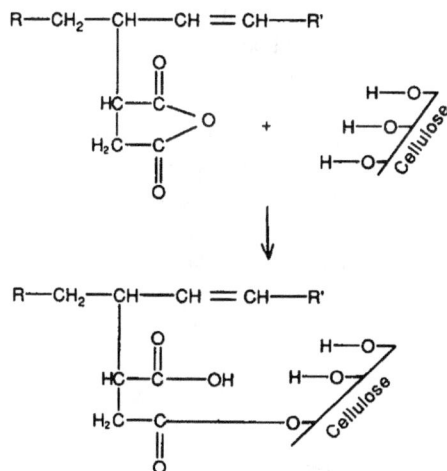

Figure 14.32. Reaction of alkenyl succinic anhydride with cellulose.

ASA's are derived from petroleum feedstock (1-alkenes or α-olefins). They are prepared from α-olefins by catalytic isomerisation, followed by reaction with maleic anhydride. The reaction is depicted in *Figure 14.33*. The commercially available ASA's are anhydrides made from alkenes with a carbon number of C-16 to C18.

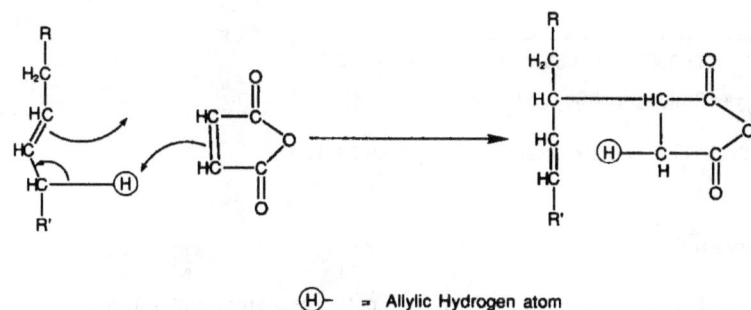

Ⓗ– = Allylic Hydrogen atom

Figure 14.33. Mechanism of the ene reaction for the formation of ASA.

The location of the double bond is important and it has been shown that internal alkenes are more effective sizing agents than α-olefins (Roberts 1991, 1997), because ASA made from α-olefins being solids at room temperature and higher temperatures are needed for dispersion, under which the ASA can hydrolyse. Secondly, ASA from α-olefins are more susceptible to hydrolysis than the internal alkene equivalents.

14.5.1 Emulsification

Emulsions are usually prepared using cationic starch as the stabilizing polymer (Chen and Woodward 1986; Roberts 1997). Potato or amphoteric starches are usually preferred, because of synergistic retention interactions, when alum is used in alkaline papermaking (Farley 1991). This is caused by the interaction between aluminium species and the phosphate group in these starch preparations.

The starch/ASA ratio is usually around 3:1. Emulsification is carried out on-site using either a low-shear venturi system or a high-shear turbine system, because of the rapid rate of hydrolysis of ASA. A small amount of a surfactant is also used to help emulsification, although higher levels may interfere with sizing. Particle size is usually between 0.5 μm to 2 μm. Particle size has a significant effect on ASA-sizing. The smaller the particle size, the better the sizing, see *Figure 14.34.*

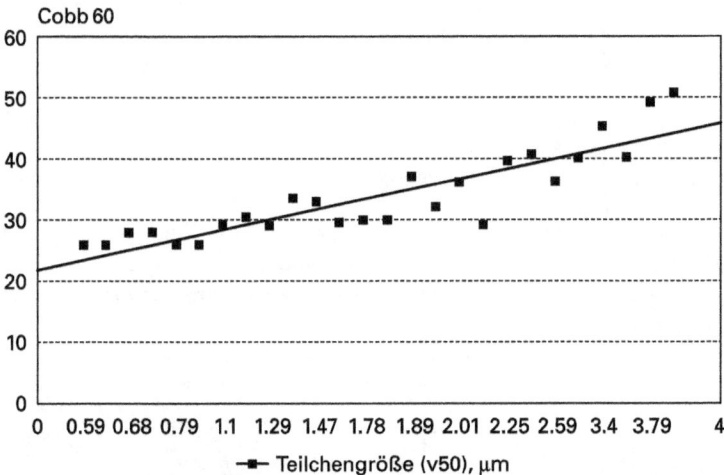

Figure 14.34. The effect of ASA emulsion particle size on sizing (Bähr 2001).

14.5.2 Hydrolytic Stability of ASA

ASA-emulsions are easily hydrolysed, particularly at high pH-values and temperatures. The effects of pH and temperature on ASA hydrolysis are shown in *Figure 14.35.*
From *Figure 14.35*, the necessity of an on-site preparation of ASA is obvious. The particle size has an effect on the rate of hydrolysis and the larger the particle size, the slowe r is the rate of hydrolysis. Optimum particle size is therefore important, in view of the above-mentioned benefits of having a smaller particle size for efficient sizing.

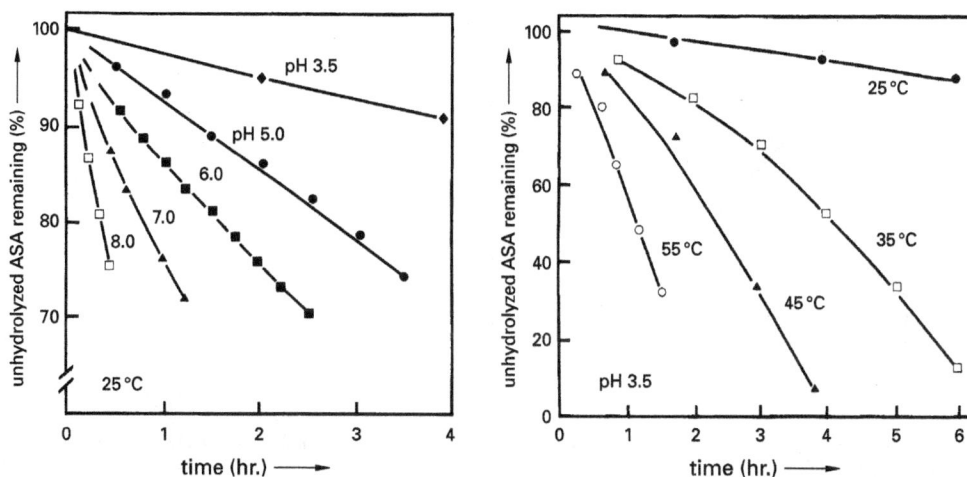

Figure 14.35. a) The effects of pH on hydrolysis of ASA at 25° C (left). b) The effect of temperature on hydrolysis at pH = 3.5 (Wasser 1987).

It is known (Wasser 1985) that the dicarboxylic acid has a desizing effect, because the hydrolysis product can overturn and expose the carboxylic acid group to the water phase, when exposed to aqueous liquids. It has also been shown that extraction of the hydrolysis product improves sizing (Roberts and Wan Daud 1988). This behaviour contrasts that of AKD, where sizing is largely unaffected by the hydrolys is product. This is the expected behaviour because the hydrolysis product of AKD, the ketone is not an amphiphatic molecule. The remedy to this effect is to lock the mobility of the dicarboxylic acid with an aluminium compound (Strazdins 1987), because the dicarboxylic acid can actually be used to size paper in a similar fashion as for rosin sizing (Hatanaka et al 1991). The Ca-soap is actually also hydrophobic (Wasser and Brinen 1998) and it is only the Na-soap or the acid form, which causes the desizing effect. It is, however also known that the Ca/Mg-soaps can cause press picking and sticky deposits in papermaking (Scalfarotto 1985) and it is therefore recommended to use aluminium compounds in conjunction with ASA-sizing. It is obvious that high single pass retention of the ASA sizing agent is critical for sizing, just as it is for AKD.

14.5.3 Reaction of ASA with Cellulose Fibres

There is direct spectroscopic evidence for the reaction between cellulose and ASA (McCarthy and Stratton 1987). The rate of reaction is strongly dependent on pH and temperature. The higher the pH and the higher the temperature, the faster is the reaction as illustrated in *Figure 14.36*. Although ASA may be used at lower pH-values than AKD, optimum pH for sizing is around 7, and the optimum pH-value will also depend on stock composition and the appropriate use of aluminium salts as discussed above.

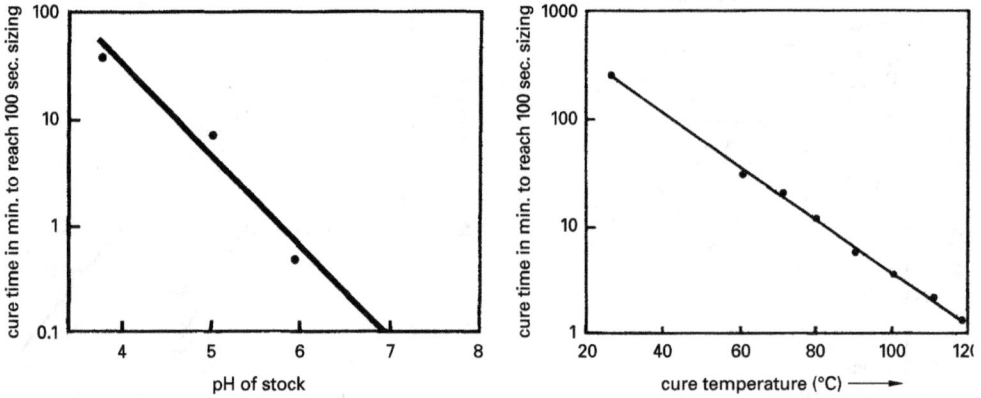

Figure 14.36. a) The effect of pH on the cure rate of ASA. Handsheets were made from $CaCO_3$ (15 %) containing stocks and were dried and cured at 90° C. b) The effect of temperature on the cure rate of ASA tub sized from toluene. (Hercules sizing test employed) (Wasser 1987).

14.6 Literature

Cited References

Arnson, T. R. (1982) The Chemistry of Aluminium Salts in Papermaking. *Tappi J.* 65(3): 125.

Arnson, T. R., and Stratton, R. A. (1983) The Adsorption of Complex Aluminum Species by Cellulosic Fibres. *Tappi J.* 66 (12): 72.

Bähr, E. (2001) Leimung mit System-ASA. *Wochbl. Papierfab.* 17: 1112.

Biermann, C. J. (1992) Rosin Sizing with Polyamine Mordants from pH 3 to 10. *Tappi* 75 (5): 166.

Bottero J.-Y., and Fiessinger, F. (1989) Aluminium Chemistry in Aqueous Solution. *Nordic Pulp Pap Res. J.*, 4(2): 81.

Bottorff, J. K. (1994) AKD Sizing Mechanism: a More Definitive Description. Tappi 77(4): 105.

Bottorff, K. J., and Sullivan, M. J. (1993) New Insights into AKD Sizing Mechanism. *Nordic Pulp Pap Res. J.* 1: 86.

Brosset, C., Biederman, G., and Sillén, L. G. (1954) Hydrolysis of Metal Ions XI. The Aluminium Ions. *Acta Chem Scand.* 8: 1917.

Cazabat, A. M. (1989) The Dynamics of Wetting. *Nordic Pulp Pap Res. J.* 4(2): 146.

Chen, G. C. I., and Woodward, T. W. (1986) Optimizing the Emulsification and Sizing of Alkenyl Succinic Anhydride. *Tappi J.* 69(8): 95.

Colasurado, A. R. and Thorn, I. (1992) The Interactions of Alkyl Ketene Dimer with other Wetend Additives. *Tappi J.* 75(9): 1430.

Cooper, C., Nicholass, J. and Thorn, I. (1995) The role of polymers in AKD-sizing. *Paper Technology* 5: 30.

Crawford, R. A., and Flood, T. A. (1989) *Preliminary NMR Study on Structure of Polyaluminium Chloride. Proc. Papermakers Conference.* Atlanta, GA, USA: Tappi Press, p 55.

Davison, R. W. (1964) The Chemical Nature of Rosin Sizing. *Tappi* 47 (10): 609.

Davison, R. W. (1988) Retention of Rosin Sizes in Papermaking Systems. *J. Pulp Pap Sci* 14 (6): J151.

Erhardt, S. M., and Gast, J. C. (1988) *Cationic Dispersed Rosin Sizes. Proc. Tappi Papermakers Conference.* Atlanta, USA: Tappi Press.

Farley C E (1991) *Use of Alum to Improve ASA Sizing Efficiency in Alkaline Paper. Tappi Proceedings, Papermakers Conference.* Atlanta, GA, USA: Tappi Press.

Hatanaka, S., Takahashi, Y., and Roberts, J. C. (1991) Sizing with Saponified Alkenyl Succinic Acid. *Tappi J.* 75(177).

Hoyland, R. W. (1978) *Swelling during the Penetration of Aqueos Liquids into Paper. Trans. 6th Fundamental Research Symp., Oxford, Vol. II.* The British Paper and Board Industry Federation, p. 557.

Jiang, H., and Deng, Y. (2000) The Effects if Inorganic Salts and Precipitated Calcium Carbonate Filler on the Hydrolysis Kinetics of Alkylketene Dimer. *J. Pulp Pap Sci.* 26(6): 208.

Kamutzki, W., and Krause, T. (1984) Neues uber Reaktionen und Mechanismen bei der Harzleimung. *Das Papier* 38 (10A): V47.

Lindström, T., and O'Brian, H. (1986) On the Mechanism of Sizing with Alkyl Ketene Dimers Part II. The Kinetics of Reaction between Alkyl Ketene Dimers and Cellulose. *Nordic Pulp Pap Res J* 1(1): 26.

Lindström, T., and Söderberg, G. (1983a) Studies on Internal Rosin Sizing. Part I: A Simple Method Based on C-14 Radioactive Labelling for Determination of Rosin Size. *Svensk Papperstidn.* 86(3): R22.

Lindström, T., and Söderberg, G. (1983b) Studies on Internal Rosin Sizing. Part II: Studies on Abietic Acid Retention in Handsheet Moulds using C-14 Fumaropimaric Acid. *Svensk Papperstidn.* 86(3): R25.

Lindström, T., and Söderberg, G. (1984) Studies on Internal Rosing Sizing. Part III. The Effects of pH, Concentration and Concentration of Calcium-lignosulfonate on Size Retention. *Svensk. Papperstidn.* 87(3): R2.

Lindström, T., and Söderberg, G. (1986a) On the Mechanism of Sizing with Alkylketene Dimers Part 1. Studies on the Amount of Alkylketene Dimer Required for Sizing of Different Pulps. *Nordic Pulp Pap Res J* 1(1): 31.

Lindström, T., and Söderberg, G. (1986b) On the Mechanism of Sizing with Alkylketene Dimers, Part III. The Role of pH, Electrolytes, Retention aids, Extractives, Ca-lignosulfonates and Mode of Addition on Alkyl Ketene Dimer Retention. *Nordic Pulp Pap Res J* 1(2): 3.

Lindström, T., and Söderberg, G. (1986c) On the Mechanism of Sizing with Alkylketene Dimers. Part IV. The Effects of HCO_3-ions and Polymeric Reaction Accelerators on the Rate of Reaction between Alkyl Ketene Dimers and Cellulose. *Nordic Pulp Pap Res J* 1(2): 39.

Lindström, T., and Söremark, C. (1977) Electrokinetic Aspects of Internal Rosin-Sizing. *Svensk Papperstidn.*, 80(1): 22.

Marton, J. (1985) *Sizing Mechanisms and the Effect of Fillers. Proc. 8th Fund. Res. Symp, Vol. 2.* Suffolk, UK: Mech. Engng. Public, p. 785.

Marton, J. (1989) Fundamental Aspects of the Rosin Sizing Process. Mechanistic Differences between Acid and Soap Sizing. *Nordic Pulp & Paper Res. J.* 4(2): 77.

Marton, J. (1990) Practical Aspects of Alkaline Sizing-on Kinetics of AKD reactions: Hydrolysis of AKD. *Tappi* 73(11): 139.

316

Marton, J. (1991) *Practical Aspects of Alkaline Sizing II. AKD in Mill Furnishes: Material Balances and Distribution Proc. Papermakers Conf., April 1991, Seattle, WA, USA.* Atlanta, GA, USA: Tappi Press, p 405.

Matijevic, E., and Stryker, L. J. J. (1966) Coagulation and Reversal of Charge of Lyophobic Colloids by Hydrolyzed Metal Ions. III. Aluminium Sulphate. *Colloid Interface Science* 22: 68.

McCarthy, W. R., and Stratton, R. A. (1987) Effects of Drying on ASA Esterefication and Sizing. *Tappi J.* 70 (12): 117.

Ödberg, L., Lindström, T., Liedberg, B., and Gustavsson, J. (1987) Evidence for ß-ketoester Formation during Sizing of Paper with Alkylketene Dimers. *Tappi* 70(4): 135.

Öhman, L-O., Wågberg, L. (1997) Freshly Formed Aluminium (III) Hydroxide Colloids-Influence of Aging, Surface Complexation and Silicate Substitution. *J. Pulp Pap Sci* 23(10): J475.

Pang, J. (1997) *Study of the Interaction of Alum with Fortified Rosin Sizing Agent and Cellulose in Papermaking Chemistry.* Dalian Institute of Chemical Physics, Chinese Academy of Sciences, May (in English).

Reynolds, W. F. (1986) *Interaction of Alum and Papermaking Fibres. Proc. Tappi Papermakers Conf.,* p. 321.

Roberts, J. C., and Garner, D. N. (1985) The Mechanism of an Akyl Ketene Dimer Sizing of Paper, Part 1. *Tappi* 68(4): 118.

Roberts, J. C., and Wan Daud, W. R. (1988) *Proc. 10th Cellulose Conf., Syracuse, NY.*

Scalfarotto, R. E. (1985) Remedies for Press Picking Boost Efficiency of ASA Synthetic Sizing. *Pulp Paper* 4: 126.

Seppänen, R., and Tiberg, F. (2000) Mechanism of Internal Sizing by Alkyl Ketene Dimers (AKD): The Role of the Spreading Monolayer Precursor and Autophobicity. *Nordic Pulp Pap Res. J.* 15(5): 452.

Shimada, K., Dumas, D., and Biermann, C. J. (1997) Properties of Candidate Internal Sizing Agents vs Sizing Performance. *Tappi* 80(10): 171.

Strazdins, E. (1977) Mechanistic Aspects of Rosin Sizing. *Tappi J.* 60(10): 102.

Strazdins, E. (1984) Critical Phenomena in Rosin Sizing. *Tappi* 67(4): 110.

Strazdins, E. (1986). The Chemistry of Alum in Papermaking. *Tappi J.* 69(4): 111.

Strazdins, E. (1989) Theoretical and Practical Aspects of Alum Use in Papermaking. *Nordic Pulp Pap Res. J.* 4(2): 128.

Subrahmenyam, S., and Biermann, C. J. (1992) Generalised Rosin Soap Sizing with Coordinating Elements. *Tappi J.* 75 (3): 223.

Thorn, I., Dart, P. J., and Main, S. D. (1993) The Use of Cure Promotors in Alkaline Sizing. *Paper Technol* 34: 141.

Traser, G., and Jayme, G. (1973) Uber die Zusammensetzung von Natriumresinat-Aluminiumsulfat-Fällungen bestimmt mittels Elementaranalyse und IR-spektroskopischer methoden. *Wochbl. Papierfab.* 101(11/12): 391.

Walkden, S. (1991) Alkaline Advance helps Liquid Packaging Board meet Rigorous Specifications. *Tappi J.* 74 (4): 103.

Wang, F., and Tanaka, H. (2001) Mechanisms of Neutral-Alkaline Paper sizing with Usual Rosin size Using Alum-Polymer Dual Retention Aid System. *J. Pulp Pap Sci* 27(1): 8.

Wang, F., Wu, Z. H., and Tanaka, H. J. (1999) Preparation and Sizing Mechanisms of Neutral Rosin Size II: Functions of Rosin Derivatives on Sizing Efficiency. *Wood Science* 45(6): 475.

Wasser, R. B. (1987) The Reactivity of Succinic Anhydride: Its Pertinence with Respect to Alkaline Sizing. *J. Pulp Pap Sci* 13(1): J29.

Wasser, R. W., and Brinen, J. S. (1998) Effect of ASA on Sizing in Calcium Carbonate Filled Paper. *Tappi J.* 81 (7): 139.

Wu, Z.-H., Chen, S.-P., and Tanaka, H. (1997) Effects of Polyamine Structure on Rosin Sizing Under Neutral Papermaking Conditions. *J. Appl Polym Sci* 65: 2159.

Zhuang, J., and Biermann, C. J. (1995) Neutral to Alkaline Rosin Soap Sizing with Metal Ions and Polyethyleimine as Mordants. *Tappi* 78(4): 155.

Zhuang, J. F., and Biermann, C. J. (1993) Rosin Soap Sizing with Ferric and Ferrous Ions as Mordants. *Tappi J.* 76 (12): 141.

Review Articles/Text Books

Davison, R. W. (1975) The Sizing of Paper. *Tappi J.* 58(3): 48.

Davison, R. W. (1990) Stock Preparation. In: Hagemeyer, R. W., and Manson, D. W. (Eds.) *Pulp and Paper Manufacture, Vol 6, 3rd Edition*. Atlanta, GA: Tappi/CPPA.

Davison, R. W., and Spurlin, H. M. (1970) In: Britt, K.W. (Ed.) *Handbook of Pulp and Paper Technology, 2nd ed*. New York: Van Nostrand Reynold Co., p. 355.

Eklund, D., and Lindström, T. (1991) *Paper Chemistry – An Introduction*. Grankulla, Finland: DT Paper Science Publications.

Erhard, S. M. (1987) *The Fundamentals of Sizing with Rosin. Sizing Short Course, Atlanta*. Atlanta, USA: Tappi Press.

Griggs, W. H., and Crouse, B. W. (1980) Wet End Sizing-an Overview. *Tappi J.* 63: 49.

Kevney, J. J., and Kulick, R. J. (1981) In: Casey, J. P. (Ed.) *Pulp and Paper Chemistry and Chemical Technology. Vol 3, 3rd ed*. New York: Wiley Interscience, p. 1547.

Roberts, J. C., Blackie (Eds.) (1991) *Paper Chemistry*. Glasgow and London, USA: Chapman & Hall, New York.

Swanson J W. (Ed.) (1971) *Internal Sizing of Paper and Board Tappi Monograph Series, No 33*. New York: Tappi Press.

Review Articles/Textbooks

Reynolds, W. F. (Ed.) (1989) *The Sizing of Paper, 2nd ed*. Atlanta GA, USA: Tappi Press

Davison, R. W. (1975) The Sizing of Paper. *Tappi J.* 58(3): 48.

Davison, R. W., and Spurlin, H. M. (1970) In: Britt, K. W. (Ed.) *Handbook of Pulp and Paper Technology, 2nd ed*. New York: Van Nostrand Reynold Co., p. 355.

Dumas, D. H. (1981) An overview of Cellulose Reactive Sizes. *Tappi J.* 64(1): 43.

Eklund, D., and Lindström, T.(1991) *Paper Chemistry – An Introduction*. Grankulla, Finland: DT Paper Science Publications.

Griggs, W. H., and Crouse, B. W. (1980) Wet End Sizing-an Overview. *Tappi J.* 63: 49.

Hodgson, K. T. (1994) A Review of Paper Sizing using Alkyl Ketene Dimer versus Alkenyl Succinic Anhydride. *Appita* 47(5): 402.

Kevney, J. J., and Kulick, R. J. (1981) In: Casey, J. P. (Ed.) *Pulp and Paper Chemistry and Chemical Technology. Vol 3, 3rd ed.* New York: Wiley Interscience, p. 1547.

Roberts, J. C., Blackie (Eds.) (1991) *Paper Chemistry.* Glasgow and London, USA: Chapman & Hall, New York.

Roberts, J. C. (1997) *A Review of Advances in Internal Sizing of Paper in Fundamentals of Papermaking Materials. 11th Fundamental Res. Symp., Cambridge, UK, Vol 1.* Leatherhead, UK: Fundamental Research Committee and Pira International, p.209.

Review Articles/Textbooks

Baes, C. F., and Mesmer, R. E. (1976) *The Hydrolysis of Cations.* New York: John Wiley & Sons, p. 112.

Dumas, D. H. (1981) An overview of Cellulose Reactive Sizes. *Tappi J.* 64(1): 43.

Eklund, D., and Lindström, T.(1991) *Paper Chemistry – An Introduction.* Grankulla, Finland: DT Paper Science Publications.

Hayden, P. L., and Rubin, A. J. (1974) *Systematic Investigation of the Hydrolysis and Precipitation of Aluminium in Aqueous Environmental Chemistry of Metals.* Ann Arbor Science, p. 317.

Hodgson, K. T. (1994) A Review of Paper Sizing using Alkyl Ketene Dimer versus Alkenyl Succinic Anhydride. *Appita* 47(5): 402.

Horoszko, W. L., and Smylie, S. E. (1997) Aluminium Compounds – Chemistry and Use. In: Gess, J. M. (Ed.) *Retention of Fines and Fillers during Papermaking.* Atlanta, GA, USA: Tappi Press, Chap. 9.

Kevney, J. J., and Kulick, R. J. (1981) *Pulp and Paper Chemistry and Chemical Technology, J P Casey, ed., Vol 3, 3rd ed.* New York: Wiley Interscience, p. 1547.

Neimo, L. (1999) Internal Sizing of Paper. In: *Papermaking Science and Technology, Vol. 4 Papermaking Chemistry.* Finnish Paper Engineers' Assoc. and Tappi Fapet Oy, Finland, p. 151.

Öhman, L. O., and Sjöberg, S. (1988) In: Kremer, J. R., and Allen, H. E. (Eds.) *Metal Speciation Theory, Analysis and Application*, Chap. 1.

Roberts, Blackie, J. C. (Eds.) (1991) *Paper Chemistry.* Glasgow, London, New York: Chapman & Hall.

Reynold, W. F. (Ed.) (1989) *The Sizing of Paper. 2nd ed.* Atlanta GA, USA: Tappi Press.

Roberts, J. C. (1997) *A Review of Advances in Internal Sizing of Paper in Fundamentals of Papermaking Materials. 11th Fundamental Res. Symp., Cambridge, UK, Vol 1.* Leatherhead, UK: Fundamental Research Committee and Pira International, p. 209.

Swanson, J. W. (1989) Mechanisms of Paper Wetting. In: Reynolds, W. F. (Ed.) *The Sizing of Paper, 2nd ed.* Atlanta, GA, USA: Tappi Press.

15 Calendering

Magnus Wikström
Korsnäs AB

15.1 Applications of the Calender Process

Most paper and board grades are calendered with the aim of producing a surface suitable for printing. The term calendering covers a large number of different operations where the web is

320

compressed in one or more rolling nips in order to improve the surface properties. The calendering can be carried out on-line in the paper machine or as an off-line operation afterwards. Depending on the position of the calender, the state and condition of the paper structure during the calendering can vary significantly. For instance, the moisture content of the paper web can be 15 % when the calendering is performed before the last dryer section in the paper machine and 5 % in off-line calendering. In the case of pigment-coated grades, the web can be calendered before and/or after the coating operation, cf. *Figure 15.1.*

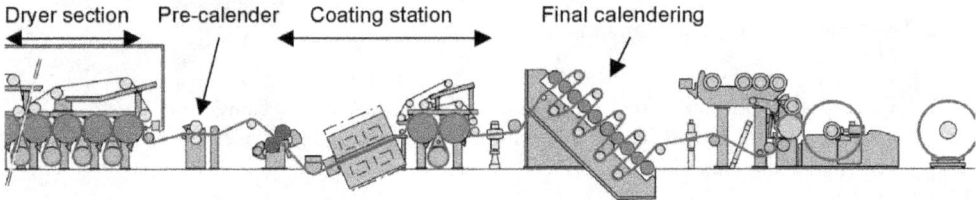

Figure 15.1. An example of calendering positions in a production line for lightweight coated wood-containing paper (LWC). The calendering is performed before and after the coating operation with a single (soft or hard) nip and with a multi nip soft calender, respectively.

Table 15.1 presents an overview of different calendering processes and configurations. The list is not complete due to the large number of local concepts or developments at the mills.

Table 15.1. Examples of different types of calenders.
Abbreviations: (H): heated roll, (P): polymer covered roll, (C): compressible soft roll made of cotton, wool/cotton or compressed synthetic fibres, (S): unheated steel or iron roll. (LWC): light-weight-coated wood-containing paper, (SC): supercalendered uncoated wood-containing paper, (WFC): woodfree coated paper.

Hard nip calenders The traditional procedure for on-line calendering (i.e. "Machine calender"). A single nip is now most common (eg. pre-calendering before coating), although multi-nip units, up to 8 rolls occur (eg. newsprint). Benefits: excellent marking and doctoring resistance, control of CD-profile.
Wet calender / Breaker stack Wet calender - water (or starch) is applied with water boxes in front of the nip entrance. Breaker stacks - calendering at high moisture content ≤15 %, on-line before the last dryer section (even soft roll covers occur in breaker stacks)
Supercalender The traditional procedure to calender paper to high smoothness and gloss, e.g. rotogravure grades. Used for WFC, LWC and SC grades. Always used off-line, up to 17 rolls. The machine speed is limited to less than 1200 m/min and the roll temperature is often less than 100 °C.
Soft calender; 1 heated soft nip Used primarily for board grades to smoothen the coated or uncoated printing side.

Soft calender; 2 heated soft nips
Two nips with the heated metal roll in the upper and the lower position, respectively, intended for paper grades that will be printed on both sides of the paper. Used for on-line calendering of LWC-offset and newsprint

Soft calender; 4 heated soft nips
Provides two heated soft nips per side in order to approach supercalendering quality on-line, e.g. SC and LWC offset grades

Soft calender; 1 soft/soft nip
The nip is formed between two polymer-covered rolls. Used for matte coated fine paper to improve the surface smoothness without increasing the paper gloss more than necessary.

Multi nip soft calenders
Combine the advantage of soft and supercalendering in order to achieve the highest supercalender quality even with on-line installations, e.g. rotogravure grades of SC, LWC, glossy WFC. OptiLoad, and Janus (*figure*) are examples of this types of calenders.

Extended soft nip / shoe-calender
Similar concept as shoe press, but generally shorter nip length (40–80 mm). Used for calendering coated and uncoated board and other packaging products.

Brush calender / brush polishing
Used to improve the gloss primarily on coated board without reducing the bulk. It is not widely used now, often replaced by soft calendering. The circumferential brush speed is 10–30 times faster than the web speed, 2–8 brushes are used.

paper roller
paper web
brush roller
paper roller

15.2 The Calendering Achievements

The surface properties and the structure of the paper product can be affected in a number of different ways by the calendering. Whether or not the achievements are desirable depends on the target product.

15.2.1 Development of Gloss and Small-Scale Surface Roughness

One of the most striking and important effects of calendering is the *enhancement of the paper gloss*. It has on several occasions been discussed whether the gloss development is a result of a shear deformation (termed micro-slip) or of a replication of the smooth roll surface. Even though some slip is likely to occur, it is today generally accepted that the latter mechanism plays a major role [1,2]. However, in somewhat different terms, high gloss can primarily be regarded as an effect of many optically parallel flat regions and it is possible that both shear and replication deformation may contribute to the forming of such a paper surface [3], but to different extents.

For paper products where a high paper gloss is aimed for it is crucial to *reduce the small-scale surface roughness* (topographic variations of a magnitude smaller than a micrometer) during the calendering, since the small-scale roughness influences the surface reflection. Calendering of matte coated grades is often carried out in such manner that the gloss increase is avoided by using a chilled iron roll mated with a soft covered roll or by having two soft-covered rolls to constitute the nip [4].

15.2.2 Reduction of Surface Porosity and in the Semi-Scale Surface Roughness

A *reduction in the porosity* is in some respects favourable since it improves the ink hold-out and thus contributes to a high and uniform print density, a low ink demand, a low print-through, a high print gloss and the possibility of using high resolution screens. On the other hand, a reduction in the porosity due to the calendering reduces the light scattering ability [5] and consequently also the brightness and the opacity. The pore size distribution is also of significant importance for the light scattering ability.

The reduction in the *semi-scale surface roughness* (topographic variations on a fibre level, i.e. the millimetre range or smaller) is to a great extent related to the decrease in the surface porosity in the case of uncoated paper. It is of importance for the numbers of missing dots in halftone prints and also for the quality of rotogravure prints [6].

15.2.3 Compression of Large-Scale Topographic Variations and Increased Porosity Variations

As described in Chapter 10, the fibres tend to form flocs when the fibre suspension is drained to form a paper web. The forming of flocs (that represent the thicker parts of the paper) and the resulting uneven mass distribution in the final sheet is described by the concept of "formation" of the dry paper. When a paper with poor formation is compressed in a hard calender nip, the stress is concentrated to the fibre flocs. Consequently an uneven reduction in the sheet porosity is obtained, due to the limited ability of the roll material to conform locally.

In the worst case, the locally very low porosity can produce almost transparent regions due to the low light scattering in these regions. This quality defect is known as blackening. Pronounced compression of the fibre flocs is reported to reduce the in-plane tensile strength of newsprint due to damage of fibres or fibre bonds [7]. In the calendering of coated grades, the local stress concentrations can cause spots with low porosity and/or high gloss in the coating layer. The spots, which are in the same order of magnitude as the fibre flocs, give raise to local variations in the ink absorption during printing. This quality defect is known as one type of print mottle.

For rotogravure printing the large-scale variations has to be reduced to some extent in order to obtain a sufficient contact between the paper surface and the printing plate. The situation is different in the cases of offset and flexography printing. Because of the deformable printing nips, a perfectly flat print surface is not necessary for flexo and offset grades [8]. The topographic unevenness on fibre floc level can therefore to some extent be preserved without reducing the print quality. On the other hand, it is important that the pore structure is homogeneous in order to avoid the local variations in ink absorption that cause print mottle.

The compression of fibre flocs during pre-calendering before pigment coating will influence the mass distribution of the coating layer.

15.2.4 Dust and Linting Control

The paper products intended for printing need a sufficient surface strength. Otherwise dust, loose fibres and poorly bonded coating pigments in the paper surface can cause serious runnability problems in the subsequent converting steps, such as deposits on the rubber blanket in the offset press. Depending on the paper product and on the calendering conditions, the calendering operation can either increase or reduce the tendency for rupture of fibres or pigment particles in the paper surface. For example, a single-nip hard-roll calendering increased the tendency for fibre rising to occur in the printing of a wood-containing paper, whereas it remained more or less unaffected by calendering in a single soft nip [9].

15.2.5 Thickness Reduction and its Consequences

When a paper is subjected to a pressure pulse in a calender nip, it can be compressed to less than 50 % of its original thickness. After the pressure is released, the paper recovers to reach its final deformation [10]. The magnitude of the permanent thickness reduction varies a lot depending on the calender conditions and the paper products (i.e. 5–30 %).

The *reduction in the average thickness* of the web is in most cases a disadvantage, especially for board and other products, which require as high a bending stiffness as possible. The bending stiffness is strongly dependent on the average thickness. A large thickness reduction can cause a significant cross machine directional (CD) expansion of the web. The expansion in CD was 0,3% when a woodfree paper exhibited a permanent thickness reduction of 20 % due to calendering [11].

When calendering certain newsgrades the thickness reduction can be beneficial as it is important to obtain a sufficient numbers of metres of paper in a reel of a specified diameter [12]. In addition, the CD thickness profile of the paper web is often adjusted by calendering in order to improve the runnability in subsequent converting steps, e.g. winding and printing. Modern calenders include sophisticated systems for CD thickness profile control [13, 14]. The progress of such systems is described below.

15.3 Paper Compression and Stress Distribution during Calendering

15.3.1 Load Profile in Cross Machine-Direction

In the old calenders at least one of the rolls was crowned to facilitate uniform load profile over the entire web width. A crowned roll implies that the roll diameter increases towards the centre of the web in order to compensate for the deflection of the rolls. The drawback of this fixed crown is that the uniform load profile is only obtained for a certain line load. To overcome this problem so-called swimming rolls were introduced where the rolls consists of a stationary beam and a rotating metal shell. The shell is supported by hydrostatic pressure chamber that acts between the stationary beam and the rotating shell. The crown could thereby be adapted to every loading case by adjusting the ratio between the loading of the bearing and the internal hydrostatic pressure in the oil-filled chamber. Later on the hydrostatic pressure chamber was replaced by hydrostatic loading elements where a thin oil film transfers the load from the stationary loading element to the rotating shell. In the latest development each loading element can be independently adjusted in order to fine-tune the thickness profile (and other properties) of the web in CD-direction, *Figure 15.2*.

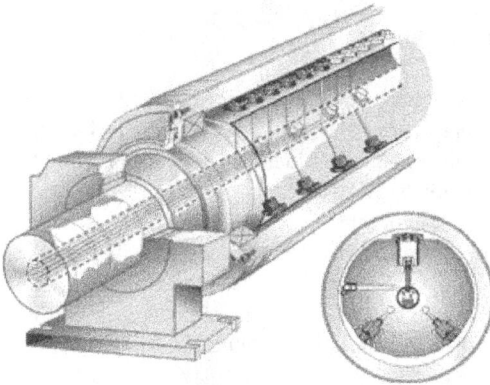

Figure 15.2. Cross-section of a calender roll with individually controlled hydrostatic loading elements.

15.3.2 Stress Distribution in Machine-Direction in the Calender Nip

The general expression for the applied load in the calender nip is the line load L_T that is the applied nip load divided with the web width. The line load cannot be used to compare different calendering situations as the nip length varies substantially for different calender configurations (from a few millimetres to several centimetres in some cases) and thus the compressive stress in the calender nip. In practice the effective (*dynamic*) nip length is difficult to establish, as it requires dynamic nip measurements that are dependent of the machine speed, the viscoelastic properties of the roll materials and the paper web. However, the *static* nip length can be estimated by applying a certain line load with a pressure sensitive film between the rolls when they not rotate. The compressive *static* stress distribution in the nip can then be estimated by the contact theory of Hertz [15]. It describes the static stress distribution in a nip between two compressible, frictionless rolls with an infinite length.

According to the Hertz equations, the maximum compressive stress σ_0, can be expressed as:

$$\sigma_0 = \frac{4}{\pi}\sigma_M = \frac{4L_t}{\pi 2a} \tag{15.1}$$

where σ_M denotes the mean compressive stress, and $2a$ the static nip length, determined by static compression tests at different line loads L_t. The compressive stress $\sigma_z(x)$ in the thickness direction at a distance x from the centre of the nip ($0 \leq x \leq a$), is then given by:

$$\sigma_z(x) = \sigma_0\sqrt{1 - \frac{x^2}{a^2}} \tag{15.2}$$

Figure 15.3 shows the static pressure distribution for nip with 6 mm nip length at a line load of 100 kN/m.

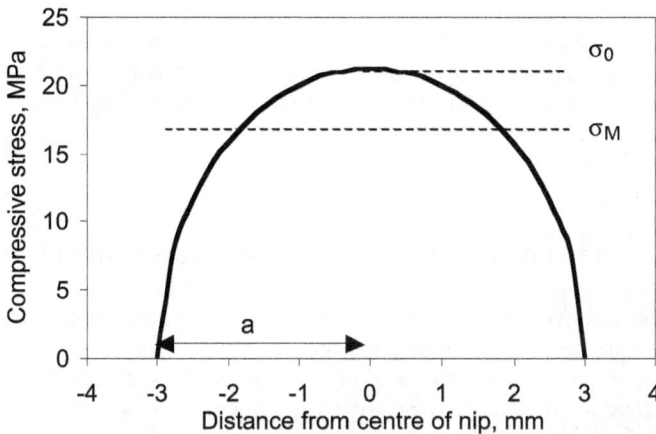

Figure 15.3. An estimate of the compressive stress distribution in a calender nip calculated according to the Hertz contact theory. The static nip length is 6 mm and the line load is 100 kN/m.

The Hertz contact theory provides also a theoretic expression for the static nip length $2a$:

$$2a = \sqrt{\frac{16L_t R^*}{\pi E^*}} \tag{15.3}$$

where E^* and R^* is given by

$$\frac{1}{E^*} = \frac{1-v_1^2}{E_1} + \frac{1-v_2^2}{E_2} \tag{15.4}$$

and

$$\frac{1}{R^*} = \frac{1}{R_1} + \frac{1}{R_2} \tag{15.5}$$

E_1, E_2, v_1, v_2 denote the Young modulus and the Poisson ratios of the roll shells (eg. polymer cover) and R_1, R_2 the radius of the rolls. An important presumption here is that the paper thickness is diminutive in relation to the thickness of a deformable roll cover and the curvature radius of the rolls.

15.3.3 Time Dependency of the Compression Behaviour of Paper

There is also a time dependence of the deformation behaviour of the paper even though it is the maximum pressure applied which has the largest influence on the resulting deformation. In practice, this time dependence is easily observed as a greater thickness reduction when the dwell time in the nip is increased, i.e. when the machine speed is reduced, at a given compressive stress [16, 17].

As paper materials have viscoelastic compression behaviour the recovery (spring back) of the paper in the calender nip will also be time dependent and thus influence the nip pressure distribution. In other words a slow spring back of the paper or an increased machine speed will shorten the dynamic nip length. However, as paper is compressed in a relatively dry state the most of the recovery occurs within a few milliseconds [18]. Though a minor slow recovery can appear afterwards for hours [19].

15.3.4 Stress-Strain Behaviour of Porous Material and Local Stress Distribution

The deformation behaviour of paper can to a great extent be ascribed to the fact that paper structures and pigment coating layers are porous materials. The paper deformation during calendering has been described in a way that stems from the field of foamed polymers [2, 20]. In this description, the compressive stress-strain behaviour of porous materials is divided into three distinct regions:

1. An initial linear elastic region where the number of contact points between the fibres (or possibly other components in the structure) increases without any yielding. The deformation is therefore not permanent, i.e. it recovers completely after unloading. This occurs only under a rather low compressive stress.
2. An elastic-plastic buckling region at not-too-large deformations where the strain increases rather rapidly although the corresponding compressive stress increase is less pronounced. This results in a permanent rearrangement of the fibre network.
3. A crushing region at higher deformations where the stress increases rapidly for very small changes in the strain. It reflects the interaction between the fibres as the fibre structure collapses to form a compact material. As a consequence, considerable damage to the fibres and the fibre network may occur.

The second region is what is aimed for in calendering. However, due to the inhomogeneous nature of paper, the first and third mechanisms may coexist locally with the second although on average the resulting deformation may be attributed to the second region.

15.4 Temperature-Gradient Calendering and other Methods for Surface Softening

During recent decades, the two most important industrial trends in calendering technology have been the development of polymeric roll covers for soft calendering and a development towards high roll temperatures (> 120 °C up to 300 °C).

The thickness reduction and the bulk deformation caused by the calendering are in most cases regarded as drawbacks. It is therefore desirable to localise the deformation to the surface regions and thus minimise the deformation of the bulk structure. A substantial part of the research on calendering has been focused on ways of achieving such effects. The general idea is to create surface regions, which are softer and thus more deformable than the interior structure of the paper. A stiffness gradient in the thickness direction is then obtained and the compressive stress necessary to achieve a certain surface property, e.g. gloss, can then be reduced. The use of lower compressive stress will partly retain the bulk structure (fibres and fibre network). Possible ways to obtaining such a stiffness gradient in the thickness direction can be divided into three approaches that can be used separately or in combination with each other; 1) temperature-gradient calendering, 2) moisture-gradient calendering and 3) material-gradient calendering.

15.4.1 Temperature-Gradient Calendering

Temperature-gradient calendering means that the paper is compressed between two rolls, one or both of which are heated to a high temperature [21, 22]. The contact time in the nip is short, which means that the surface regions become hotter than the bulk material. The surface material (e.g. fibres or coating binders) will therefore become comparatively more compressible. This approach is the most common way to obtain a stiffness gradient in the thickness direction.

Several positive effects can be obtained from temperature-gradient calendering even though it does not show benefits in every application. The usefulness of temperature-gradient calendering tends to be strongly dependent on the fibre composition of the web, e.g. on the type of wood and the degree of delignification [23]. The calendering temperature had no significant effect on the final result when pre-calendering a coated woodfree grade [24]. On the other hand, when paper based on thermomechanical pulp was pre-calendered the best surface properties were obtained by coating sheets when the pre-calendering was performed at a high temperature [25]. Moreover, the surface roughness, the paper gloss and the tear index at a given bulk level were improved for newsprint calendered between two iron rolls heated to 210 °C instead of to 50 or 70 °C [22].

The softening necessary for a temperature-gradient effect to develop is often discussed in terms of the glass transition temperature (T_g) of the wood polymers or coating binders [26, 27]. However, the wood polymers can exhibit a substantial softening even below T_g that can be important in this context [23].

On a coated paper surface, the conditions for obtaining thermal softening differ. An almost immediate thermal softening was found, indicated by a weak time-dependence of the gloss development in the calendering of coated woodfree and LWC papers [17]. The component of the coating layer that can exhibit thermal softening in a heated calender nip and thus affect the mechanical properties of the coating layer is primarily the binder. Different types of latices used as binders in coating layers have different thermal softening behaviours [28]. The positive effects of temperature-gradient calendering can therefore to some extent be promoted by adapting the type of coating binder to calendering at high temperatures by changing the type of coating latex (see Section 15.5).

The actual shape of the temperature gradients in the paper during calendering is however not yet fully understood. Several studies points to a rapid decrease of the roll temperature only a few microns from the paper surface contacted by the heated roll [29, 30, 31]. Parameters that have an important influence on the shape of the temperature gradient are for example the thermal conductivity of the paper during compression, the diffusion distances within the paper structure and the heat transfer coefficient between the paper and the roll. Therefore the surface roughness, the moisture content and the nip pressure will certainly influence the shape of the temperature gradient. The heat conduction equation, that describes the thermal flux between two infinite plates, has been used in several studies as a rough estimate of the temperature distribution in the thickness direction:

$$\frac{\partial^2 T}{\partial z^2} = \frac{1}{\alpha} \frac{\partial T}{\partial t} \tag{15.6}$$

where its simplest solution is:

$$T(z,t) = T_0 + (T_s - T_0) \cdot \left[1 - erf\left(\frac{-z}{2 \cdot \sqrt{\alpha \cdot t}} \right) \right] \tag{15.7}$$

where:
$T(z,t)$ = temperature at a distance z from the heated roll surface at time t
T_0 = temperature of the paper web as it enters the nip
T_S = temperature of the heated roll surface
α = thermal diffusivity of the paper
$erf(x)$ = error function (a kind of distribution, cf. normal distribution)

α is given by:

$$\alpha = \frac{k}{\rho \cdot C_p} \tag{15.8}$$

where
k = thermal conductivity
ρ = density
Cp = specific heat

The temperature gradient develops simultaneously with the pressure build up when the paper is compressed in a heated calender nip. *Figure 15.4* shows the temperature profile at different depths in the paper as function of time. The results, such as the thermal diffusivity values, are based on platen press experiments [31]. It can though be regarded as an illustration of the heating of certain segments of the paper when it passes through a heated calender nip.

A more complex analysis of temperature gradient calendering can be carried out by modelling of simultaneous compression and thermal softening using finite element analysis [32].

The high-temperature calender rolls, i.e. thermo rolls, are normally made of chilled cast iron or hardened steel [11]. Forged steel also occurs, as well as chromium plated cast iron rolls. The thermo rolls should be very smooth, provide the right thermal properties for sufficient heat transfer, good corrosion resistance and a low adhesion to the surface material of the web. Moreover, the rolls surface have to be hard in order to impart good abrasive wear resistance and enable doctoring of the roll surface. A thermal spray coating, eg. by plasma treatment, can improve the properties of the roll surface. The achievements can be a harder or a smoother roll surface depending of the type of thermal spray coating material used. Aluminium oxide, chromium oxide and tungsten carbide are some examples.

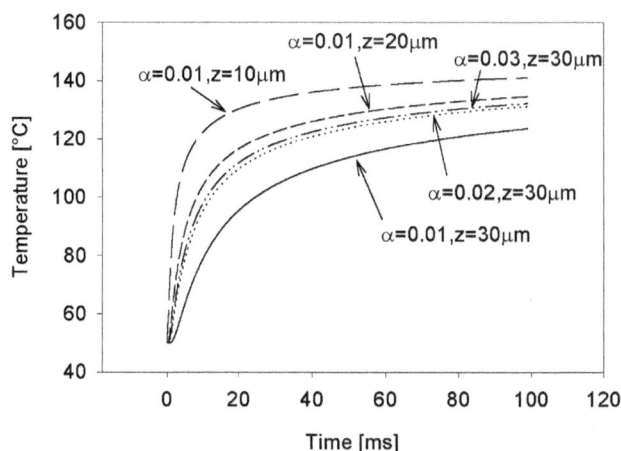

Figure 15.4. Temperature as a function of contact time for different thermal diffusivities, α, and at different distances, z, from the surface. The temperature profiles were calculated by using the heat conduction equation with a surface temperature of 150 °C. The thermal diffusivity is given in mm²/s.

The thermo rolls are normally heated by oil, water or steam flowing in peripherally drilled pipes through the shell of the roll. To obtain the highest calendering temperatures internal or external induction heating is also possible, i.e. heating generated by an electromagnetic field [33]. This technique is not yet so common in mill scale, but several new industrial installations have been completed in the recent years.

15.4.2 Moisture-Gradient Calendering

Spraying steam or water onto the web immediately before the calender nip is another way to selectively soften the paper surface. Since the time for water penetration is very short, a moisture gradient develops [34]. The presence of a moisture gradient may be especially effective when it is combined with high roll temperatures, since the moisture reduces the softening temperature of the wood polymers, lignin, hemicellulose and cellulose [35]. The possibility of achieving a positive effect of the moisture gradient depends upon the composition of the paper. For instance, higher gloss and lower ink demand was found for newsprint when applying water with spray nozzles before the calender nip [34]. Moreover, applying a water film to a woodfree uncoated paper before calendering with a soft nip reduced the gloss variations at a given average gloss level. The average gloss, surface roughness and bending stiffness were not however improved by the water application [36]. A water box in the calender stack in the pre-calendering of solid bleached board gave a coated product with better surface smoothness [37]. Moistening of the web after the coating operation is not so common due to the risk for enhanced coating deposits on the calender rolls.

As an example of applications, steaming of the web is common for uncoated wood-containing supercalendered paper (i.e. SC paper). A number of steam showers are then placed before the nip entrances in the calender stack. Despite the risk for deposits, this concept has also been applied to LWC-grades. *Figure 15.5* shows a schematic cross section of a steam shower. Modern steam showers are zone controlled in order to enable adjustment of the CD gloss profile.

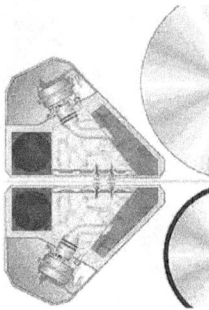

Figure 15.5. A cross-section of a steam shower.

When the web temperature is high (> 60 °C), for instance when the paper will be calendered right after the dryer section, the steam cannot condensate on the web surface. There is then two alternative ways to moisten the web; either to reduce the web temperature by using cooling cylinders before the calender or to transfer water instead of steam by using a water spray box or a water box. In the former case the water is sprayed on the web with nozzles and in the latter case the web passes a roller nip with a water pond.

15.4.3 Multi Layers Products Adapted to Promote Surface Deformation

Multi-layer board (and perhaps in the future stratified formed woodfree paper) offers another possibility to produce a surface more deformable than the inner structure. The basic idea is to direct the most deformable fibres to the surface region. The fibres can be more deformable for geometrical reasons (wall thickness, length to width ratio etc.) or because of their composition (e.g. lignin content). The requirements regarding the surface and other properties in a wider sense, e.g. whiteness and surface strength limit the possibilities of specifically adapting the surface fibres with respect to their response to the calendering operation. Material-gradient calendering may be a realistic possibility for some high-quality coated products where the coating operation is performed in two or more steps. The coating formulation can then be adapted to the position of the layer in the thickness direction. The pre-coating ensures a proper interaction with and coverage of the base sheet. In the top coating, either the binder or the pigment can be used to enhance the surface compression. For instance, hollow plastic pigments are more deformable than inorganic pigments and are therefore of interest as a component in the formulation of the top coating [38].

15.5 Calendering with Soft Deformable Nips

15.5.1 Principal Features of Soft Deformable Calendering Nips

The characteristic of a soft calender nip, here regarded as a nip formed between a rigid roll with a metal surface and a polymer-covered backing roll, is that both the paper and the backing roll are compressed. The rigid roll surface is replicated in the contacting paper surface, through plastic deformation during the compaction [39]. The paper surface in contact with the polymer-covered backing roll deforms correspondingly with the cover. The paper surface deformation is normally limited on this side of the paper; e.g. the gloss increases only slightly on the paper surface contacted by the deformable roll. The magnitude of the roll cover deformation depends on the line load, machine speed, roll diameter, stiffness and thickness of the cover material etc. The deformation of the available covers varies over a broad range. For instance, rolls filled with synthetic or natural fibres are reported to shrink in the circumferential direction during loading, whereas polymeric unfilled rolls are rather incompressible and therefore expand in the nip [2].

Soft calendering is often performed with the rigid roll heated to rather high temperatures. It is then clear that the characteristics of the backing roll will influence both the pressure and the temperature pulse experienced by the paper in the calender nip. The effects of a more deformable backing roll on the nip conditions are here divided into four characteristics (which are of course more or less interrelated):

1. An increase in the nip length at a given line load will *decrease the maximum pressure level* and produce a *more evenly distributed pressure in the machine direction* (even if the paper exhibits no thickness variations). A certain pressure is required to achieve a permanent deformation of the surface. Since the applied pressure does not vary in the thickness direction, a change in the nip pressure cannot be used in order to localise the deformation to the surface regions.

2. An increase in the nip length will *prolong the nip dwell time* at a given machine speed. This is important since paper products exhibit to some extent a viscoelastic character in the thickness direction [40, 41]. By varying the machine speed with a constant line load, with one type of nip and no heating, the effect of a prolonged dwell time can be estimated. For instance, when an uncoated woodfree paper was calendered, both the gloss and the thickness decreased linearly with decreasing machine speed [9]. However, the line load had a substantially greater influence on the gloss and on the thickness reduction than the dwell time [16].

3. As described in the previous section the *heat transfer* in the hot calender nip is time-dependent. Therefore, the roll deformation that influences the dwell time in the nip will also affect the heat transfer to the paper.

4. The paper surface replicates the smooth surface of the rigid heated roll, whereas the deformable backing provides a flexible support. The deformation of the backing roll can therefore to some extent be adapted to *conform to local variations of the paper structure,*

such as fibre flocs, cf *Figure 15.6*. The deformable rolls thus give a more uniform pressure distribution in the nip, and this lead to a paper structure with a more even density distribution [16]. On the other hand, this limits the possibilities of adjusting the CD thickness profile of the paper web.

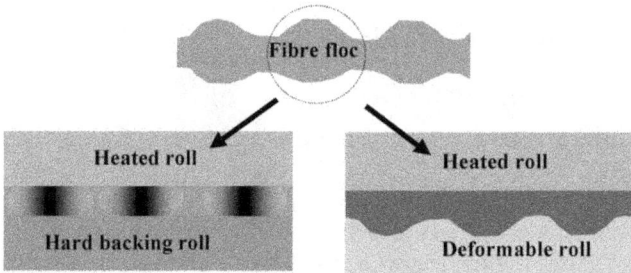

Figure 15.6. A schematic drawing of the pressure distribution and the deformation of the paper structure when using a hard and a deformable backing roll, respectively. In the case of the hard backing roll the shading illustrates stress concentrations directed towards the fibre flocs.

15.5.2 The Influence of Soft Nips on the Calendering Result

The effects on the calendering result of using a soft instead of a hard nip (metal roll surfaces only) have been studied extensively. The advantages of soft calendering have often been attributed to the more gentle compression where the stresses are not only localised to the thicker parts of the paper structure. The strength properties are therefore better preserved [2]. Compared to a hard nip, the more even compression in soft calendering causes less porosity variations and the tendency for calender blackening and print mottle is therefore reduced [42]. In another study the in-plane density distribution in uncoated woodfree paper calendered with soft or hard nips was characterised [9]. The soft calendering enhanced the density variations less than calendering with the hard nip. On the other hand, hard nip calendering produces a paper web with a more uniform thickness. When soft calender nips are used, the CD thickness profile of the web has to be of acceptable evenness before the web enters the calender nip, e.g. by adding steam in different positions in CD using zone-controlled steam showers [43].

15.5.3 The Evolution of the Soft Calendering

Soft-nip calendering comprises a number of different calendering concepts. The common feature is that at least one of the two rolls constituting the nip is more deformable than a roll with a steel or iron surface. Supercalendering and soft nip calendering have in a historic perspective been distinguished as two different types of calendering process in which deformable rolls are used.

The main features of the two calendering techniques have differed substantially. The *supercalender* has, with very few exceptions, been an off-line installation with a roll stack of 8 to 14 rolls. Most often two supercalenders are required to keep up with the production from one paper machine. The *soft calendar* has mainly been installed on-line in the paper machine or coating station, and consists of 1 × 2 rolls, 2 × 2 rolls, 1 × 3 rolls or 2 × 3 rolls.

The character of the deformable rolls has also been different. The supercalender rolls have a typical diameter of 400 mm and have traditionally been made of compressed natural fibres (cotton, wool/cotton or asbestos/cotton fibres), even though rolls of compressed synthetic fibres (i.e. filled rolls) and polymer-covered rolls are now more frequently used. Due to the extensive numbers of different concepts of soft calendering, the diameter of the rolls varies over a broad range from 400 to 1200 mm. The soft calender roll covers are always made of different types of polymers, such as polyurethane.

In the latest calendering concepts, the supercalender and the soft calender appear to be combined into a single technique. The design of these new concepts (e.g. Janus, OptiLoad and Pro-Soft) is often reminiscent of traditional calender stacks, but a number of improvements make it possible to produce supercalendered quality with on-line installations [44], *Figure 15.7*. For example, the filled rolls are being replaced by polymer-covered rolls, the rolls with metal surfaces can be heated to higher temperatures (170 °C or more), the dead load of the rolls can be relieved so that the line load can be equal in each nip and the calenders are equipped with high-speed web-threading systems [45]. Cost effectiveness is the major driving forces for replacing the supercalender operation, normally two supercalenders are required two keep up with the production from one paper machine.

Figure 15.7. A multi nip soft calender installed on-line on a LWC paper machine (Haindl Papier) that started up 2000.

It was reported already in the 1960's that an increase in roll temperature up to 200 °C substantially improved the gloss and smoothness at the same bulk when soft calender coated board [46]. Obviously, even though the temperature-gradient calendering concept was originally developed for hard nips, i.e. for a nip formed between two iron or steel rolls, it has long been known that high-temperature calendering combined with soft nips can be very favourable. This early technique for hot soft calendering, known as gloss calenders, was carried out with heated iron rolls and rubber covered backing rolls. The maximum line load was however rather limited due the rubber covers sensitivity for shear deformation that caused heat build. Today's soft rolls with synthetic polymer covers have been progressively improved over the last decades and are now therefore used in an increasing number of application areas. Examples of such improvements are a higher marking resistance, a lower mechanical hysteresis (resistance to thermal overstress in order to avoid local heating-up) and stronger bonding between the cover and the roll shell. The soft rolls can therefore act as backing rolls under more demanding conditions than before; i.e. higher compressive nip stress or together with metal rolls heated to 250 °C or higher.

A modern deformable roll cover is built up in layers of materials with different mechanical properties. Examples of materials used for elastic calender covers are polyurethane, epoxy resin, aromatic polyamides, mineral fillers and synthetic or mineral fibres. The outer layer must have a sufficient wear and marking resistance. The thermal properties are also important such as low thermal expansion, tolerance for external heat (eg. transferred from the thermo roll) and low heat built-up due to the subsequent compression. The built up of the outer layer is usually epoxy resin reinforced by polyester or aramide fibres and sometimes filler added in order to improve the wear resistance, *Figure 15.8*. The intermediate layer and the base layers prevent stress concentrations between the deformable outer layer and the stiff metal shell and consist of different types of woven and non-woven synthetic or mineral fibres. The hardness of soft rolls is most often described as °Shore D, °Shore A or P&J (Pusey and Jones). °Shore D is applicable for the most common hardness range. A comparatively hard soft roll used for soft calendering of newsprint can have a hardness of 92–95 °Sh D and a softer one (eg. intended for coated board) can have a hardness of 60–82 °Sh D (or 3–5 P&J). Note that for P&J a lower number means a harder material.

functional layer

base layer

Figure 15.8. Soft deformable roll covers consist of a smooth outer layer and winded intermediate and base layers of synthetic fibres (left) and (right) a cross section of a polymer cover.

Another direction of the development of the soft calendering technique is an extended nip length. For conventional soft calendering the dynamic nip length rarely exceeds 20 mm. In the last decade a calendering method with an extended soft nip has been introduced, so called shoe calender or long nip calender. The nip is formed between a normal heated roll and a deformable belt supported by a shoe. Oil is added between the belt and the shoe in order to perform hydrostatic lubrication, cf. shoe presses for wet pressing. With this technique the dynamic nip length is extended to up to 80 mm. With shoe calendering the principal features of soft calendering (see above) has been taken further. Due to the very uniform pressure distribution the calendering technique will not enhance porosity variations that can cause print mottle and gloss variations for example. When the technique is combined with high calendering temperature an intense calendering can be carried out at comparatively low compressive stress [47, 48]. The technique is preferentially used for grades where preserving the bulk as much as possible is especially important, such as different types of board, and also for paper grades where compression localised to the fibre flocs is an critical issue (for example due to a poor formation). It is less suitable for grades that require multi nip calendering.

15.6 Calendering of Coated Grades

15.6.1 Pre-Calendering before Coating

The most coated grades are subjected to some kind of pre-calendering before the coating operation. Both hard and soft nips occur as well as single and multi nip configurations. The objective of the pre-calendering differs depending on paper grades and coating technique eg;

- Levelling of the thickness profile of the base paper in CD before the coating. This is of importance for controlling the difference in the properties between the centre and the edges of the rolls caused by for example the shrinkage during drying.
- Sealing of the surface pore structure of the base paper in order to reduce the penetration of coating colour into the base paper.
- Compression of fibre flocks and single loose bond fibre in the base paper surface in order to improve the coating coverage of the base paper.

However, the effect of the pre-calendering is partially gone strayed during the coating operation. When the base paper comes into contact with the liquid phase of the coating colour during the coating operation, a roughening of the surface of the base paper can occur. This effect can be quite pronounced for instance for pre-calendered LWC paper [49]. It has been suggested that the roughening related to pre-calendering is primarily an effect of fibre shape recovery [25]. The pre-calendering introduces stresses in the paper structure. The water absorbed by the base paper during the coating process can release these stresses, and this causes out-of-plane dimensional changes in the base paper during the consolidation of the coating layer. Fibre swelling and bond rupture may also contribute to the roughening.

The paper structure is compressed at rather low moisture content when it is calendered. As a consequence, it is improbable that any new bonds will be formed due to the re-organisation of the fibre structure. The amount of water present when the smooth paper surface is formed seems to be of great importance. It has been showed for example that fibres that have collapsed by press-drying are more stable when wetted than those collapsed by temperature-gradient-calendering [25].

15.6.2 Calendering after the Coating Operation

Temperature-gradient calendering is commonly used also for calendering of coated grades. It is clear that a high roll temperature makes the coating layer more deformable. Therefore, in the most cases, the targeted surface smoothening and gloss can be obtained at a lower compressive stress or with fewer calender nips by increasing the calendering temperature. However, an important differences compared to hot calendering of uncoated grades is that main component of the surface, the pigments do not soften. A typical coating layer contains 80–95 weight-% inorganic pigments (e.g. kaolin clay and calcium carbonate) that are not greatly affected by the temperature increase in the range for a heated calender nip. In contrast, the synthetic latex binders normally exhibit rather marked softening in the temperature range for calendering and it constitutes in the most cases a substantial proportion of the coating layer (5–10 weight-%). It is therefore a clear relation between the thermomechanical properties of the synthetic binders (i.e. latex) and the compressive stiffness of the coating layer, which affects the deformation of the coating layer in the calender nip [27].

It becomes evident in *Figure 15.9* that the temperature dependent softening behavior of different styrene-butadiene latex binders can vary significantly. The storage modulus was characterised for pure latex films at different temperatures [50]. The influence of the latex composition on their mechanical properties is more discussed in Chapter 16.

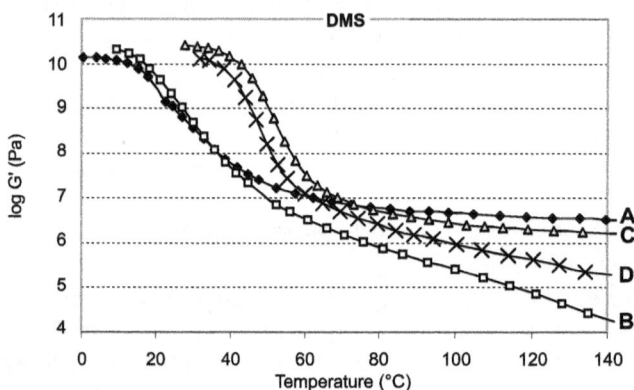

Figure 15.9. The storage modulus at a frequency of 1 rad/sec of pure latex films at different equilibrium temperatures.

When calendering coated grades there is risk for coating deposits on the calender roll surface that can be formed at a rather low rate but create a problem due to accumulation over time. At low calendering temperatures the primary reason for the deposit formation is insufficient binding of pigment particles in the coating layer, i.e. dusting. At high temperature calendering the softening of the latex binder is necessary for attaining the benefits of temperature gradient calendering, but on the other hand an extensive softening may increase the adhesion of the coating layer to the calender surface to a point where the cohesive strength of coating layer is exceeded and sticking occurs. The moisture content of the web is very critical for the cohesion of the coating components and affects the deposit formation both at high and low calendering temperatures.

Additionally, the properties of the pigment particles (such as the shape, particle size distribution and abrasive properties) influence the calendering ability of the coating layer. For example, platy clay particles in the uppermost surface of the coating layer can be redirected and paralleled by the calendering operation and this will promote both a proper smoothness and gloss [51]. However, the most coating formulations of today contain both platy and more spherical particles, thus the relevance of this effect may be limited (see Chapter 16). In general, the smoothens and gloss of the coating layer after the calendering operation is to a great extent determined by the structure uniformity of the coating layer before the calendering [52, 53]. The use coating pigments where the individual pigment particles deform by the calendering operation is however an exception from this. An example of such particles is hollow sphere polystyrene pigments [38, 54]. However, due to the high material costs their industrial relevance has so far been very limited.

15.7 References

[1]   Crotogino, R. H., and Gratton, M. F.: Hard-nip, and soft-nip calendering of uncoated groundwood papers. *Pulp Paper Can.* 88,12(1987)T461.

[2]   Rodal, J. J. A.: Soft-nip calendering of paper, and paperboard. *Tappi J.* 72,5(1989)177

[3]   Peel, J. D.: *Recent developments in the technologhy, and understanding of the calendering processes, Trans. 9th Fundam. Research Symp., Cambridge, Vol. 2.* Mech. Eng. Publ. Ltd, London, UK, 1989, p. 979.

[4]   Millington, K.: Development in calendering technology for printing papers. *Paper Technology* 34,6(1993)38.

[5]   Johnson, R. W., Abrams, L., Maynard, R. B., and Amick, T. J.: Use of mercury porosimetry to characterize pore structure, and model end-use properties of coated papers – Part I: Optical, and strength properties. *Tappi J.* 82,1(1999)239.

[6]   Bristow, J. A., and Ekman, H.: Paper properties affecting gravure print quality. *Tappi J.* 64,1(1981)115.

[7]   Moffat, J. M.: *Newsprint calendering: constraints, and possibilities.* Proc. UMIST Symp.Calend. Supercalend., Manchester, UK, 1975.

[8]   Saarelma, H. J., and Oittinen, P. T.: *Paper, and print noise as limiting factors of information capacity, Trans. 10th Fundam. Research Symp., Oxford, Vol. 1.* Pira Int., Leatherhead, UK, 1993, p. 363.

[9]   Granberg, A.: *Fukt- och temperaturinverkan vid kalandrering av papper (Lic. Eng Thesis).* Royal Institute of Technology, Stockholm 1996 (in Swedish).

[10] Steffner, O., Nylund, T., and Rigdahl, M.: Influence of the calendering conditions on the structure, and the properties of woodfree paper – a comparison between soft nip calendering, and hard nip calendering. *Nordic Pulp Paper Res. J.* 13,1(1998)68.

[11] Lif, J.O., Wikström, M., Fellers, C., and Rigdah, , M.: Deformation of paper structure during calendering as measured by electronic speckle photography. *J. Pulp Paper Sci.* 23,1(1997)J481.

[12] Peel, J. D.: *Developments in calendering technologies, PIRA reviews of Pulp, and Paper Technology.* Pira Information Centre, Leatherhead, UK, 1990.

[13] Knecht, W.: *Fast response CD caliper control, Proc. TAPPI Finishing, and Converting Conference.* Tappi Press, Atlanta, USA, 1988, p. 199.

[14] Brendel, B.: The Küsters Hydro-vario roll – a new effective tool for the papermaker. *Das Papier* 42,7(1988)325.

[15] Pav, J., and Svenka, P.: The Soft Compact Calender – The answer to the challenge for smoothness, and level surface at high speeds. *Das Papier* 39,10A(1989)V178.

[16] Wikström, M., Nylund, T., and Rigdah, M.: Calendering of coated paper, and board in an extended soft nip. *Nordic Pulp Paper Res. J.* 12,4(1997)p. 289.

[17] Keller, S. F.: *Calendering variables affecting coated paper properties, Proc. TAPPI Coating Conference.* Tappi Press, Atlanta, USA, 1992, p. 71.

[18] Brown, T. C., Crotogino, R. H., and Douglas, W. J. M.: Measurement of paper strain in the nip of a calender. *J. Pulp Paper Sci.* 20,9(1994)J266.

[19] Steffner, O.: *Strukturförändringar hos papper – några effekter av kalandrering och bestrykning (Lic. Eng Thesis).* Royal Institute of Technology, Stockholm, 1993 (in Swedish).

[20] Rusch, K. C. J.: Load-compression behaviour of flexible foams. *J. Appl. Polym. Sci.* 13(1969)2297.

[21] Kerekes, R. J., and Pye, I. T.: Newsprint calendering: an experimental comparison of temperature, and loading effects. *Pulp Paper Can.* 75,11(1974)T379.

[22] Crotogino, R. H.: Temperature-gradient calendering. *Tappi J.* 65,1(1982)97.

[23] Salmén, L.: Temperature softening of the components of paper: Its effect on mechanical properties. *Trans. Techn. Section CPPA* 5,3(1979)TR 66.

[24] Steffner, O., Nylund, T., and Rigdahl, M.: *Influence of precalendering on the properties of a coated woodfree paper, and the covering ability of the coating, Proc. TAPPI Coating Conference.* Tappi Press, Atlanta, USA, 1995, p. 335.

[25] Skowronski, J.: Surface roughening of pre-calendered basesheets during coating. *J. Pulp Paper Sci.* 16,3(1990)J102.

[26] Vreeland, J. H., Jewitt, K. B., and Ellis, E. R.: *Substrata thermal molding – a breakthrough in the understanding, and practise of the hot calendering of paper, Proc. TAPPI Coating Conference.* Tappi Press, Atlanta, USA, 1989, p. 179.

[27] Wikström, M., Mäkelä, P., and Rigdahl, M.: *Influence of temperature, and type of coating latex on the out-of-plane compression behaviour of coating layer. Proc. TAPPI Advanced Coating Fundam. Symp..* Tappi Press, Atlanta, USA, 1999, p. 201.

[28] Parpaillon, M., Engström, G., Pettersson, I., Fineman, I., Svanson, S. E., Dellenfalk, B., and Rigdahl M.: Mechanical properties of clay coating films containing styrene-butadiene copolymers. *J. Appl. Polym. Sci.* 30(1985)581.

[29] Kerekes, R. J.: A simple method for determining the thermal conductivity, and contact resistance of paper. *Tappi* 63,1(1980)137.

[30] Luong, C. H., and Lindem, P. E.: *A method to determine heat conduction in paper during calendering. Proc. Int. Paper Phys. Conf., CPPA, Montreal, Canada.* 1995, p. 1.

[31] Rättö, P., and Rigdahl, M.: Temperature distribution a paper sheet subjected to a short pressure pulse from a heated plate. *Nordic Pulp Paper Res. J.* 13,2(1998B)101.

[32] Rättö, P.: *On the compression properties of paper – implications for calendering (Doctor of Technology Thesis).* Royal Institute of Technology, Stockholm2001.

[33] Okamoto, K.: *New aspects on temperature gradient calendering. Proc. TAPPI Finishing, and Converting Conference.* Tappi Press, Atlanta, USA, 1989, p. 168.

[34] Lyne, M. B.: *The effects of moisture, and moisture gradients on the calendering of paper. Trans. 6th Fundam. Research Symp., Oxford, Vol. 2.* British Paper Board Fed., London, 1977, p. 641.

[35] Salmén, L. and. Back E. L.: Moisture dependent thermal softening of paper, evaluated by its elastic modulus. *Tappi* 63,6(1980)117.

[36] Granberg, A., Nylund T., and Rigdahl M.: Calendering of moistened woodfree uncoated paper. *Nordic Pulp Paper Res. J.* 11,3(1996)132.

[37] Steffner, O., and Rigdahl, M.: *Moisture-gradient calendering of board.* Submitted to 1999 TAPPI Coating Conference1999.

[38] Hemenway, C. P., Latimer, J. J., and Young, J. E.: *Hollow-sphere polymer pigment in paper coating. Proc. TAPPI Coating Conference.* Tappi Press, Atlanta, USA, 1984, p. 20.

[39] Rodal, J. J. A.: Paper deformation in a calendering nip. *Tappi J.* 76,12(1993)63.

[40] Van Haag, R.: Physikalische Grundlagen des Glätten. *Wochenbl. Papierfabr.* 125,18(1997)872.

[41] Rättö, P., and Rigdahl, M.: The deformation behaviour in the thickness direction of paper subjected to a short pressure pulse. *Nordic Pulp Paper Res. J.* 13,3(1998A)180.

[42] Stevens, R. K., Mihelich, W. G., and Neill, M. T: *On-line soft calender for uncoated groundwood grades. Proc. TAPPI Coating Conference.* Tappi Press, Atlanta, USA, 1989, p. 191.

[43] Begemann, U.: *Module jet headbox concept. Operating experience with single, and multi-layer headboxes processing different paper grades. Proc. first ECOPAPERTECH Conf., Helsinki, Finland.* 1995, p. 315.

[44] Grant, R.: Fresh approaches to calendering technology. *Pulp Paper Int.* 39,11(1997)25.

[45] Gamsjäger, N., and Bader, B.: *Innovations on elastic calender covers. Proc. 5th Int. Conf. New Available Tech., Stockholm, Part 2.* SPCI, Stockholm, Sweden, 1996, p. 1014.

[46] Brecht, W., and Müller, G.: Test performed with a gloss calender. *Tappi* 51,2(1968)61A.

[47] Wikström, M.: *Influence of temperature, and pressure pulses on the calendering result (Doctor of Technology Thesis).* Royal Institute of Technology, Stockholm 1999.

[48] Moreau-Tabiche, S., Maume, J. P., Morin, V., Piette, P., Gurein, D., and Chaussy, D.: *Wood containing paper pilot calendering: obtaining results through super calendering, soft nip calendering, and extended nip concept. Proc. 1999 PTS Coating Symposium.* PTS, München, Germany, 1999, 43E.

[49] Engström, G., and Lafaye, J. F.: Precalendering, and its effects on paper–coating interaction. *Tappi J.* 75,8(1992)117.

[50] Wikström, M., Carlsson, R., and Salminen, P.: *Einfluss von viskoelastischen Eigen-schaften von Latex auf die Laufegeinschaften des Kalenders bei Satinage bei hohen Tem-*

340

peraturen. Proc. 2001 PTS Coating Symposium. PTS, München, Germany, 2001, C42 (in German).

[51] Elton, N. J., Gate, L. F., and Hooper, J. J.: *Texture, and orientation of clays in coatings. Mineralogical Society Clay Minerals Group Golden Jubilee Meeting, Aberdeen, UK.* 1997.

[52] Lee, D. I.: *A fundamental study on coating gloss. Proc. TAPPI Coating Conference.* Tappi Press, Atlanta, USA, 1974, p. 183.

[53] Wikström, M., Bouveng, M., Lindholm, J., and Teirfolk, J.-E.: The influence of different coating latices on printing, and surface properties of a paper calendered at high temperatures. *Wochenbl. Papierfabr.* 128,19(1999)1321.

[54] Haskins, W. J., and Lunde, D. I.: Hollow-sphere pigment improves gloss, printability of paper. *Pulp & Paper* 65,5(1998)53.

16 Pigment Coating

Gunnar Engström
Department of Surface Treatment, Karlstad University

16.1 Introduction

Coated paper and coated board are printing substrates used for printed matters, printed in multi-colour, on which high demands are directed on sharpness and brilliance in the printed image. Examples, of such printed matters are magazines, catalogues, brochures and consumer packages of different kind. In magazines, catalogues, and brochures the advertisements must be exposed in such a way so they catch the attention of the reader. Analogous is true for consumer packages; they must expose the packed article so that they catch the attention of the buyer. More than 90 % of the production of coated paper and coated board is used for printed matters of this kind. The coating is applied to the paper or the board in the form of a coating colour, consisting of pigments, binders and additives dispersed in water. There are different methods to apply the coating. After forming the coating layer, the water is evaporated with a dryer. The coating can be performed both in-line and off-line and in one or in multiple steps. Usually 10 g/m^2 is applied in each step, which corresponds to a thickness of approximately 7 µm. In paper coating the machine speed is higher than in board coating. This is due to the higher grammage of board, and besides of the grammage, the machine is also controlled by the coating technique used. The fastest coaters currently in operation run at machine speeds approaching 2000 m/min. A brief summary of the definitions of coating colour, coated paper/board, and coating layer is given in *Table 16.1* The surface treatment of paper and board briefly described above is called pigment coating or just coating.

Table 16.1. Brief summary of the definitions of coating colour, coated paper/board, and coating layer.

Coating colour	Coated paper/board	Coating layer
Pigment, binder, additives, and water	Paper is coated on both sides (C2S), board on one side (C1S). Single coating Double coating Triple coating (board)	Coat weight of each layer 8–12 g/m^2, corresponding to a thickness of 6–9 µm.

Surface sizing is another type of surface treatment. In this treatment the paper or board is applied with a thin surface layer, generally consisting starch, in order to improve the surface strength. The size is applied in the form of a water solution using similar techniques as for pigment coating. This chapter is focused on pigment coating and surface sizing is discussed just briefly.

16.2 Coated Products

In the classification system used for paper and board, coated products are divided into coated mechanical paper, coated wood free paper and coated board. Below are some of these grades described.

16.2.1 Coated Mechanical Paper

LWC (Light Weight Coated): The base paper consists of 50 % mechanical pulp and 50 % chemical pulp and usually it does not contain any filler. The paper is coated on both sides and the coat weight is approximately 10 g/m^2 per side. Coated grammage is to be found within the range 50–65 g/m^2. The final coated paper is usually supercalendered.

ULWC (Ultra Light Weight Coated): The base paper contains more chemical pulp then that for LWC and it is usually added with filler in the form calcined clay. The paper is single coated on both sides and the coat weight use to be 6–8 g/m^2. The final coated paper is usually soft-calendered in-line. Coated grammage is to be found within the range 39–42 g/m^2.

16.2.2 Coated Wood-Free Paper

Coated wood free paper, which also is named coated fine paper, is coated on both sides and both single coating and double coating exist. In single coating the coat weight is 10–12 g/m^2 and in double coating approximately 10+10 g/m^2. Coated grammage is to be found within a relatively wide grammage range, 80–400 g/m^2, with the centre of gravity within the range 115–150 g/m^2. The higher grammages are called "board weights", but in contrast to board, which is formed in multiple plies, the base paper for the board weights is form in one ply.

The fibre composition of the base paper, which is produced at a neutral pH-value, is usually 40 % fully bleach chemical soft wood pulp and 60 % fully bleached chemical hard wood pulp. The opacity of sheets formed of chemical pulp is low and therefore the base paper is pigment filled and ground calcium carbonate (GCC) or precipitated calcium carbonate (PCC) are usually used for that purpose. The base paper is sized and usually also surface sized. Base papers for coated wood free papers are characterised by high brightness.

16.2.3 Coated Board

There are five types of coated board: solid bleached sulphate board (SBS), folding box board (FBB), white lined chip board (WLB), carrier board (CB) and liquid board (LB). The baseboard differ for the different board products, but common for all is that they are coated on one side in two steps approximately 10+10 g/m^2, or in three steps, approximately 10+10+10 g/m^2, and that the baseboard is built up in plies, usually 3–5. Counted from the side to be coated the names of the plies are top ply, centre ply (which can be more the one), and bottom ply.

SBS: The top and bottom plies consist of fully bleached hard wood pulp and the centre ply or plies of fully bleached soft wood pulp. The baseboard, which is produced at a neutral pH, is sized and usually also surface sized, but it contains no filler. The baseboard for SBS is characterised by high brightness.

FBB: The top ply consists of fully bleached chemical hard wood pulp and the centre and bottom plies of mechanical pulp. The top ply, which is produced at a neutral pH, is sized in contrast to plies based on mechanical pulp. None of the plies contains filler.

WLC: The top ply is based on fully bleached non-printed recycled paper and the centre and bottom plies of non-de-inked printed recycled paper, which gives the two latter plies a grey tone with an ISO brightness of approximately 30 %. The side to be coated must have high brightness and therefore it is important that the top ply has high opacity so that it covers the low brightness ply beneath. Bleached recycled paper contains pigment in the form of filler and coating pigments and therefore the top ply will be pigment filled, which increases its opacity. Filler in the form of calcined clay or titanium dioxide are sometimes added in order to increase the opacity further. Normally none of the plies are sized.

CB and LB: The top ply of these products consists of fully bleached chemical hard wood fibre and the centre and bottom plies of unbleached chemical soft wood pulp. In order for the top ply to cover the centre ply beneath, which has low brightness, the top ply must have high opacity. Therefore filler is usually added to the top ply in the form of calcined clay or titanium dioxide. All plies are produced and sized at a neutral or an acid pH.

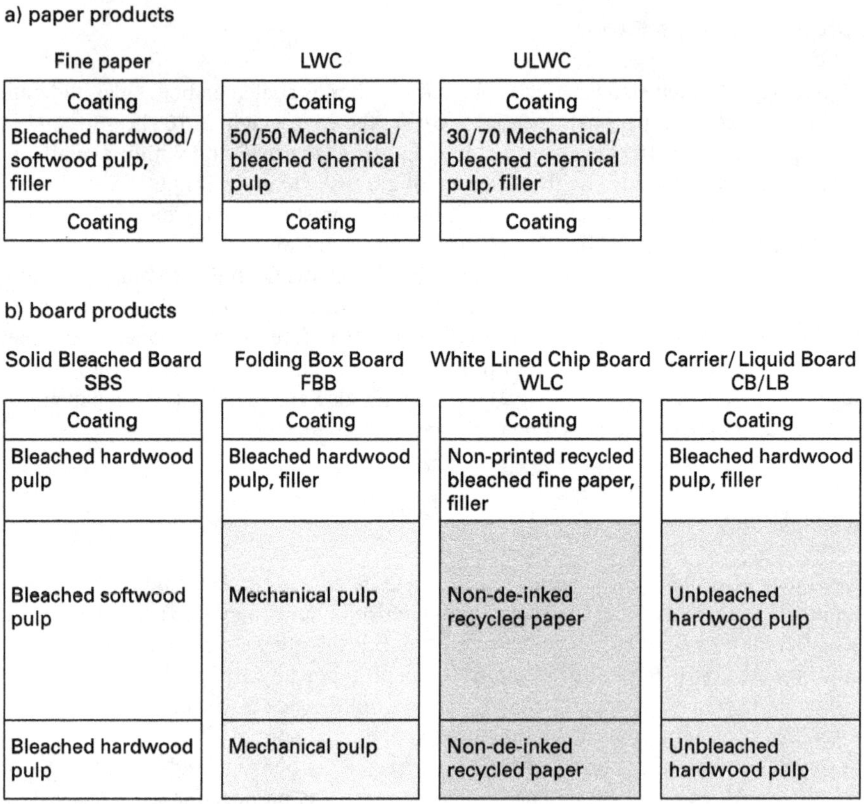

a) paper products

Fine paper	LWC	ULWC
Coating	Coating	Coating
Bleached hardwood/ softwood pulp, filler	50/50 Mechanical/ bleached chemical pulp	30/70 Mechanical/ bleached chemical pulp, filler
Coating	Coating	Coating

b) board products

Solid Bleached Board SBS	Folding Box Board FBB	White Lined Chip Board WLC	Carrier/ Liquid Board CB/LB
Coating	Coating	Coating	Coating
Bleached hardwood pulp	Bleached hardwood pulp, filler	Non-printed recycled bleached fine paper, filler	Bleached hardwood pulp, filler
Bleached softwood pulp	Mechanical pulp	Non-de-inked recycled paper	Unbleached hardwood pulp
Bleached hardwood pulp	Mechanical pulp	Non-de-inked recycled paper	Unbleached hardwood pulp

Figure 16.1. Build up of some coated a) paper and b) board products.

16.3 Demands on Coated Paper and Coated Board

Coated paper and coated board are printing carriers, to be printed in lithographic offset and roto-gravure, and in the case of board also in flexography. The demands on the coating are placed by the printing method used. However, common for all printing methods is, that high print quality prerequisites that the coating layer covers the substrate and that it is homogenous in all respects; in porosity, in chemical composition, in topography, and in reflectance. High brightness and usually also high gloss are other desired properties.

Base paper and base board for coating, like all paper products, are rough and therefore diffi-cult to cover with a thin coating layer. *Figure 16.2* shows schematic cross sections of such a substrate coated to the same coat weight but with two different coating colours, one yielding poor coverage, *Figure 16.2a*, and the other yielding good coverage. The coverability of the coating layer is governed by its density or porosity. At the same coat weight a coating with low density coating (porous) is thicker and, evidently, yields better coverage than one with high density (dens).

Figure 16.2 illustrates the topographical or physical coverage of the base paper. However, not only the physical coverage is important but also the optical coverage. High optical coverage is especially important when coating substrates of low brightness, e.g. LB or WCB or when the coat weight is low, e.g. LWC and ULWC, because the substrates must not show through the coating. If it does the coated material will have a mottled appearance, because the coat weight varies, *Figure 16.2*. High optical coverage is promoted by a high light scattering coefficient (s-value) of the coating layer. Usually high topographical coverage is accompanied with high opti-cal coverage and vice versa.

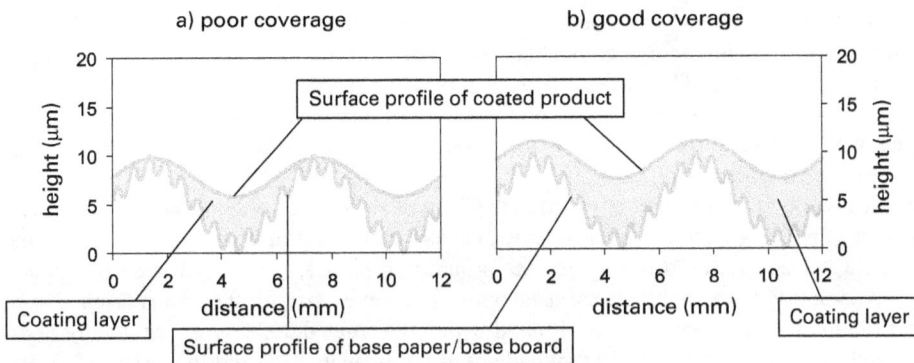

Figure 16.2. Schematic cross sections of coated papers/boards illustrating a) poor coverage and b) good coverage of the base.

The print quality is to a great extent determined by the way the ink interacts with the pore system of the coating layer, and therefore the porosity and the pore size distribution of the coat-ing layer are important parameters. As already indicated these parameters also govern the cov-erability of the coating layer. The pore system is built up to meet the demands placed by the specific printing method by selecting suitable pigments, binders, and additives. Lithographic offset and flexography demand a strong coating, because the inks used in these methods are tacky. The tackier the ink is the higher are the forces, which the coating surface is exposed to during the printing ink transfer. The strength of the coating for rotogravure paper is not critical.

Coated board is not only a printing carrier, but also a packaging material. This means that additional demands are directed on that material, such as good runnability in converting machines, good glueability and low bacteria counts. High stiffness, which gives good runnability, is a key parameter of coated board. Good glueability is demanded because coated board is used for glued capsules and low bacteria counts because it is used for food packages.

16.4 Surface Sizing

Paper and board to be printed in lithographic offset must have high surface strength. This is due to the fact that the surface of the paper or board to be printed in this printing method is exposed to high tensile forces in the printing nip. These in turn arises from the tacky inks used. However, the demand on high surface strength does not only apply for for coated products but also for non-coated ones. To meet the demand on high surface strength non-coated products are therefore surface sized. Base paper for coated fine paper and all types of baseboard are also most often surface sized prior to the coating in order to avoid the coating to penetrate into the base paper/base board, which has a detrimental effect on the coating coverage.

When printing in lithographic offset the non-printed areas of the paper or board to be printed are applied with fountain solution, which contains water. In multicolour printing these areas are printed with ink in the subsequent printing units. The water releases stresses and breaks fibre bonds. These causes lose bonded fibres in the surface to rise, and during printing these fibres are picked up by the ink and accumulated on the rubber blanket in the printing press. Fibres accumulated on the rubber blanket has a detrimental effect both on the print quality on the runnability in the printing press and to overcome these problems the printing press must be stopped and the rubber blanket cleaned. The surface sizing improves the surface strength by bonding the surface fibres. This reduces the accumulation of fibres on the rubber blanket. (Rubber blanket is the name of the cylinder from which the ink is transferred to the substrate to be printed. This cylinder is covered with rubber blanket.).

The size is applied to the paper or board in the form of an aqueous polymer solution, generally based on starch. Pigmented sizes also exist, usually consisting of equal parts of starch and pigment. These sizes are used to upgrade the printing properties, e.g. of uncoated fine paper. Improved newsprint can also be surface sized with a pigmented size for the same reason. A typical coat weight in surface sizing is to be found within the range 0.5–2.0 g/m^2 per side for regular sizes and 2.0–4.0 g/m^2 per side for pigmented sizes. Raw materials used for surface sizes and machines used for applying the size are the same as for regular pigment coating and are therefore not commented separately in the following text.

16.5 Coating Raw Materials

16.5.1 Pigments

Pigments are the main components in coatings and, as already mentioned, the properties of these govern the properties of pore system as well as the printing properties and the brightness and the gloss of the final coated material. The binder does also affect these properties but to a lesser extent.

In Europe approximately 72 % of the pigment consumption consists of GCC, 20 % of kaolin clay and 8 % of other pigments. In the latter group talc is the largest single pigment, followed by titanium dioxide and plastic pigment. In the US 73 % kaolin clay, 15 % GCC, and 12 % other pigment are used.

The particle size distribution is a parameter, which affects the pigment properties, and this is measured usually using the Sedigraph instrument. The measuring principle is sedimentation, and the particle size distribution curve obtained shows the cumulative mass percentage as a function of the equivalent spherical diameter. *Figure 16.3* shows Sedigraph-curves for two GCC;s used in coating. The percentage below 2.0 μm is usually used to indicate the particle size of the pigment. The slope of the particle size distribution curve gives information whether the distribution is broad or narrow. Other techniques, which can be used for the measurement of particle size distribution, are quasi-dynamic light scattering and laser diffraction.

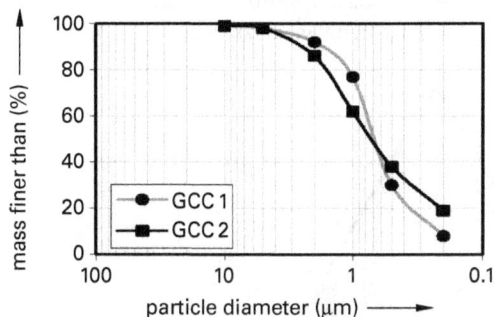

Figure 16.3. Particle size distribution curves for GCC. (From Knappich et al. 1999. Reproduced from 1999 Tappi Coating Conference Proceedings by the permission of TAPPI Press, Atlanta).

The particle size expressed as the percentage < 2.0 μm affects most properties of the coating layer. When the percentage < 2.0 μm increases the smoothness and gloss of the coating layer increases. The light scattering index passes through a maximum, due to that the light scattering coefficient attain a maximum when the diameter of pores in the coating layer is equal with half the wavelength of light, *Figure 16.4*. The pore diameter, but not the porosity, is governed by the percentage < 2.0 μm.

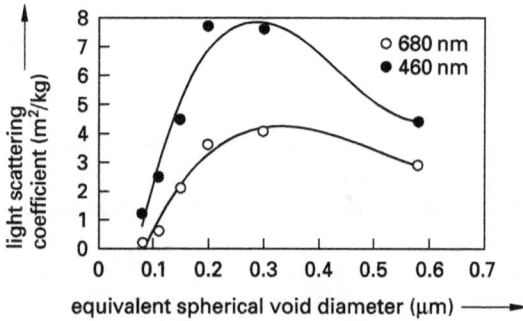

Figure 16.4. Light scattering coefficient as a function of equivalent void diameter. 680 nm and 480 nm denote the wavelength of the light used for the measurement. (From Alince and Lepoutre 1980. Reproduced by the permission of Journal of Colloid and Interface Science).

The width of the particle size distribution governs the pigment packing and the porosity of the coating layer. A broad distribution gives a dense coating because the fine particles are located to the space in between the coarser particles. A narrow distribution with low percentage of fines gives a porous coating. Thick porous coatings exhibit better coverage than dens and thin ones. Usually a high porosity is associated with a high light scattering coefficient.

The effect of width of the particle size distribution on the pigment packing is schematically illustrated in 2-D for discs in *Figure 16.5a* and *Figure 16.5b*. In *Figure 16.5a* the discs of equal size occupy a larger area than the discs in *Figure 16.5b*, which vary in size.

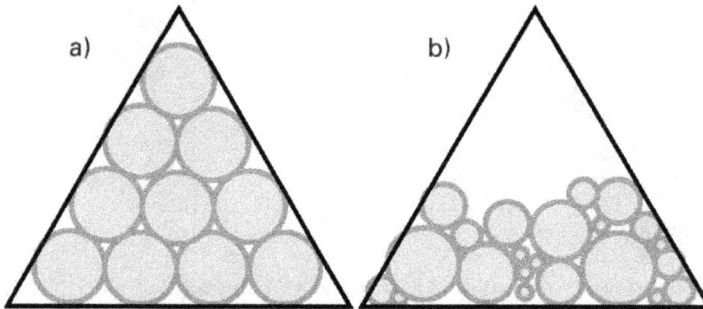

Figure 16.5. Schematic illustration of pigment packing.

The surface area of the pigment increases when the percentage < 2.0 μm increases, and due to that the binder demand for a given surface strength (IGT-pick strength) and the chemical interaction with other components in the coating colour increase.

Coating colours for industrial purposes must exhibit good rheology and be mechanically stable. Moreover the solids content must be high. To meet those demands electrostatically stabilised slurries are prepared of the pigments. Non-stabilised pigments aggregate in an uncontrolled way.

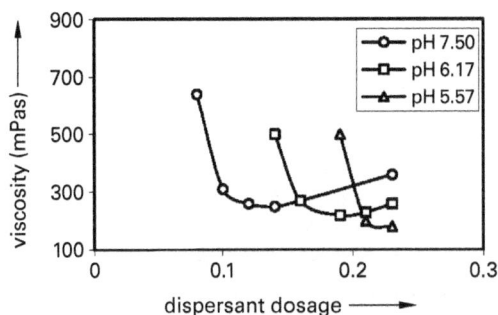

Figure 16.6. Dispersant demand curves for kaolin clay at pH 7.50, 6.17, and 5.57. (From Snyder, Coating Additives. Reproduced by the permission TAPPI Press, Atlanta).

A dispersant is used for the stabilisation and this usually consists of a sodium salt of a low molecular weight (ca. 30000) polyacrylic acid (NaPA), which when adsorbed on the pigment gives this an anionic character. For maximum stabilisation, which coincides with minimum viscosity of the slurry, an optimum dosage of dispersant is demanded. *Figure 16.6* shows dispersant demand curves for kaolin clay at pH 7.50, 6.17 and 5.57, which in turn shows that the dispersant demand for kaolin clay decreases when the pH increases.

Kaolin clay is a weathering product of feldspar in gneiss and granite. Kaolin is the name of a family of minerals and clay the name of mineral particles within the fraction 1–2 μm. There are two types of kaolin clay or clay, which is the term generally used in coating context: primary clay and secondary clay. Primary clay is found where it once was formed. This is not the case for secondary clay. Secondary clay has been transported with creeks and rivers from the place it was formed, to the shoreline, where it settled and where it now is found. However, the water transport does not only mean that the clay has been moved from one place to another, but also a purification process. Therefore secondary clay contains fewer impurities in the form of quarts and feldspar and has a more homogenous mineral composition than primary clay.

The main mineral in clay is kaolinite, which is a 1:1 phyllosilicate mineral. Kaolinite is built up in layers, every second consisting of $Al(OH)_3$-octahedral sheets and every second of SiO_2-tetraeder sheets. The sheets are bonded to each other with hydrogen bonds. This is a weak bond, which is easily broken by the thermal forces the clay is exposed to in nature, and therefore clay is a platy mineral. Primary clays are platier than secondary ones. Delaminated clays is a type of clay with high aspect ratio, *Figure 16.7*, produced by grinding the coarse fraction rejected when classifying the crude clay.

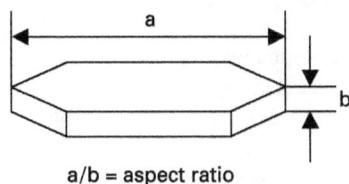

Figure 16.7. Definition of aspect ratio.

Significant for clay is their good coverability and clay is therefore used when this property is critical, for instance when the coat weight is low. Therefore coatings for LWC contain a significant amount of clay. Coatings for ULWC usually contain clay as the solely pigment. High aspect ratio gives good printing properties in rotogravure.

The rheological properties of the coating colour are affected by the mineral composition and by the aspect ratio of the clay. English clays, which contain mica and montmorillonite, the latter a swellable kaolin mineral, exhibits poorer rhelogy than US-clays which mainly contain kaolinite. Increasing the aspect ratio has a detrimental effect on the rheology.

Calcined clay is produced by heat treatment of clay at 1100 °C. At that temperature the clay is chemically modified simultaneously as the pigment particles are sintered to porous aggregates. The porous character of the pigment gives excellent coverabilty but also poor rheology. Calcined clay, which also is called titanium dioxide extender, is used when coating substrates of low brightness.

Ground calcium carbonate (GCC) originates from three different rocks: chalk, limestone, and marble. All three rocks are based on the mineral calcite. The difference between the rocks is how tight the calcite crystals are bonded to each other. In chalk they are loosely bonded, and chalk used to be classified as loose limestone. Marble is crystalline and formed from limestone, which has been exposed to high temperature and high pressure. The GCC-grades used in coating constitute of wet grinded products.

GCC has significantly higher pigment brightness then clay but it does not give the same good coverage as clay, which is seen in both a higher density and a lower light scattering coefficient of the coating, *Figure 16.8*. GCC also gives lower gloss than clay. Therefore GCC is used in coatings where coverage is not critical, i.e. when coating high brightness substrates and when the coat weight is high and a matt coating is desirable. GCC is not used in coatings for rotogravure.

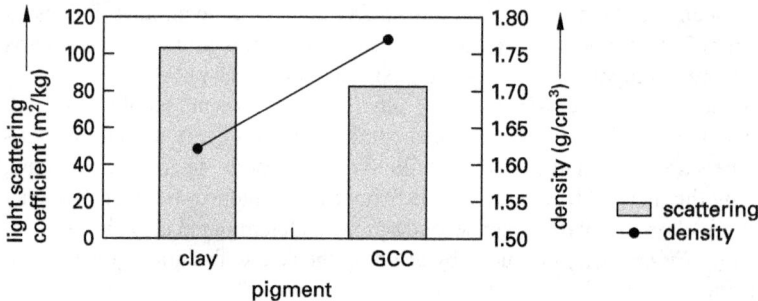

Figure 16.8. Density and light scattering coefficient for coating layers based on GCC and clay.

The pigment slurry can be prepared to significantly higher solids than is possible with clay; up to 78 % compared with 70 %, and that difference is reflected in higher solids content and better rheology and runnability of the coating colour. Rheology is dealt with in Paragraph 16.10 and runnability in Paragraph 16.8.

Other pigments. Talc is, like clay, a phyllosilocate mineral and it is extremely platy. It is mainly used for coating of LWC intended for rotogravure. The mineral is hydrophobic and the pigment is due to that difficult to wet. Therefore a special make down procedure and special dispersants are needed for the slurry preparation.

Titanium dioxide exists in two crystal forms: rultile and anatase both having high refractive index, which gives the pigment excellent coverability. Titanium dioxide is used for coating of low brightness substrates, e.g. WLC and CB.

Plastic pigments are latices with high glass transition temperature, which do not form film. The pigment shape is spherical and the diameter is small and is found within the range 0.1–0.2 μm. Plastic pigments are used in contexts where high gloss is demanded.

Pigment blends. There are many demands made on the coated sheet; it should be bright and glossy and it should exhibit superior printing properties. In order to meet these demands the coating colour used is usually based on blends consisting of two or three pigments. Coating formulations and pigment blends are discussed in Paragraph 16.12.

16.5.2 Binders

The binder system in a coating colour consists of two types of binders: latex and water-soluble binder. The function of these differs. The only function of the latex is to bind the pigment particles to each other and to the substrate, whereas function of the water-soluble binder is to be as well a pigment binder as a thickener and a water retention agent. Most often the latter properties are the important ones and therefore water-soluble binders are often also called thickener or water retention agent. The term co-binder also exists.

Latex binders. The latices used in coating colours are synthetic polymer dispersions. They are mono disperse and the particles are spherical with a diameter usually within the range 0.15–0.20 μm. Latex is produced using emulsion polymerisation and the particles are stabilised with a tenside.

There are three types of polymers used: 1) styrene-butadiene (SB-latex), 2) styrene-butylacrylate (acrylate-latex), and 3) polyvinylacetate (PVAc-latex). In order to modify the properties of the latex, functional groups are linked to the polymer backbone, e.g. – COO, which improves the mechanically stability of the dispersion.

The glass transition temperature, T_g, of the polymer is an important parameter of the latex and in production of SB-latex T_g is controlled by the ratio styrene to butadiene in the polymer. Styrene makes the polymer stiff and butadiene makes it soft. For commercial latices the ratio styrene/butadiene varies within the range 40/60 to 80/20 which corresponds to a range in T_g of minus 25 °C to plus 50 °C. For acrylate latices T_g is controlled in the same way, i.e. by the ratio styrene to butylacrylate. Corresponding ranges are 40/60 to 60/40 and minus 10 °C to plus 40 °C. Butylacrylate makes the polymer soft. The polymer in PVAc-latices, has a T_g of approximately 30 °C.

T_g affects the porosity and the stiffness of the coating layer and latices with a low T_g are therefore used when a low porosity or a soft and compressible coating is desired. The latter is important for rotogravure. On the contrary is true, that a high T_g-latex is chosen if a porous and a stiff coating is desired. High T_g also promotes light scattering and gloss. It should be pointed out that the strength of the coating decreases with increasing porosity. Latices with high T_g are therefore generally considered as poorer pigment binders than latices with low T_g.

The polymers in SB and acrylic latices are rather alike and they usually do not interact with the pigments commonly used in coating colours. However, this does not apply for PVAc latex, because this is hydrophilic in contrast to the polymers in the two other latices just mentioned.

The interaction gives the coating colour high viscosity and the coating layer high porosity. PVAc-latices are rather uncommon in Europe but are popular for board coating in the US.

In order for the latex to work as a binder it must form a film. The film formation process is schematically illustrated in *Figure 16.9*. When water is transported away from the latex, the latex spheres are approaching each other successively. When they are brought in contact with each other they flow together and form a film if the surrounding temperature exceeds the minimum film formation temperature, T_{film}, which attain approximately the same as value T_g. If the surrounding temperature is less than T_{film}, discrete latex particles are formed.

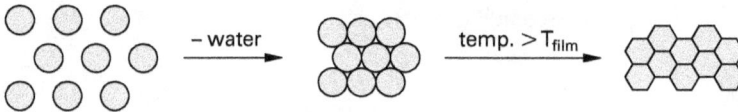

Figure 16. 9. Schematic illustration of film formation of latex.

Water-soluble binders. The two most commonly used water-soluble binders are starch and the sodium salt of carboxymethylcellulose (Na-CMC). Other less used water-soluble binders are protein, polyvinyl alcohol (PVA), alkali swellable polyacrylates, and associative thickeners. **Starch** is a natural polymer, built up by glucose monomers, *Figure 16.10*, and it is created in plants by photosynthesis. In the plant, starch is present in the form of granules consisting of coiled polymers. There are two types of starch: 1) amylose, which is a linear polymer, and 2) amylopectin, which is a branched polymer. The ratio amylose to amylopectin varies between different plants, *Table 16.2*. The starch used for coating originates from potato or corn.

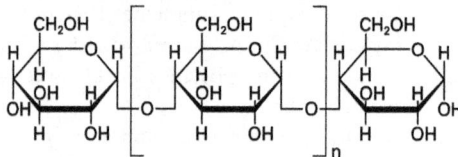

Figure 16.10. Structure formula of starch; a) amylose and b) amylopectin.

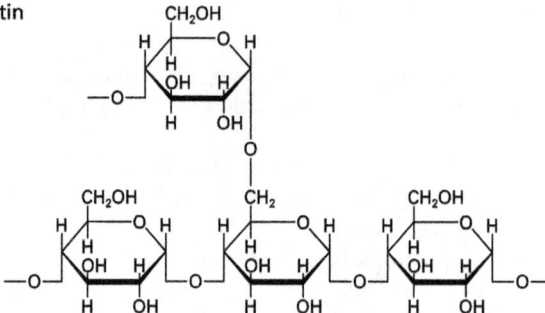

Table 16.2. Amylose and amylopectin in different plants.

	Amylose, %	Amylopectin,%
Potato	21	79
Corn	28	72
Wheat	28	72
Tapioca	17	83
Waxy maize	0	100

The molecular weight of native starch is too high to be able to prepare a water solution of it at a solids content suitable for coating. Therefore the native starch must be broken down before it can be used, which can be performed either at the supplier or at the mill. During this process it is common to simultaneously oxidise the starch, using hypoclorite, in order to avoid retrograda-tion, which is an irreversible association of starch molecules. The retrogradation gelatinises the starch solution, and makes it useless.

Besides oxidised starch, which is the most commonly starch type used in coating, chemically modified starches are also used in special cases in order to obtain certain unique properties. The modification consists of a functional group bonded via ether or ester bonds to the methyl-group on the glucose monomer.

In order to work as binder the starch must be brought into solution which is achieved by cooking. During the cooking, when the temperature is raised, the starch granules swell, and at 70 °C the molecules in the granule begin to solvate and at 95 °C they are completely dissolved. The cooking can be performed batch wise at 95 °C or continuously using jet cookers at elevated pressure and short residence time at 120–140 °C.

Starch is mainly used as water-soluble binder in offset formulations for LWC, and character-istic for starch is that it imparts stiffness to the coating. As already mentioned starch is also used for surface sizing of paper and board, both in non-pigmented and pigmented sizes.

Sodium carboxymethylcellulose. NaCMC or CMC as it is usually called, is a polyelectro-lyte produced from cellulose by reaction first with NaOH and then with $ClCH_2COOH$ (mono-chloroacetic acid). The monomer, which CMC is built up of, is the same monomer that builds up starch. The difference between CMC and starch is the bond between the glucose monomers, *Figure 16.11.* In CMC every second glycoside bond is upwards and every second is down-wards, whereas in starch all bonds are in the same direction. This makes CMC to an extended molecule and starch to a coiled one. CMC can be characterised as an all-purpose water soluble binder. CMC is also used for surface sizing but is not as common as starch.

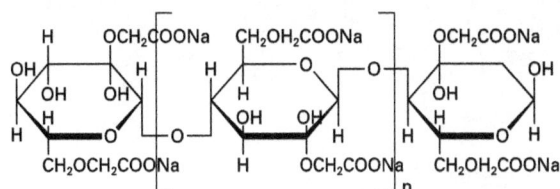

Figure 16.11. Structure formula of Na-CMC.

16.5.3 Additives

Additives is the name of a group of chemicals, which are added to the pigment slurry or coating colour in order to attain certain desirable properties, which can be functional or non-functional. They are added in small amounts, usually <1 pph (parts per hundred). Some common additives are: dispersants, biocides, wet strength agents (hardeners), lubricants, and optical brighteners.

- Dispersants are used for the preparation of pigment slurries. The most commonly used dispersant is NaPA. Slurry preparation is dealt with in Paragraph 16.5.1.
- Biocides are added to the slurry, as well to the final coating colour, in order to prevent bacterial growth in these. Bacteria in the final coated paper or board are not tolerated intended for packaging of food. Common biocides are glutharaldehyde and isothiazolin.
- In offset printing, water (fountain solution) is applied to the coating surface, and the water may reduce the strength of the coating, which is not acceptable. Therefore wet strength agents are used with the purpose to harden the coating (cross-link the binders). Wet strength agents used are: ammonium zirconium carbonate (AZC), glyoxale and melamine formaldehyde resins.
- The pigment particles in coatings are hard and wear on the cover of the calender rolls during calendering. To restrict the wear lubricants may be used: usually ammonium stearate.
- Optical brighteners are used in order to enhance the brightness of the coated material. Incident UV-light is reflected in the visible wavelength region and increases in that way the brightness. Chemically, optical brighteners are sulphonated stilbenes derivatives.

16.6 Coating Machines

There are two principles for coating of paper or board: 1) an excess of the coating colour is applied to the substrate to be coated, which in a later stage of the process is metered to the desired coat weight. 2) a pre-metered amount corresponding to the desired coat weight is applied to the substrate to be coated.

16.6.1 Excess Processes

Applicator systems. There are three methods for the application of coating colours: 1) roll application, 2) jet application, and 3) short dwell time application (SDTA), *Figure 16.12*.

applicator
coating pan

jet

coating inlet

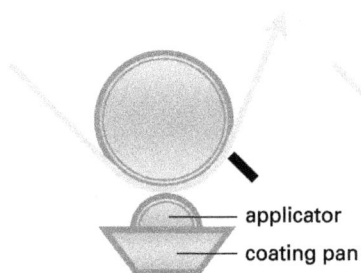

Roll application
Autoblade (Metso)

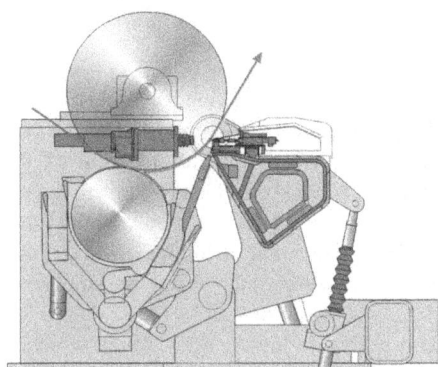

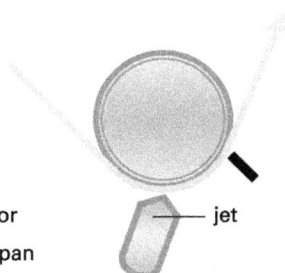

Jet application
OptiCoat Jet (Metso)

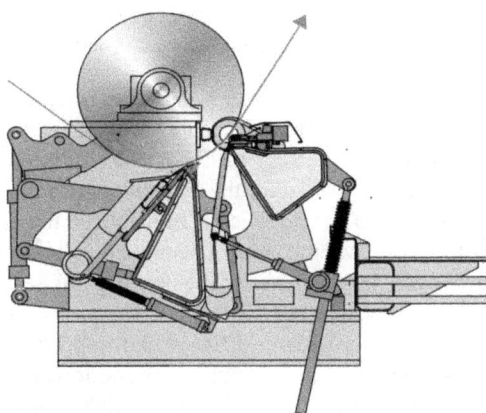

Short Dwell Time Application (SDTA)
OptiBlade (Metso)

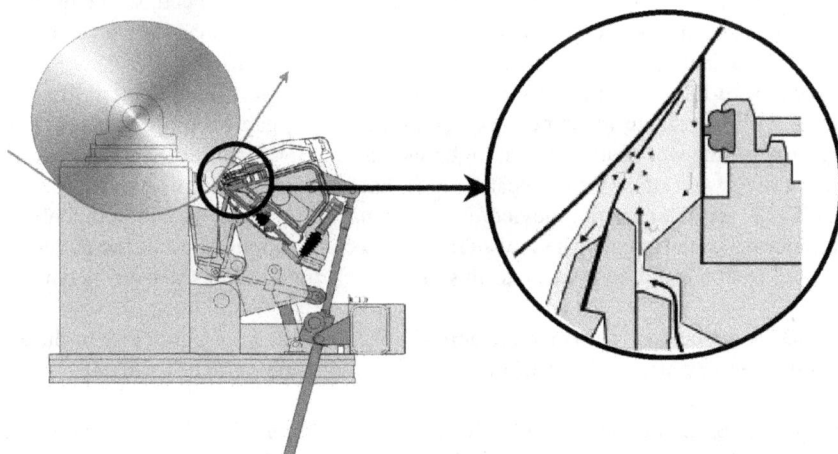

356

■ SEALING BLADE

- **Seals air off from the chamber**
- **Constant coating color acceleration distance =>**
  even impulse
- **Controlled uniform wetting line**

■ CHAMBER LENGTH < 20 mm

- **Undisturbed premetering layer**
- **Short acceleration distance =>**
  no disturbing vortices

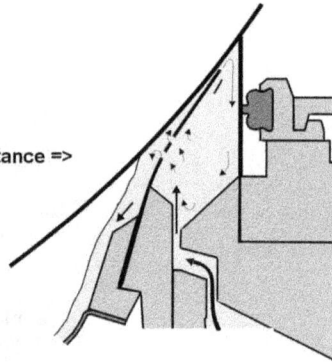

Figure 16.12. Different application systems. (Courtesy of Metso Paper Oy).

In the coaters, *Figure 16.12*, the paper is led over a rubber-covered backing roll and when using roll or jet applicators the coating colour is applied right under or slightly in front of the vertical axis of the backing roll. For SDTA the coating colour is applied just in front of the metering device, which for this type of applicator generally is a steel blade. The metering device is mounted on a beam, which is movable so that the metering device can be brought in contact with the paper when coating and moved away from it for replacement of the metering device, cleaning etc. The diameter of the backing roll is found within the range 0.5–1.5 m and it is larger at higher machine speeds.

The applicator systems differ in two respects: 1) the pressure during the application, which is high, 0.5–4.0 bar, for roll applicators and low, close to atmospheric pressure, for jet and SDTA applicators, and 2) the dwell time, which is long, 15–75 ms for roll and jet applicators, and short, 1.25–6.25 ms for SDTA applicators.

The applicator roll is rubber-covered like the backing roll, and its diameter is usually 25 % of that of the backing roll. It rotates in the machine direction with a perimeter speed, which usually is 25 % of the machine speed. The gap between the applicator roll and the substrate to be coated is usually 0.5 to 2.0 mm. Both the rotational speed and the gap can be varied.

As already mentioned the pressure is high in the applicator nip in roll application and the pressure increases with increasing machine speed and diminishing gap opening. The pressure forces the dewatering of the wet coating and at high machine speeds, for instance when coating of LWC, the pressure dewatering in the applicator nip is that high that it controls the dewatering between the applicator and the metering device. At lower machine speeds, as in board coating, the pressure dewatering is margin and the dewatering between the applicator and the metering device is controlled by the dwell time prior to this. The dwell time decreases with increasing machine speed.

With jet and SDTA applicators, in which the application occurs at a pressure close to the atmospheric pressure, the dewatering is controlled by the dwell time, which, as already mentioned, is short.

The most frequently used coating method is **blade coating**, *Figure 16.13*, and as the name suggests the metering device consists of a steel blade. More then 90 % of all coated material produced today is blade-coated. Blade coating is a rational process operating at high machine speeds and at high solids contents of the coating colour. The fastest coaters are those for LWC and they operate at speeds, which approach 2000 m/min. The solids content is in the range 62–67 % depending on coating colour formulation. The blades used are 0.4–0.5 mm thick and 75 mm wide.

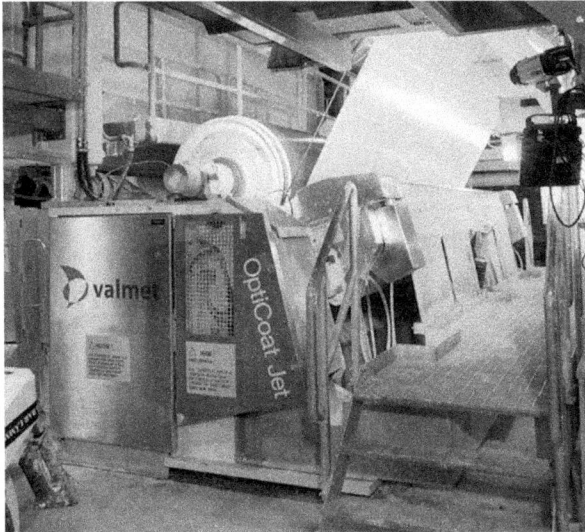

Figure 16.13. Blade coater. (Courtesy of Metso Paper Oy).

The blade is fixed in a holder mounted on a beam, *Figure 16.14*, which can be moved towards the paper so that the blade can be pressed against it when coating. In connection with blade change, cleaning, web breaks, the blade beam is moved away from the paper. A blade is normally wearied out after eight hours. Blade with ceramic tip lasts considerably longer.

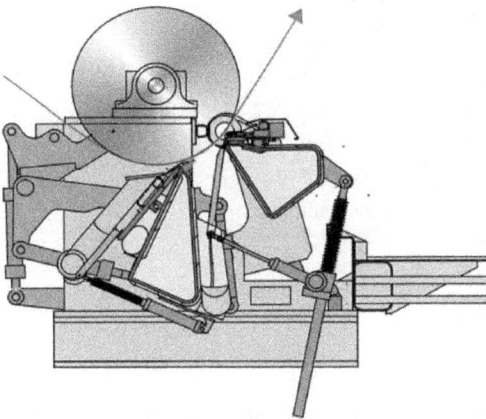

Figure 16.14. Schematic picture of a blade coater. (Courtesy of Metso Paper Oy).

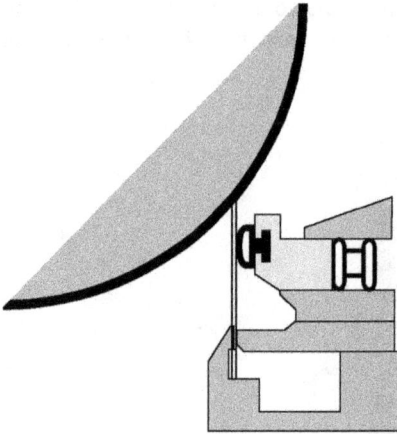

Figure 16.15. Blade pressure control with hose.

The coat weight is controlled by applying a pressure on the blade approximately 10 mm from the tip. Previously this was done with a rubber hose behind the blade, which was blown up to increase the pressure, *Figure 16.15*. When increasing the blade pressure, which has to be done during production to compensate for the blade wear, the blade angle decreases, and because the blade angle affects the sheet properties it is desired to coat at a constant blade angle. In modern blade coaters, therefore, the hose is replaced with a strip of steel and a construction, which, simultaneously as the blade pressure is increased or decreased, adjusts the blade angle so that it remains constant, *Figure 16.16*.

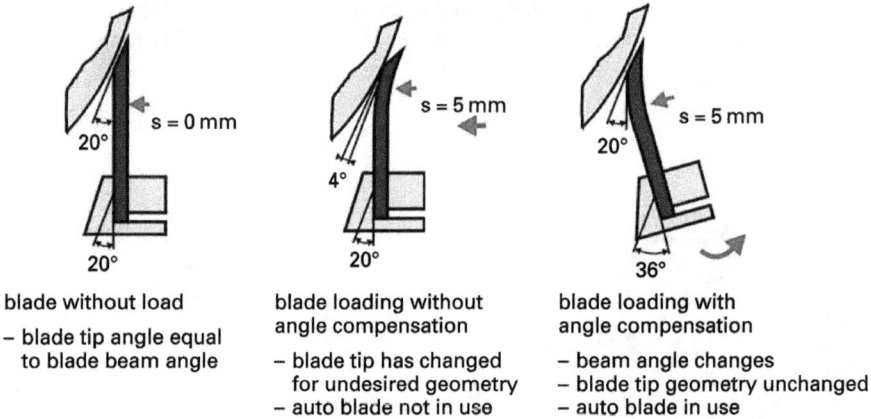

blade without load

– blade tip angle equal to blade beam angle

blade loading without angle compensation

– blade tip has changed for undesired geometry
– auto blade not in use

blade loading with angle compensation

– beam angle changes
– blade tip geometry unchanged
– auto blade in use

Figure 16.16. Blade pressure control at constant blade angle. (Courtesy of Metso Paper Oy).

There are two modes of blade coating: 1) stiff or bevelled blade and 2) bent blade, *Figure 16.17*. When coating using bent blade the tip angle is approximately 45°, and when coating using bent blade approximately 10°. The relationship between coat weight and blade pressure is not the same for stiff and bent blade. For stiff blade the coat weight decreases with increasing pressure whereas the opposite is true for bent blade, *Figure 16.18*. The fact that the coat weight increases when the blade pressure is increased in bent blade coating, is due to that the blade angle simultaneously is reduced. This leads to an increased hydrodynamic pressure, which pushes the blade away from the paper more than the increased blade pressure moves it closer. SDTA allows coating only using stiff blade.

Stiff blade is the most commonly used mode for high speed coating e.g. for LWC, and when the target coat weight is low or moderate. At higher coat weights, 12–14 g/m² or higher, or at lower machine speeds, bent blade is usually used.

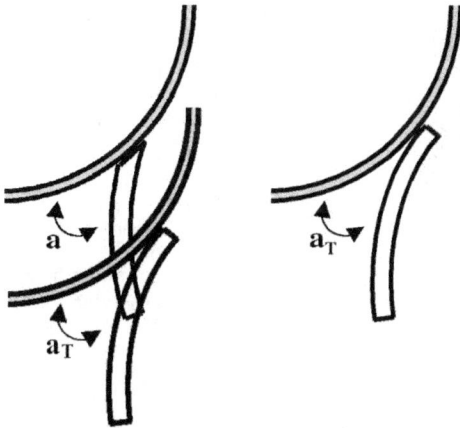

Figure 16.17. Stiff and bent blade.

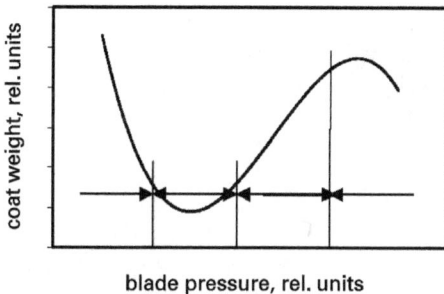

Figure 16.18. Coat weight as a function in blade pressure for stiff and bent blade.

Roll blade, *Figure 16.19* an alternative to blade for pre-coating of board. The roll blade consists of a wire-wounded rod, approximately 5 mm in diameter, in a support mounted in a frame. The one end of the rod is attached to a motor, which brings it in rotation. The rod is lubricated and cleaned with water, which is pumped through the support. The roll blade can replace the blade in a blade coater and it is clamped in the blade holder in the same way as the blade.

The coat weight is controlled as in blade coating with a rubber hose behind the rod blade, which presses it against the substrate to be coated. However, the coat weight is not that sensitive to changes in hose pressure as is the case for blade coating. Mainly, the coat weight is governed by the diameter of the wire, which is wounded around the rod.

The roll blade gives more of a contour coating than the blade, which is considered as a filling in coating method. The roll blade reduces also the risk for scratches and streaks in the coating surface, which is a common problem in blade coating. If coarse particles get stuck underneath the blade tip they give rise to streaks. The risk for particles to get stuck underneath the rod is considerably less because the rod rotates. The solids content of the coating colour is usually 3–5% units lower than for blade coating.

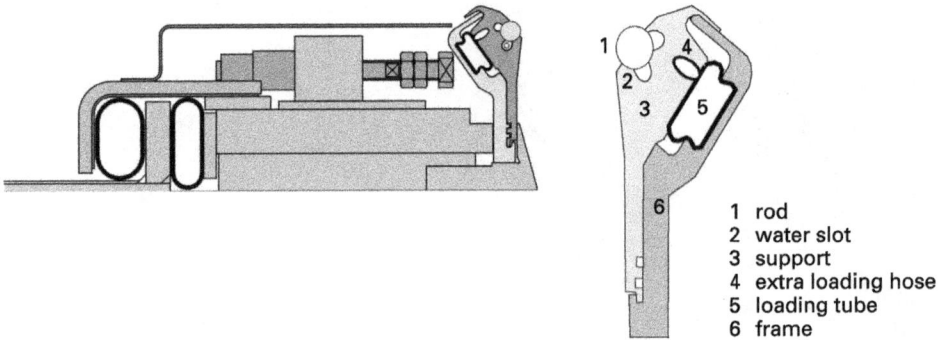

1 rod
2 water slot
3 support
4 extra loading hose
5 loading tube
6 frame

Figure 16.19. Roll blade. (Courtesy of Metso Paper Oy).

Air knife coating, *Figure 16.20*, was previously a common method for top coating of board. In air knife coating the excess of coating colour is metered with an air jet, directed against the substrate to be coated. The excess of coating colour blown off is collected in a vacuum box after which it is de-aerated and re-circulated.

Air knife coating is a contour coating method. The solids content of the coating colour is low, 40–45 %, and the method works at machine speeds below 400 m/min. The speed limitation is the main reason why the air knife coater is replaced by blade coater for top coating in modern coaters for board operating at higher machine speeds than 400 m/min.

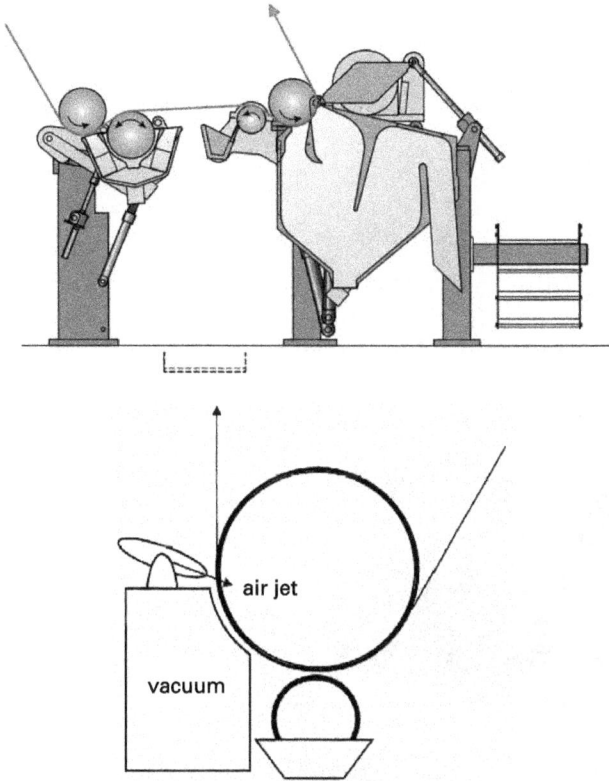

Figure 16.20. Schematic picture of an air knife coater. (Courtesy of Metso Paper Oy).

Two-side coating with blade. In blade coating one side at the time of the substrate is coated. Mirroblade is a new type of blade coater, which coats both sides simultaneously, *Figure 16.21*. The substrate to be coated is passing in between two blades standing on each other's tips. The coating colour is applied with jet applicators, one for each side of the paper.

The Mirroblade coater works with special steel blades with a soft polymer covered tip. With these blades a contour coating is obtained. *Figure 16.22* shows burnout images on Mirroblade coated material using a) polymer covered blades and b) regular steel blades. The more even reflectance in the image of the material coated with the polymer covered blades shows that the mass distribution of the coating layer is more uniform, and that it therefore has the character of a contour coating.

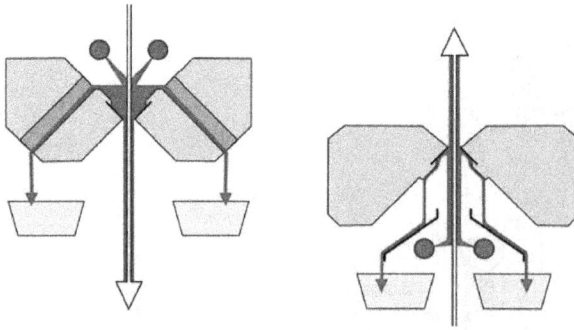

Figure 16.21. Mirroblade coater. (Courtesy of BTG AB).

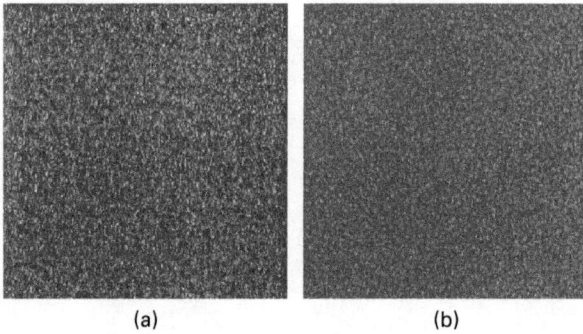

(a) (b)

Figure 16.22. Burnout images on paper coated using a) conventional blade and b) blade with polymer coated tip. (Courtesy of BTG AB).

16.6.2 Pre-Metering Methods

In coating using **Metered Size Press** (MSP) a pre-metered amount of the coating colour is dosed on a transfer roll, *Figure 16.23*. The transfer to the substrate takes place in the nip in between the transfer roll and the backing roll, and usually both sides of the substrate is coated simultaneously. The transfer roll is covered with rubber or polyurethane with a hardness of 10–20 P&J. Its diameter is approximately 1 m.

The dosage of the coating colour to the transfer roll is usually performed using a SDTA unit equipped with roll blade. Both wire wounded and smooth rods are used and usually the diameter of the rods is larger, 10–25 mm, than that for the rods used in board coating. It is easier to control the dosage and the coat weight with the hose pressure when using smooth rods.

The coating colour can also be dosed on the transfer roll volumetrically using a wire-wounded rod of large diameter, approximately 300 mm, *Figure 16.24*. The open area in between the dosage roll and the transfer roll, which is formed when these are brought in contact with each other, gives the amount of coating colour transferred. The transfer roll, which has a soft cover, approximately 100 P&J, rotates at somewhat lower circumference speed than the web does.

Figure 16. 23. Metered size press. (Courtesy of Metso Paper Oy).

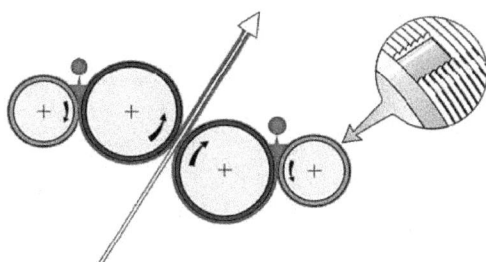

Figure 16.24. Volumetric coating transfer. (Courtesy of BTG AB).

The MSP coater was developed for surface sizing and within this application area it has replaced the size press. In the size press the size is applied to the paper from a pond in the nip between two rolls, *Figure 16.25*. However, the MSP is not only used for surface sizing but also for coating at lower coat weights in the range 6–10 g/m², and both LWC and coated wood free paper are presently MSP coated. The solids content of the coating colour is lower than that in blade coating. MSP coating is considered as a contour coating method.

The web may get stuck in the nip between the transfer rolls and tear off. Web breaks causes production losses and is considered as a serious runnability problem. Coating with the MSP yields less web breaks than coating with the blade. This makes the MSP well suited for coating of low grammage papers, e.g. LWC and ULWC. Runnability in blade coaters is discussed in Paragraph 16.8.

For coat weights exceeding 8–10 g/m² the coating becomes patterned and therefore coating with MSP is restricted to lower coat weights. Because the pattern reminds of that of an orange peel it is called "orange peel pattern". The pattern is caused by a sub-atmospheric pressure, developed in the coating film in the exit of the transfer nip, causing cavitation. The cavitation in turn leads to filament and droplet formation when the coating film splits during the transfer to the substrate, *Figure 16.26*. If the filament ends on the paper side do not level out the prior to the immobilisation, the coating layer of the surface becomes patterned.

The droplets formed are thrown out by the centrifugal force and form a mist in exit of the transfer nip. The mist constitutes an environmental problem and causes also a significant loss of coating colour and is considered as a serious runnability problem. The problem is called "misting".

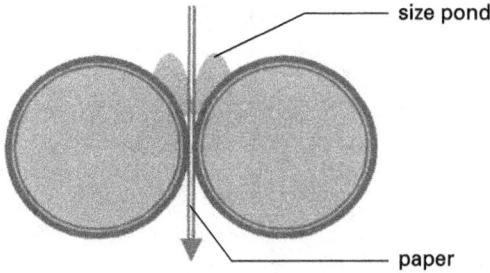

Figure 16.25. Size press.

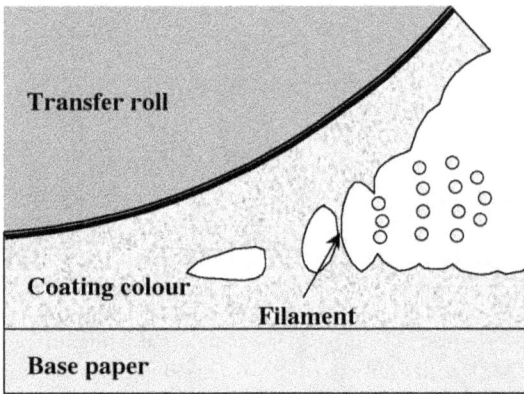

Figure 16.26. Formation of filaments and associated orange peel pattern and misting.

16.6.3 New Coating Techniques

In the coating methods discussed above the coating layer is formed, with the assumption of the air knife coater, using a metering device in contact with the paper. Since the trend towards ever higher machine speeds is assumed to continue, the forces acting on the paper during the forming of the coating layer will also increase. This increases the risk for web breaks. In order to reduce that risk contact free coating methods for high speed coating are currently under testing.

16.7 Dryers

The water applied to the paper together with coating colour is evaporated with a dryer. A lay out of a typical dryer for coated paper is shown in *Figure 16.27*. A common drying strategy begins with infrared (IR) drying, continuous with hot air drying and finishes with cylinder drying. Other combinations also exist, e.g. IR+hot air-drying.

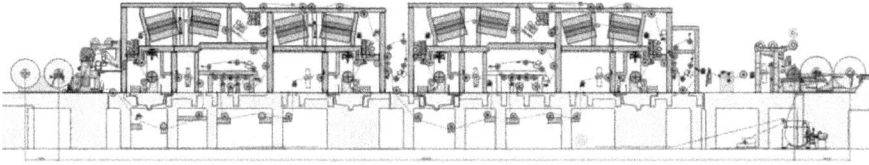

Figure 16.27. Dryer lay out. (Courtesy of Metso Paper Oy).

In hot air drying the coating layer must be heated up to a temperature identical with the wet bulb temperature of the impingement air before the evaporation commences. The heating progress considerable faster with IR than with hot air dryers and therefore the drying usually begins using IR-dryers. Cylinder drying demands that the coating surface is sticky free before it can be brought in contact with the cylinder, and therefore this dryer is solely used for the final drying.

There are two types of IR-dryers: 1) electrically heated and 2) gas heated. The electrically heated dryers emit radiation of somewhat shorter wavelength than the gas heated ones; 800–2000 nm and 1500–2000 nm respectively. A certain amount of the radiation penetrates through the paper and in order to increase the yield with respect to evaporation, a reflector is placed behind the paper web. The evaporation rate of IR-dryers at maximum effect is 40–50 kg H_2O/m^2h. For drying of coated paper are air born hot air dryers used, consisting of a number of blow boxes, arranged zigzag on each side of the paper web. The air impingement and the geometry of the blow boxes are such that paper is supported by an air cushion when passing through the dryer, *Figure 16.28*. For drying of coated board hot air dryers are used, with impingement nozzles just on the coated side, blowing the impingement air perpendicular towards the board web. Rolls support the board web when passing through the dryer.

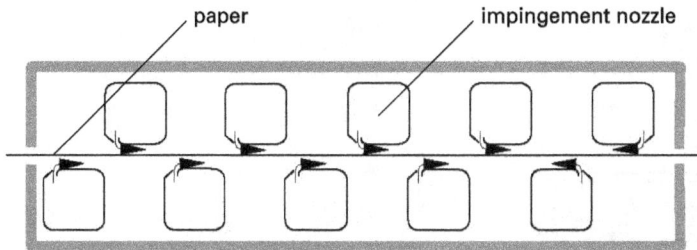

Figure 16.28. Air born hot air dryer.

The hot air is heated with steam or gas. Temperatures of approximately 200 °C is reached with steam and approximately 400 °C with gas. For the air born dryers the air velocity is linked to the temperature and the velocity increases when the temperature is increased and vice versa. At 400 °C the velocity can be that high as 60 m/s and at that setting the evaporation rate is approximately 60 kg H_2O/m^2h.

16.8 Runnability in Blade Coaters

Good runnability means that the coating colour can be applied to the paper and metered without any problems. The concept of runnability includes several different problems. However, three main types can be sorted out: 1) web breaks, 2) defects in the coating surface, and 3) material build-up on the blade tip.

There are two causes for web breaks related to runnability. The web may break because it is weakened that much by the water, taken up from the coating colour between the applicator and the blade, so that it does not withstand the forces exposed to when passing beneath the blade tip. Alternatively, the web may break because the rheological properties of the coating colour are such that a high blade pressure is needed to meter the coating to the desired coat weight. Due to the high blade pressure the paper may get stuck beneath the blade tip and tear off. It should be pointed out that runnability related causes for web breaks are more common for low grammage papers like LWC than for products of higher grammage like board.

Streaks and scratches in the coating surface, *Figure 16.29*, belong to the group of runnability related defects in the coating surface. They are caused by disturbances in the flow beneath or in the vicinity of the blade tip. The causes of the streaks and scratches on fundamental level are not clarified, but flow instabilities in the coating colour during the forming of the coating layer and air in the coating colour are possible causes. Common for these are that they are related to the rheology of the coating colour, and that dilatant colours or colours, which exhibit high viscosity at high shear rates, are prone to give rise to streaks and scratches.

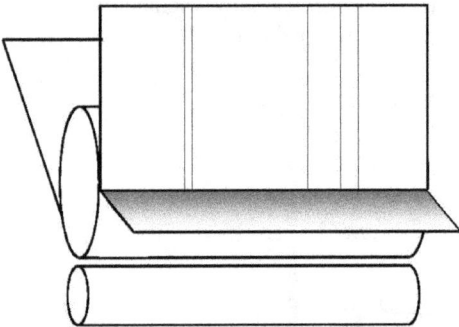

Figure 16.29. Streaks and scratches in blade coating.

The material built-up on the blade, *Figure 16.30*, is termed bleeding, whiskers or stalagmites. It is a common problem at high speed coating and coating at high solids contents. The material build-up is, analogous with streaks and scratches, caused by the rheology of the coating colour, but it does not exist a direct link to any defined rheological parameter. A hypothesis is that coating colours with strong interaction between the components may give rise to material build-up.

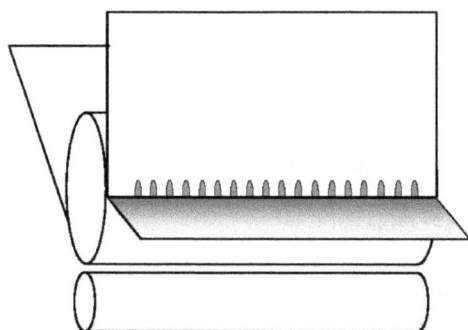

Figure 16.30. Material-build-up on the blade tip.

16.9 Forming and Consolidation of Coating Layers

16.9.1 Consolidation

During the consolidation of the coating layer, when water is released, either through evaporation or absorption, the pigment particles will gradually approach each other to finally be brought contact and form a pigment matrix. Watanabe and Lepoutre (1982) have studied the consolidation of coating layers by performing draw down on polyester substrate and measuring the gloss, reflectance over black background, and solids volume concentration on these during the consolidation process. The result of their study is summarised in *Figure 16.31*, where the gloss and the reflectance over black background (a measure of opacity) is plotted against the solids volume concentration.

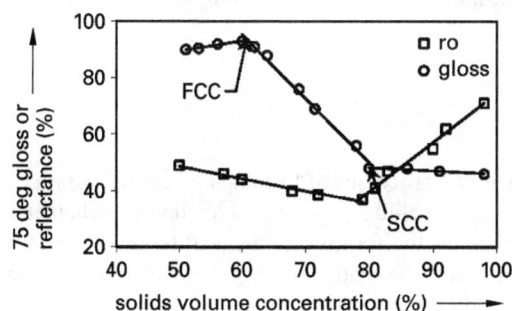

Figure 16.31. Gloss and reflectance over black background as a function of solids volume concentration for draw downs on polyester film. (From Watanabe and Lepoutre 1982. Reproduced by the permission of *Journal of Applied Polymer Science*).

The inflection points on the curves respectively in *Figure 16.31* are called the first critical concentration, FCC, and the second critical concentration, SCC. Up to FCC the coating surface is wet and optically smooth and therefore the gloss is high. At the FCC a pigment matrix is formed, which breaks through the surface and makes it rough. This reduces the gloss because gloss is governed by roughness with a length scale equal with the dimension of the pigment particles.

Between the FCC and the SCC, the pigment matrix shrinks due to the capillary forces that are developed in the matrix during this process when it is emptied on water. The shrinkage is seen in decreasing reflectance over black background (opacity) and decreasing gloss. The size and shape of the pigment particles, as well as the type and amount of binder, affect the shrinkage.

At SCC the final structure is fixed and air voids begin to appear and the over black background increases. The gloss is just marginally affected after the SCC.

The glass transition temperature (T_g) of the latex used is an important parameter in this context, because it governs the stiffness of the pigment matrix. A stiff matrix, which is obtained using stiff latex with a high T_g, shrinks less than a soft matrix, which is obtained using soft latex with a low T_g. The drying temperature also affects the stiffness of the pigment matrix, if containing latex, and thus the shrinkage, which increases when the drying temperature is increased.

16.9.2 Water Pick-up

The paper picks up water from the wet coating layer from the time of application until the FCC, which is reached somewhere in the beginning of the dryer. The water is picked up both by the pore in between the fibres and by the fibres themselves. The water taken up by the fibres is called fibre swelling. After the FCC the water is removed by evaporation.

The fibre swelling plasticizes the paper and releases stresses in it. This in turn causes fibre flocks to expand and loose bonded fibres to lift in the base paper beneath the coating layer. The associated surface movement in the base paper roughens the coated paper which affects the texture and the appearance of the coated surface, as well as the printing properties of the coated paper in a detrimental way.

The water pick up also increases the solids content in the wet coating layer metered by the metering device. This has an adverse effect on the rheology of the coating colour and on the runnability in the coater.

16.9.3 Forming of the Coating Layer

Figure 16.32 illustrates schematically the steps in the processes of forming and consolidation of the coating layer; forming, recovery, shrinkage and levelling. *Figure 16.32a* shows a schematic cross section of the base paper, *Figure 16.32b* shows the forming of the coating layer beneath the blade tip, *Figure 16.32c* shows the recovery after the blade, and *Figure 16.32d* shows the shrinkage and levelling during the consolidation after the blade.

As already mentioned base papers for coating, like all papers, are rough, and the roughness or surface profile can be described in terms of different length scales. Two length scales, which are rather easy to define, are determined by the dimensions of the fibre flocks and of the fibres building up the flocks. A typical diameter of a fibre flock is 4–8 mm and of a fibre is 20–40 μm. The surface profile of the base paper in *Figure 16.32* and *Figure 16.33* is for the sake of simplicity described as the sum of two sine functions having wavelengths equal to mid points of these intervals.

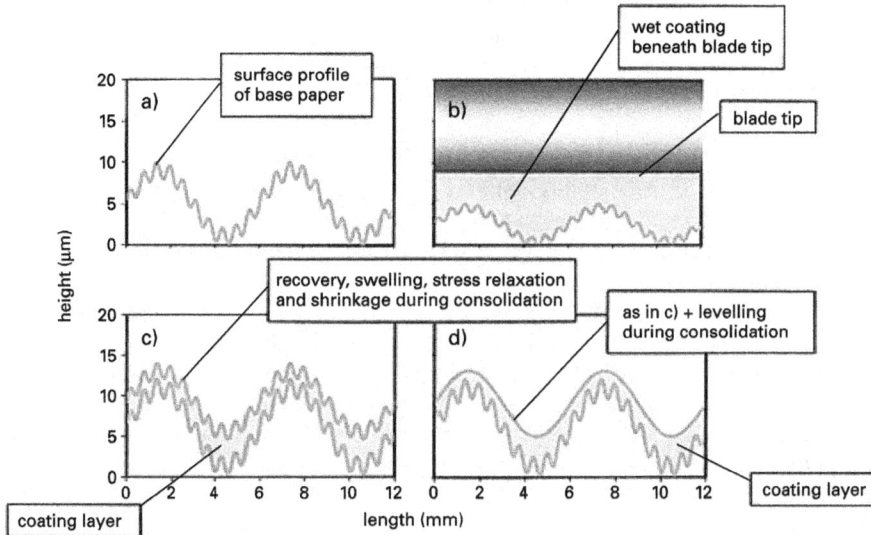

Figure 16.32. Surface profile of the base paper a) prior to the coating, b) in compressed state beneath the blade, and the surface profiles of the base paper and coated paper respectively c) after recovery and moisture expansion and d) ditto + levelling.

In blade coating the base paper is compressed beneath the blade tip during the forming of the coating layer, and the thickness distribution of the coating layer is given by the upper boundary surface in contact with the blade and the lower boundary surface in contact with the paper, *Figure 16.33*. If the upper boundary surface is smooth, which is a reasonable assumption in blade coating because the blade is straight, then the thickness distribution of the coating layer is determined solely by the surface profile of the base paper in a compressed state during the forming of the coating layer. In order to obtain a coating layer with even thickness, which is desirable, the amplitude of the surface profile must be low. This is obtained if the base paper is compressible and if the coating layer is formed at high blade pressure.

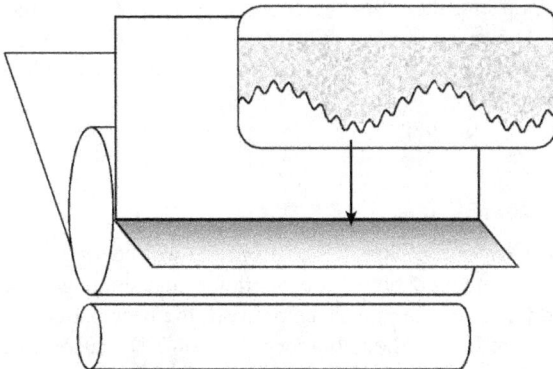

Figure 16.33. The mass distribution of the coating layer is governed by the surface profile of the base paper in a compressed state.

The compressibility of the base paper at the time of forming the coating layer is affected by the amount of water picked up between the application and the blade. The water plasticize the fibre network. A high water pick up is promoted by roll application, long dwell time, and low water retention of the coating colour.

The blade pressure needed for a given coat weight is governed by the rheology of the coating colour used. Rheology is discussed in paragraph 10. One important factor in this context is the solids content, and the higher the solids contents is, the higher is the blade pressure needed. High blade pressure yields a coating layer with even thickness distribution. However, it should be added that high blade pressure and high water pick up may have a negative impact on the runnability, cf. Paragraph 16.8.

After the passage beneath the blade the base paper recovers when the pressure on it is released. The recovery, the fibre swelling, and the stress relaxation, the two latter released by the water which the base paper is exposed, roughens the base paper. The manner in which these factors affect the surface profile of the base paper is illustrated in *Figure 16.32c*. In *Figure 16.32c* is also illustrated how the shrinkage of the coating layer during the consolidation affects the surface profile of the final coated paper.

The shrinkage referred to here is not the shrinkage of the filter cake discussed in Paragraph 16.9.1 but the shrinkage from wet to dry coating. Of course, the shrinkage depends of the coating composition and clay based coatings shrinks less then GCC-based ones. Important is also the solids content of the coating colour. High solids coatings shrink less than coatings of lower solids. For normal coating colour the shrinkage is to be found within the range 50–60 %.

If the recovery, the fibre swelling, the stress relaxation, and the shrinkage are the only factors that control the surface profile of the final coated paper, then the surface profile of the base paper should always show through, and stamp the character of the coating surface to the same extent at both low and high coat weights, *Figure 16.32c*. However, at a high coat weight the base paper texture is not seen in the coating surface to the same extent as at a low coat weight. This is due to the fact that the coating levels out after the forming and erases the texture of the base paper in the coating surface, *Figure 16.32d*. The length scale of the levelling corresponds to the fibre dimension and the length scale of the recovery to that of the fibre flocks.

The higher degree of levelling at higher coat weights is, of course due to the fact that the coating then remains fluid for a longer period of time. However, it should be pointed out that the smoother surface obtained at higher coat weights are not solely caused by the levelling. Because of the lower blade pressure, the compression and therefore the subsequent recovery are also less. This also leads to a smoother surface. The compression of the base paper beneath the blade tip is governed, besides of the fibre composition, also of the water take off between the application and the blade, which plasticizes the base paper.

16.9.4 Binder Migration

A certain fraction of the binder follows the water that is removed from the wet coating layer during the consolidation. Since the water is removed both by evaporation and drainage, the binder will be enriched to the coating surface and to the interface between the base paper and the coating layer. This redistribution of the binder is called "binder migration". Because more water is to be removed at high coat weights, more binder will thus be enriched in the surface and the interface of the coating layer where the coat weight is high. An uneven mass distribution thus leads to an uneven binder distribution. The general opinion is that uneven binder distribution in the coating surface may cause mottle (back trap mottle) in the printed image when printing in lithographic offset.

The drying strategy affects the binder migration and the tendency for mottling in the offset print. High evaporation rate up to FCC and low evaporation rate between FCC and SCC reduces the risk for mottling. The evaporation rate after FCC is not critical.

16.10 Rheology

Rheology is described as the science of deformation and flow of materials and how these properties are affected by stresses that the materials are exposed to. There are two types of deformation 1) reversible or elastic deformation during which the structure of the material remains intact and 2) irreversible deformation or flow during which the structure of the material is broken down. For solid materials the elastic deformation is the most important one, whereas the flow properties (viscous properties) are the most important ones for liquids. Materials, which exhibit both elastic and viscous properties, are termed viscoelastic materials and pigment slurries and coating colours are examples of such materials.

The ratio of the elastic properties to the viscous properties is dependent on the time during which the deformation works. At short times (pulses) the elastic character is stronger than at longer times. The opposite is of course true for the viscous properties.

The rheology of coating colours determines their behaviour when pumping, screening, and coating. The properties of the final coated paper are also affected by the rheology. Rheology is an important tool for studying chemical interaction in coating colours.

16.10.1 Viscosity

Figure 16.34 shows schematically a liquid, which is sheared between a stationary and a moving plate. The shear stress, τ, is defined in a normal manner as the shear force per unit area with the unit Pa (Pascal) and the shear rate, γ, as the shear rate gradient dv/dh with the unit s^{-1} (reciprocal seconds). The shear rate is a measure of the deformation, which the fluid is exposed to, and the deformation increases when the shear rate increases. The viscosity η is given by the ratio τ/γ and has the unit Pas. An older unit for viscosity is Pois. 1 Pois = 10 Pas.

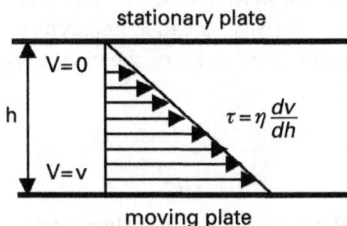

Figure 16.34. A liquid sheared between a stationary and a moving plate.

When rheologically characterising pigment slurries and coating colours (or any liquid), it is customary to record a flow curve, which shows the shear stress as a function of the shear rate. *Figure 16.35* shows flow curves for the four basic types of flow: Newtonian, Bingham, dilatant and shear thinning (pseudoplastic).

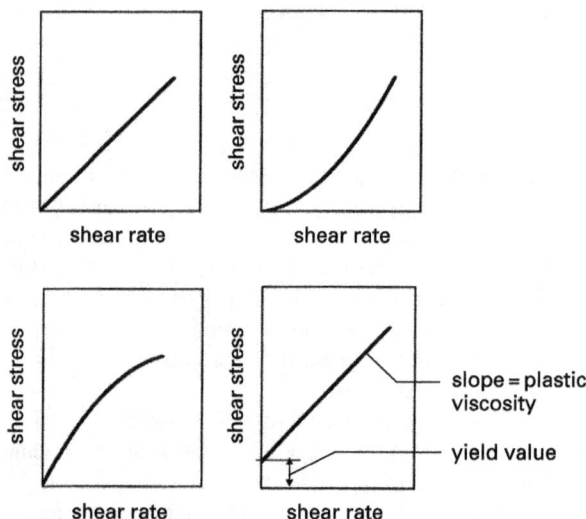

Figure 16.35. Flow curves.

A Newtonian liquid is characterised by a linear flow curve, Equation (16.1), which starts at the origin. The viscosity, η, which is equal with the slope of the flow curve, which in this case is a straight line, is constant for Newtonian liquids.

$$\tau = \eta \dot{\gamma} \tag{16.1}$$

A Bingham liquid is also characterised by a linear flow curve, Equation (16.2), but it starts a bit up on the shear stress axis. The intercept, τ_o, is called yield value and is a measure of the network strength of the fluid which must be exceeded before the liquid starts to flow. In coating colours τ_o is generally taken as a measure of chemical interaction between the components in the colour. The ratio τ/γ, which by definition is equal with the viscosity, is called apparent viscosity and is not constant but decrease with increasing shear rate. The slope of the curve, η_{PL}, which is constant, is called plastic viscosity.

$$\tau = \tau_o + \eta_{PL} \dot{\gamma} \tag{16.2}$$

When the logarithm for the shear rate is plotted against the logarithm of the shear stress a straight line is in many cases obtained. Liquids, which exhibit such flow curves, are called power law liquids and the flow curves for these can be written:

$$\log \tau = k + n \log \dot\gamma \qquad (16.3)$$

where k denotes consistency factor and n denotes flow index. For $n = 0$ the liquid is Newtonian, for $n < 1$ it is shear thinning, and for $n > 1$ it is dilatant. Shear thinning means that the viscosity decreases with increasing shear rate and dilatant that the viscosity increases with increasing shear rate.

The viscosity of shear thinning or dilatant liquids is not only dependent of the shear rate, but also of the time during which the shear works. The time dependency is usually investigated with a shear ramp, *Figure 16.36*. If the up curve shows higher shear stress (viscosity) at a given shear rate the liquid is said to be thixotropic. If the opposite is true the liquid is said to be rheopectic.

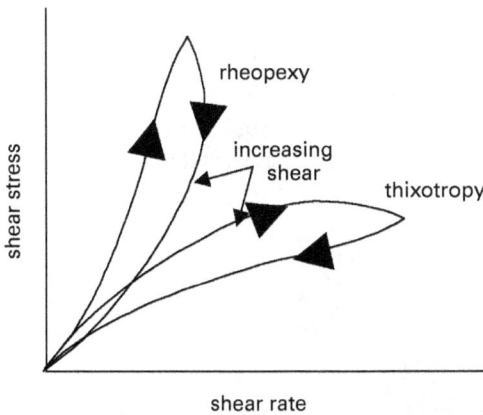

Figure 16.36. Shear ramps.

16.10.2 Viscosity of Suspensions

Pigment slurries and coating colours are suspensions, i.e. particles suspended in water or water based liquid phase. The presence of particles increases the viscosity as well as the shape of the flow curve.

For diluted suspensions of mono dispersed hard spheres Einstein has derived an expression, equation 4, which describes the relationship between the viscosity of the suspension, η, the viscosity of the liquid phase, η_s, and the volume fraction of the spheres, φ.

$$\frac{\eta}{\eta_s} = 1 + 2{,}5\phi \qquad (16.4)$$

For concentrated suspensions, like pigment slurries and coating colours, whose flow curves can be described with the help of the Newton or Bingham equations, the viscosity of the suspension can be predicted using the Dougherty-Krieger equation, which is a semi-empirical equation based on the Einstein equation:

$$\frac{\eta}{\eta_s} = \left(1 - \frac{\phi}{\phi_{max}}\right)^{-[\eta]\phi_{max}}$$

(16.5)

where φ_{max} denotes the maximum volume fraction and $[\eta]$ the intrinsic viscosity, which describes the shape of the pigment particles. For spheres $[\eta]$ is equal with 2.5 (cf. the Einstein). The Dougherty-Krieger equation predicts the hydrodynamic viscosity, i.e. it does not consider chemical interaction between the components and their effect on the viscosity. When the Dougherty-Krieger equation is applied on pigment slurries $[\eta]$ denotes a mean value of the intrinsic viscosity.

16.10.3 Viscoelasticity

As already said pigment slurries and coating colours are viscoelastic materials, which exhibit both viscous and elastic properties. The elastic properties can be determined by subjecting the liquid for an oscillating shear, in the form of a sinusoidal strain or stress, and measuring the response of these, *Figure 16.37*.

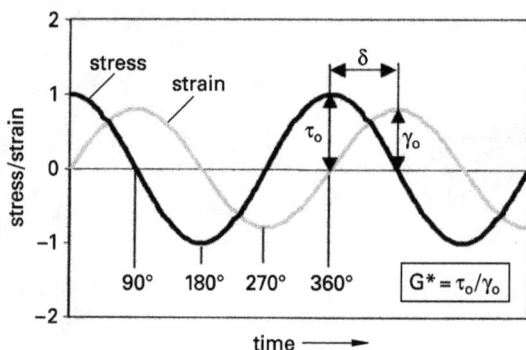

Figure 16.37. Respond as a function of sinusoidal stress or strain.

The response curve is phase shifted with an angle, δ, in relation to the imposed curve. The complex shear modulus G^* is defined by the ratio τ_0/γ_0. Based on G^* and δ the storage modulus G' and the loss modulus G'' can be calculated. G' describes the elastic properties and G'' the viscous properties of the liquid and δ the degree of elasticity. $\delta = 0°$ indicates that the liquid is 100 % elastic and $\delta = 90°$ that it is 100 % viscous.

A simple model used to describe viscoelasticity of fluids is the Maxwell element, which consists of a spring and a dashpot in series, *Figure 16.38*. When the element is loaded, the spring is stretched and as soon as the stress has exceeded a sufficiently high level to overcome the resistance of the dashpot (viscosity) the piston begins to move. When the element is reloaded the spring returns to it original length whereas the dashpot remains in its final position. The load and reload of the Maxwell element explains in a clear way the phase shift between stress and strain illustrated in *Figure 16.38*.

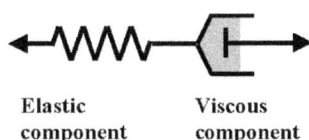

Elastic   Viscous
component   component

Figure 16.38. Maxwell element.

16.10.4 Measurement of Viscosity and Viscoelasticity

When pigment slurries and coating colours are prepared, as well as when coating with the latter one, these are exposed to shear rates within a wide shear rate range. During pumping of the coating colour to the coater the shear rate is approximately 10^2 s^{-1}, during the forming of the coating layer beneath the blade tip 10^6 s^{-1}, and during the levelling after the blade 10^{-2} s^{-1}. The shear rates which the pigment slurries and coating colours are exposed to in connection with the coating colour preparation and coating are summarised in *Table 16.3*.

Table 16.3. Shear rates in coating.

Process	Shear rate, s^{-1}
Pumping	10^0–$5 \cdot 10^2$
Screening	10^4–10^5
Application (roll)	10^4–10^5
Forming beneath the blade tip	10^6–10^7
Levelling after the blade	10^{-1}–10^{-2}

There is no rheometer, which can measure the viscosity within such a wide shear rate range as that existing in coating. Therefore several different types of rheometers must be used for a complete characterisation of the rheology of pigment slurries and coating colours, *Figure 16.39*. For routine measurements at low shear rates, < 30 s^{-1}, the Brookfield viscometer is used and for semi high shear rates within the range 10^2–10^5 s^{-1} the Hercules High Shear Viscometer is used. In research and development contexts rheometers such as Bohlin, Rheologica or similar are used for measurements within the range 10^{-2}–10^3 s^{-1}. These rheometers are also used for viscoelastic measurements. Capillary viscometers are used for measurement within the high shear range 10^4–10^7 s^{-1}.

The Brookfield viscometer is a standard instrument, which is commonly used at the mills for the measurement of the viscosity of pigment slurries and coating colours. The measuring device consists of a spindle, which is immersed in the liquid to be measured. The Hercules-viscometer is a so-called Couette instrument, which means that the measuring device consists of a cylinder and a cup (two concentric cylinders). For these instruments the shear rate is defined and given by the rotational speed and the clearance between the cylinder and the cup. The rheometers used for research and development work are either Couette or cone plate instruments.

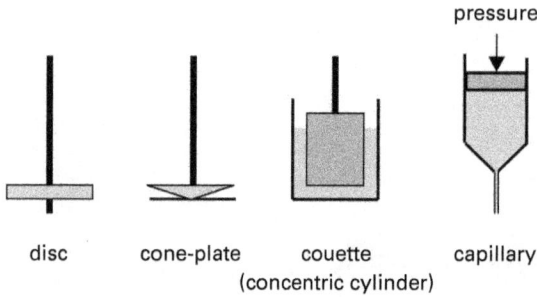

Figure 16.39. Rheometers used for characterisation of coating colours.

16.11 Chemical Interaction

Pigment slurries and coating colours can be regarded as colloidal systems, and in these weak interaction forces of van der Waal's type act between the components. These interaction forces build up a network structure. Coating colours having a network structure are often called "structured" colours.

The yield value, τ_o, of the flow curve, Equation (16.2), and the storage modulus G' are indicators of a network structure. The network structure is weak and shear forces easily break it down. An example is the shear forces beneath the blade tip, which the coating colour is exposed to during the coating layer is formed in blade coating. The capillary forces, which the coating layer is exposed to during the consolidation, may also break down the network structure. If the network structure is reversible after shearing and if it withstands the capillary forces, a porous coating with superior coverage is obtained.

It should be pointed out that all components used in ordinary coating colours exhibit an anionic character and that these components interact with each other. Components with cationic character interact stronger and are only used when the purpose is to create a strong network structure.

16.11.1 Salt

The stability of pigment slurries is governed by the amount of dispersant adsorbed, its conformation, and the thickness of the electrostatic double layer. The thickness of the latter, $1/\kappa$, is given by the expression

$$\kappa = \left(\frac{2e_o^2 N_A Z^2 c_3}{\varepsilon \varepsilon_o kT} \right)^{0,5} \tag{16.6}$$

where e_o denotes the electronic charge, N_A the Avagadros constant, Z the valency of the counterion, c_3 the electrolyte concentration, ε the relative permittivity, ε_o the permittivity of free space, k the Boltzmann constant, and T the absolute temperature.

As is evident from Equation (16.6), the thickness of the electrical double layer decreases when the salt concentration and the valence of the cation increases. However, the salt concentration and the valence do not only affect the thickness of the electrical double layer but also the conformation of the dispersant, which govern its dispersing effect. In salt free environmental, NaPA is an extended molecule because the acid groups along the backbone repel each other. In presence of salt, the molecule coils and losses in dispersing ability.

Addition of salt to pigment slurries causes a destabilisation and a formation of a network structure, partly due to that the thickness of the electrical double layer reduces and partly due to that the dispersant looses in efficiency. The network formation increases the viscosity, and in *Figure 16.40* it is shown that the viscosity of a clay slurry increases marginally when added with NaCl, whereas it increases considerable when added with $CaCl_2$.

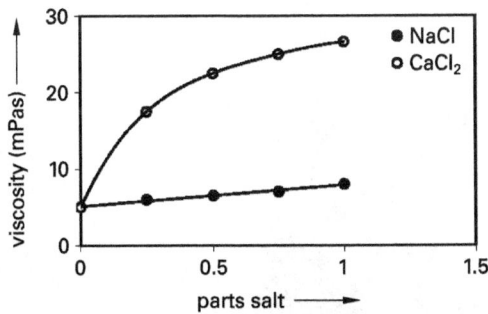

Figure 16.40. Viscosity of clay slurry as a function of salt content. (From Fadat and Rigdahl 1986. Reproduced by the permission of Nordic Pulp & Paper Journal).

Clay could be regarded as an inert pigment with no solubility in water. GCC in contrast, is slightly soluble at the pH-values normally used in coating and GCC slurry therefore contains a certain amount of Ca^{2+}. When clay slurry is mixed with GCC-slurry Ca^{2+} from the GCC-slurry will destabilise the clay slurry and form a network. This means that the viscosity of the pigment mix will be higher than that obtained from the volume fraction of the pigments respectively. Maximum network formation coincides with maximum viscosity and for the example shown in *Figure 16.41* this was obtained at a ratio clay/GCC of 40/60. *Figure 16.42* shows that the maxima for the porosity and the light scattering coefficient for dry coating layer were obtained at the same ratio.

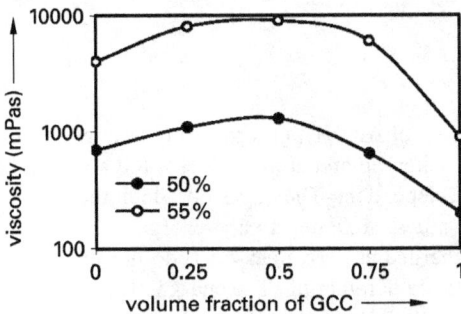

Figure 16.41. The viscosity as a function of the ratio clay/GCC in the coating colour. (From Alince and Lepoutre 1983. Reproduced by the permission of Tappi Journal).

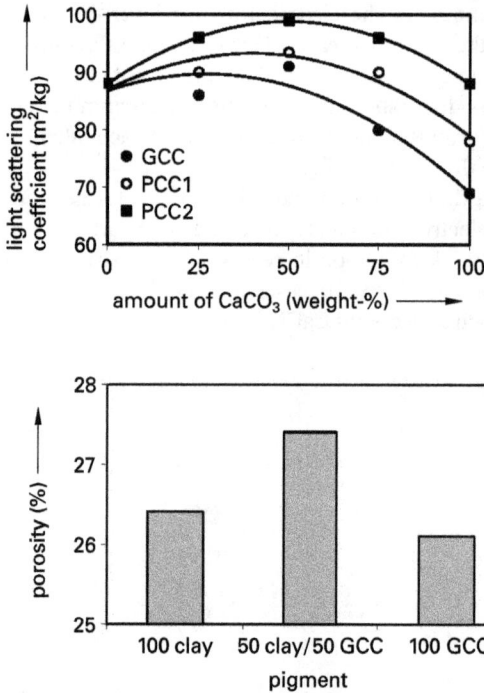

Figure 16.42. a) Light scattering coefficient (From Engström and Rigdahl 1992. Reproduced by the permission of Nordic Pulp & Paper Journal) b) porosity as a function of the amount of clay and GCC in the coating.

Ca^{2+} react in the same way with Na-CMC and similar water-soluble binders as with NaPA. CMC is, like NaPA, an extended molecule in salt free solution. When Ca^{2+} is adsorbed on the negative groups on the NaCMC backbone these groups do not repel each other any longer. Since the repulsion keeps the molecule extended it coils when Ca^{2+} is adsorbed. The viscosity of water solutions of CMC is higher when the CMC molecules exist in extended form. The viscosity of the latex products most frequently used is only marginally affected by presence of salt.

16.11.2 pH

The edges and the one face of the clay particle consists of Al(OH)$_3$ groups, which can take up or release one proton. At pH-values < 8.0 a proton is taken up and at pH-values > 8.0 a proton is released. The charge of the clay particle is thus pH-dependent. The edges on which the dispersant is adsorbed are positively charged at pH-values > 8,0 negatively charged at pH-values < 8.0. The fact that the edges are negatively charged at pH-values > 8.0 do not mean that positive sites are missing, but that the negative sites are dominating. The charge of the Al-octahedral face, however, does not change at the pH-value 8.0, because it is governed by isomorphic substitution. The charge of the face is always negative, but it is more negative at higher pH-values. The charge of the Si-tetrahedral face is not affected by pH and is always negative.

The fact that the charge of the clay particle is pH-dependant affects its interaction with the dispersant. At low pH-values more dispersant is needed for optimum dispersion, as is evident in *Figure 16.4*. In addition. the slurry attains higher viscosity. For GCC the solubility increases with decreasing pH. This was discussed above in Paragraph 16.11.1.

16.11.3 Water Soluble Binders

The shear modulus, G', is a parameter used to quantitatively describe the network structure in pigment slurries and coating colours, and as already mentioned the structure is the result of interaction between the components in these. Pigment slurries exhibit low values of G' like also water solutions of water-soluble binders as CMC and starch. The structure is developed when mixing the pigment slurry with the water-soluble binder. The latex normally has a minor effect on the structure formation. The development of G' when mixing clay slurry with CMC and starch respectively, is shown in *Figure 16.43*. GCC-based coating colours are usually less structured than the corresponding clay based ones.

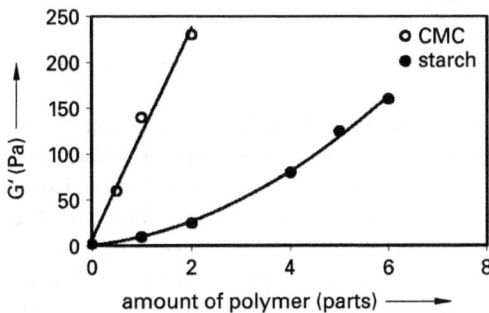

Figure 16.43. G' for clay slurry as function of amount water-soluble binder. (From Ericsson et al. 1991. Reproduced by the permission of Deutcher Fachverlag GmbH, Frankfurt am Main).

Other parameters used to quantify formation of network structure is the relative plastic viscosity, η_{PL}/η_o, cf. Equation (16.2), and the yield stress τ_o. The relative plastic viscosity is obtained by first measuring the plastic viscosity of the coating colour, η_{PL}, centrifuging the coating colour and then measuring the plastic viscosity of the supernatant, η_o. The relative plastic viscosity for clay-NaCMC and clay-starch colours (without latex) as a function of the amount water-soluble binder is shown in *Figure 16.44*.

Values of the relative plastic viscosity exceeding the relative plastic viscosity at zero amount water-soluble binder indicate interaction between the pigment and the water-soluble binder, and values below this value that the water soluble has dispersing effect. In *Figure 16.44* it is shown that NaCMC interacts stronger with the clay than starch at normal amounts of water-soluble binder, which for NaCMC is 0.5–1.0 parts, and for starch 4.0–6.0 parts. The stronger interaction that NaCMC give rise to explains the higher porosity and better coverage obtained with that water-soluble binder.

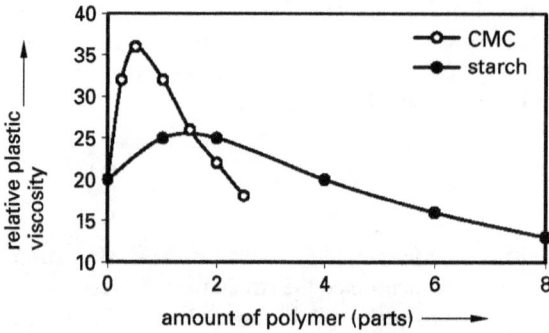

Figure 16.44. The relative plastic viscosity as a function of parts NaCMC or starch for a clay based coating colour. (From Ericsson et al. 1991. Reproduced by the permission of Deutcher Fachverlag GmbH, Frankfurt am Main).

The adsorption of NaCMC and starch on NaPA-stabilised clay is very small but it is anyhow larger for NaCMC than for starch. In *Figure 16.45* the adsorption of NaCMC and starch respectively on clay is shown at the salt content 0.1 M NaCl. At the salt content of 0.2 M NaCl NaCMC and starch are not adsorbed at all. The adsorption of NaCMC and starch on non-stabilised clay is significant.

For systems in which the water soluble binder is not adsorbed on the pigment, or if the adsorption is that weak that the adsorbed binder is not able to overlap between two pigment particles, the interaction mechanism is called "depletion flocculation" and this mechanism is suggested for the interaction between clay and NaCMC and starch respectively. Depletion flocculation can be explained in terms of concentration gradients. If the binder concentration is lower around two pigment particles than in between them, the concentration gradient will cause binder to diffuse from the area in between the particles to the area outside. This causes an osmotic pressure between the particles, which holds these together and in this way forms a network.

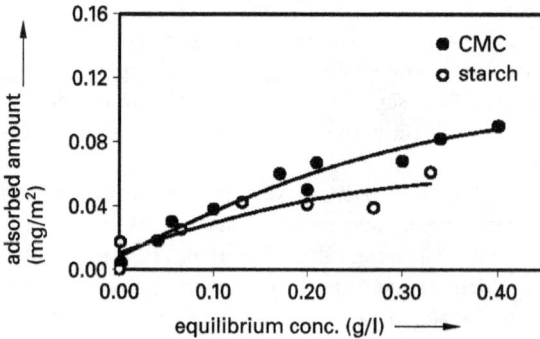

Figure 16.45. Adsorption of NaCMC and starch on clay. (From Ericsson et al. 1991. Reproduced by the permission of Deutcher Fachverlag GmbH, Frankfurt am Main).

16.12 Coating Formulations and Coating Colour Preparation

16.12.1 How Coating Formulations are Written

Coating formulations are written in parts per hundred of the pigments used, and therefore the pigment fraction is always equal with 100 parts. An example of a formulation consisting of three pigments, one water soluble binder, one latex, and one additive is shown in *Table 16.4*, together with the corresponding amounts, dry and wet, needed to prepare a coating colour based on 2.00 kg pigment. The solids contents used for the conversion of kg (dry) to kg (wet) are also included in the table.

Table 16.4. Principles for writing coating formulations.

	Parts	Kg (dry)	Solids, %	Kg (wet)
Pigment 1	50	1.000	72.0	1.389
Pigment 2	45	0.900	70.0	1.286
Pigment 3	5	0.100	50.0	0.200
Water soluble-binder	1	0.020	10.0	0.200
Latex	10	0.200	50.0	0.400
Additive	0.5	0.010	40.0	0.025
Total		2.230		3.500
Solids (calculated), %			63.68	

16.12.2 Coating Formulations

Table 16.5 shows examples of coating colour formulations used for coating of some grades of paper and board. LWC and coated board are both printed in offset and rotogravure while coated wood free paper solely is printed in offset. Offset and rotogravure do not direct the same demands on the coating and this is considered when formulating the coating colours. Other factors to consider are converting of coated board into capsules, runnability, coat weight, and brightness of the substrate.

The formulation of the pre-coating colour for board, Formulation No.1, is based on coarse GCC and the reason for that is that this pigment gives a rough surface, which is suitable for top coating using blade without problems with steaks and scratches. If the baseboard is of low brightness, which usually is the case if the baseboard contains unbleached pulp or recycled paper in the centre plies, the coverability of the top coating is particularly important. Therefore calcined clay and titanium dioxide are found in the top coating Formulation No. 2. Formulation No. 3 is a top coating formulation for SBS. The binder content is higher for board than for offset paper, which is due to that the surface strength of coated board is controlled by demand on glueability and that of paper by demand on printabilty.

Formulations for coated wood free papers and SBS, which Formulation No. 3 and No. 4 are example on, use to be rich on GCC, because the demand on opacity of the coating layer is mod-

Table 16.5. Typical coating formulations for paper and board.

	Board			Coted wood-free	LWC		ULWC
	Precoating	Top-coating			Offset	Roto-gra-vure	
	1	2	3	4	5	6	7
Coarse GCC	100						
Fine GCC		30	90	90	50		
Delaminated clay		60			50	50	100
Ultra-fine clay			10	10			
Calcined clay		5					
Titanium dioxide		5					
Talc						50	
Na-CMC	1	1	1	1		1	
Oxidised starch					6		6
Latex	16	16	16	12	6	4	6

erate. This is due to the fact that the baseboard or base paper for these grades exhibit high brightness and therefore can be allowed to show through the coating. The ultra fine clay in Formulation No. 3 and No. 4 imparts gloss.

Formulation No. 5 is a typical formulation for LWC intended for offset. GCC and starch are not used for rotogravure because these components give a stiff and incompressible coating layer, which is detrimental for printability in rotogravure. Therefore these components are not included in the rotogravure formulation no. 6. Starch in the offset Formulation No. 5 has been replaced with CMC and latex, and GCC has been replaced with delaminated clay and talc. The demand on surface strength of the coating layer is significantly lower in rotogravure than in offset because the ink printed with is low viscous and not tacky. Therefore the binder content is lower in the rotogravure Formulation No 6. It should be pointed out that the binder content is the same in both offset and rotogravure formulations for board because the surface strength of coated board, as already mentioned, is controlled by the demand on glueability and not by the printability. The clay in Formulation No 7 improves the coverage of the coating layer, which is especially important for this type of paper because of the low coat weight.

16.12.3 Coating Colour Preparation

In industrial preparation of coating colours, each component in the colour must exist in liquid form. The pigments used are delivered to the mill either in dry form as a powder or ready dispersed as slurry. If delivered dry, slurries are prepared which are stored in tanks, one for each pigment. Most water-soluble binders are delivered dry, e.g. starch and NaCMC, and these are dissolved and pumped to storage tanks. The latex and most of the additives are delivered in liquid form, and like the pigments and the water-soluble binders, stored in storage tanks.

The coating colour is usually prepared batch wise in a mixer. When preparing a colour the pigment slurry is first added to the mixer, then the water-soluble binder and after some 20 minutes agitation, the latex and the additives. Finally water is added to adjust the solids content to the target value. In industrial preparation it takes approximately 45 minutes to prepare one batch. The pigment slurries are usually screened before pumped to the mixer. The ready coating colour is also screened. A schematic illustration of the coating colour preparation is shown in *Figure 16.46*.

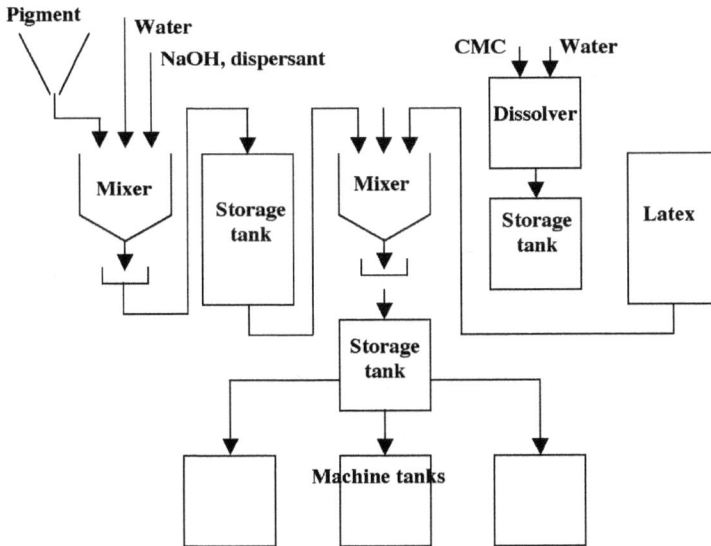

Figure 16.46. Schematic illustration of the coating colour preparation.

Index

388

390

HOLZFORSCHUNG

International Journal of the Biology, Chemistry, Physics, and Technology of Wood

Editor-in-Chief: Oskar Faix, Germany

Publication frequency: bi-monthly (6 issues per year).
Approx. 700 pages per volume. 21 x 29.7 cm
ISSN (Print) 0018-3830
ISSN (Online) 1437-434X
CODEN HOLZAZ
Language: Englisch

Holzforschung is an international scholarly journal that publishes cutting-edge research on the biology, chemistry, physics and technology of wood and wood components. High quality papers about biotechnology and tree genetics are also welcome. Rated year after year as the number one scientific journal in the category of Pulp and Paper (ISI Journal Citation Index), *Holzforschung* represents innovative, high quality basic and applied research. The German title reflects the journal's origins in a long scientific tradition, but all articles are published in English to stimulate and promote cooperation between experts all over the world. Ahead-of-print publishing ensures fastest possible knowledge transfer.

Indexed in: Academic OneFile (Gale/Cengage Learning) – Aerospace & High Technology Database – Aluminium Industry Abstracts – CAB Abstracts – Ceramic Abstracts/World Ceramic Abstracts – Chemical Abstracts and the CAS databases – Computer & Information Systems Abstracts – Copper Data Center Database – Corrosion Abstracts – CSA Illustrata – Natural Sciences – CSA / ASCE Civil Engineering Abstracts – Current Contents/Agriculture, Biology, and Environmental Sciences – Earthquake Engineering Abstracts – Electronics & Communications Abstracts – EMBiology – Engineered Materials Abstracts – Engineering Information: Compendex – Engineering Information: PaperChem – Journal Citation Reports/Science Edition – Materials Business File – Materials Science Citation Index – Mechanical & Transportation Engineering Abstracts – METADEX – Paperbase – Science Citation Index – Science Citation Index Expanded (SciSearch) – Scopus – Solid State & Superconductivity Abstracts.

W
DE
G
de Gruyter
Berlin · New York

www.degruyter.com

www.ingramcontent.com/pod-product-compliance
Lightning Source LLC
Chambersburg PA
CBHW051115200326
41518CB00016B/2508